METHODS IN MOLECULAR BIOLOGY

For further volumes:
http://www.springer.com/series/7651

For over 35 years, biological scientists have come to rely on the research protocols and methodologies in the critically acclaimed *Methods in Molecular Biology* series. The series was the first to introduce the step-by-step protocols approach that has become the standard in all biomedical protocol publishing. Each protocol is provided in readily-reproducible step-by step fashion, opening with an introductory overview, a list of the materials and reagents needed to complete the experiment, and followed by a detailed procedure that is supported with a helpful notes section offering tips and tricks of the trade as well as troubleshooting advice. These hallmark features were introduced by series editor Dr. John Walker and constitute the key ingredient in each and every volume of the *Methods in Molecular Biology* series. Tested and trusted, comprehensive and reliable, all protocols from the series are indexed in PubMed.

Chromatin Immunoprecipitation

Methods and Protocols

Edited by

Franziska Greulich

Metabolic Programming, TUM School of Life Sciences, Technical University of Munich, Freising, Germany

Humana Press

Editor
Franziska Greulich
Metabolic Programming
TUM School of Life Sciences, Technical University of Munich
Freising, Germany

ISSN 1064-3745 ISSN 1940-6029 (electronic)
Methods in Molecular Biology
ISBN 978-1-0716-4070-8 ISBN 978-1-0716-4071-5 (eBook)
https://doi.org/10.1007/978-1-0716-4071-5

This Humana imprint is published by the registered company Springer Science+Business Media, LLC, part of Springer Nature.
The registered company address is: 1 New York Plaza, New York, NY 10004, U.S.A.

Preface

The genome stands as the master regulator of our body, encoding all essential information in its DNA. It plays a pivotal role in orchestrating tissue- and stimulus-specific gene regulatory networks. These networks, in turn, govern both our fundamental cellular processes and our intricate responses to environmental challenges. It is hardly surprising, then, that both scientists and the general public are captivated by the way a fixed DNA sequence can direct the vast array of functional behaviors within our bodies, and how these encoded instructions are executed at the cellular level. Central to these regulatory mechanisms are DNA-binding factors, which serve two main functions: translating the DNA sequence into specific biological actions or recording the cellular history of an individual cell through various means, such as histone marks, DNA compaction, and spatial positioning of nucleosomes.

Our comprehension of the functional genome has been profoundly advanced by the development of Chromatin Immunoprecipitation (ChIP), as well as next-generation sequencing (NGS) techniques [1]. These technologies form the cornerstone of the methodologies detailed in this volume. ChIP, a method for identifying the interactions between proteins and DNA within living cells or tissues, becomes exponentially more powerful when combined with next-generation sequencing. This combination, known as ChIP-seq, facilitates the genome-wide mapping of transcription factors, co-regulators, histone modifications, and nucleosomes [2].

Since its development by DS Gilmour and JT Liz in 1984, initially employing UV cross-links to study RNA polymerase in *E. coli* [3], ChIP has revolutionized our understanding of numerous biological processes, including transcription, replication, and DNA repair, as well as the specifics of cell type differentiation in both health and disease contexts. Large-scale projects, such as the Encyclopedia of DNA Elements (ENCODE) initiative, have further enriched our knowledge by attributing biological functions to the human genome and supplying researchers with a wealth of high-quality data [4].

Moreover, in tandem with other genome-wide profiling techniques such as RNA-seq, ATAC-seq, and 3D genome interaction mapping, ChIP-seq has provided sufficient data for the application of machine learning algorithms that are now capable of unraveling gene regulatory networks and assessing the biological significance of non-coding genetic variants [5–8], underscoring the immense impact of ChIP and its related methodologies.

ChIP protocols and their derivatives are highly diverse, recapitulating the necessity to adjust the protocol based on tissue type and the specific factor under investigation, not just during the sample preparation phase but also throughout data analysis. This book aims to offer a glimpse into the myriad protocols and strategies used to experimentally and computationally (Chapters 4, 5, and 13) dissect gene regulation, whether through profiling DNA-binding proteins, exploring DNA footprinting, or studying histone modifications.

ChIP can be performed under native conditions (N-ChIP, Chapter 1) to map strong and abundant interactions with DNA [9], or via cross-linking (X-ChIP) to capture more transient associations [10]. Especially, two-step crosslinking procedures (Chapters 2 and 3) employing additional protein-protein cross-linkers allow us to profile non-DNA binding chromatin interactors and indirect interactions of transcription factors with DNA [11]. Each approach, from the choice of tissue or cross-linking conditions to the properties of the protein of interest, significantly influences the efficiency and specificity of ChIP [12, 13].

Beyond traditional ChIP-seq, we explore high-resolution techniques like ChIP-exo [14] and its derivative ChIP-nexus [15], designed for mapping protein-DNA interactions at base pair resolution (Chapter 6). The advent of single-cell technologies has highlighted the variability in gene expression and DNA accessibility across cellular populations and within tissues, necessitating the development of methods like Cleavage Under Targets and Release Using Nuclease (CUT&RUN, [16], Chapters 9, 10, and 11) and Cleavage Under Targets and TAGmentation (CUT&TAG, [17], Chapter 12). These methods allow for the profiling of DNA-binding proteins with significantly reduced input material and enhance the signal-to-noise ratios, reducing sequencing costs and avoiding biases inherent in traditional ChIP. Those and other developments enable researchers to aim for single-cell resolution (Chapter 14) [18].

As we delve into these topics, this book hopes to serve as a valuable resource on the evolving field of chromatin immunoprecipitation techniques, marked by rapid advancements and exciting developments. I extend my deepest gratitude to all the contributing authors for their dedication to refining these techniques and sharing their expertise with the next generation of genome biologists.

Freising, Germany *Franziska Greulich*

References

1. Mardis ER (2008) Next-generation DNA sequencing methods. Annu Rev Genomics Hum Genet 9:387–402
2. Mardis ER (2007) ChIP-seq: welcome to the new frontier. Nat Methods 4(8):613–614
3. Gilmour DS, Lis JT (1984) Detecting protein-DNA interactions in vivo: distribution of RNA polymerase on specific bacterial genes. Proc Natl Acad Sci U S A 81(14):4275–4279
4. Consortium EP, Dunham I, Kundaje A, et al (2012) An integrated encyclopedia of DNA elements in the human genome. Nature 489(7414):57–74
5. Avsec Z, Weilert M, Shrikumar A, et al (2021) Base-resolution models of transcription-factor binding reveal soft motif syntax. Nat Genet 53(3):354–366
6. Klie A, Laub D, Talwar JV, et al (2023) Predictive analyses of regulatory sequences with EUGENe. Nat Comput Sci 3(11):946–956
7. Zhou J, Troyanskaya OG (2015) Predicting effects of noncoding variants with deep learning-based sequence model. Nat Methods 12(10):931–934
8. Avsec Z, Agarwal V, Visentin D, et al (2021) Effective gene expression prediction from sequence by integrating long-range interactions. Nat Methods 18(10):1196–1203
9. Kasinathan S, Orsi GA, Zentner GE, et al (2014) High-resolution mapping of transcription factor binding sites on native chromatin. Nat Methods 11(2):203–209
10. Schmiedeberg L, Skene P, Deaton A, et al (2009) A temporal threshold for formaldehyde crosslinking and fixation. PLoS One 4(2):e4636
11. Nowak DE, Tian B, Brasier AR (2005) Two-step cross-linking method for identification of NF-kappaB gene network by chromatin immunoprecipitation. Biotechniques 39(5):715–725
12. Teytelman L, Thurtle DM, Rine J, et al (2013) Highly expressed loci are vulnerable to misleading ChIP localization of multiple unrelated proteins. Proc Natl Acad Sci U S A 110(46):18602–18607
13. Gavrilov A, Razin SV, Cavalli G (2015) In vivo formaldehyde cross-linking: it is time for black box analysis. Brief Funct Genomics 14(2):163–165

14. Rhee HS, Pugh BF (2012) ChIP-exo method for identifying genomic location of DNA-binding proteins with near-single-nucleotide accuracy. Curr Protoc Mol Biol Chapter 21:Unit 21.24
15. He Q, Johnston J, Zeitlinger J (2015) ChIP-nexus enables improved detection of in vivo transcription factor binding footprints. Nat Biotechnol 33(4):395–401
16. Skene PJ, Henikoff S (2017) An efficient targeted nuclease strategy for high-resolution mapping of DNA binding sites. Elife 6:e21856
17. Kaya-Okur HS, Wu SJ, Codomo CA, et al (2019) CUT&Tag for efficient epigenomic profiling of small samples and single cells. Nat Commun 10(1):1930
18. Bartosovic M, Kabbe M, Castelo-Branco G (2021) Single-cell CUT&Tag profiles histone modifications and transcription factors in complex tissues. Nat Biotechnol 39(7):825–835

Contents

Preface . . . . *v*
Contributors . . . . *xi*

1 Native ChIP: Studying the Genome-Wide Distribution of Histone Modifications in Cells and Tissue 1
Sandra Nitsch and Robert Schneider

2 Glucocorticoid Receptor-Dependent Binding Analysis Using Chromatin Immunoprecipitation and Quantitative Polymerase Chain Reaction 17
Denis Nalbantoglu, Jonathan M. Preuss, and Sabine Vettorazzi

3 Chromatin Immunoprecipitation in Adipose Tissue and Adipocytes: How to Proceed and Optimize the Protocol for Transcription Factor DNA Binding 35
Céline Jouffe, Kenneth A. Dyar, and Nina Henriette Uhlenhaut

4 Bioinformatics Core Workflow for ChIP-Seq Data Analysis 47
Tamas Schauer

5 Differential Analysis of Protein-DNA Binding Using ChIP-Seq Data 63
Christiane Boeckel, Xavier Pastor, Matthias Heinig, and Thomas Walzthoeni

6 An Optimized High-Resolution Mapping Method for Glucocorticoid Receptor-DNA Binding in Mouse Primary Macrophages 91
Suhail A. Ansari and Nina Henriette Uhlenhaut

7 Sequential ChIP-Seq 109
Joseph D. Schinderle and Irina M. Bochkis

8 Reverse Chromatin Immunoprecipitation (R-ChIP) 123
Xuejing Wen and Yucheng Wang

9 Estrogen Receptor Chromatin Profiling by CUT&RUN 133
Bruno Gegenhuber and Jessica Tollkuhn

10 Cleavage Under Targets and Release Using Nuclease (CUT&RUN) in Macrophages 151
Antonia Babl and Franziska Greulich

11 Cleavage Under Targets and Release Using Nuclease (CUT&RUN) of Epigenetic Regulators 169
Andrew D. McCray and Xiaofeng Wang

12 CUT&Tag for Efficient Epigenomic Profiling of Frozen Tissues 181
Qi Yin, Yang Li, and Yuxin Yin

13 Integrative Analysis of CUT&Tag and RNA-Seq Data Through Bioinformatics: A Unified Workflow for Enhanced Insights 191
Niccolò Liorni, Alessandro Napoli, Marta Adinolfi, Manlio Vinciguerra, and Tommaso Mazza

14 Single-Cell Histone Modification Profiling with Cell Enrichment Using sortChIC 215
Helena Viñas Gaza, Vivek Bhardwaj, and Peter Zeller

15 XL-DNase-Seq: Footprinting Analysis of Dynamic Transcription Factors 243
Kyu-Seon Oh, Mohammad Aqdas, and Myong-Hee Sung

16 ChEC-Seq: A Comprehensive Guide for Scalable and Cost-Efficient Genome-Wide Profiling in *Saccharomyces cerevisiae*........ 263
Tamar Gera, Divya Krishna Kumar, Gilad Yaakov, Naama Barkai, and Felix Jonas

Index *285*

Contributors

MARTA ADINOLFI • *Bioinformatics Unit, IRCCS Casa Sollievo della Sofferenza, San Giovanni Rotondo, Italy*

SUHAIL A. ANSARI • *Institute for Diabetes and Endocrinology (IDE), Helmholtz Zentrum Muenchen, Neuherberg, Germany; German Center for Diabetes Research (DZD), Neuherberg, Germany*

MOHAMMAD AQDAS • *Laboratory of Molecular Biology and Immunology, National Institute on Aging, National Institutes of Health, Baltimore, MD, USA*

ANTONIA BABL • *Metabolic Programming, TUM School of Life Sciences, Technical University of Munich, Freising, Germany*

NAAMA BARKAI • *Department of Molecular Genetics, Weizmann Institute, Rehovot, Israel*

VIVEK BHARDWAJ • *Hubrecht Institute-KNAW (Royal Netherlands Academy of Arts and Sciences), Oncode Institute, Utrecht, The Netherlands; University Medical Center Utrecht, Utrecht, The Netherlands*

IRINA M. BOCHKIS • *Department of Pathology, University of Pittsburgh School of Medicine, Pittsburgh, PA, USA; Pittsburgh Liver Research Center, University of Pittsburgh School of Medicine, Pittsburgh, PA, USA*

CHRISTIANE BOECKEL • *Core Facility Genomics, Helmholtz Zentrum München Deutsches Forschungszentrum für Gesundheit und Umwelt (GmbH), Neuherberg, Germany*

KENNETH A. DYAR • *Institute for Diabetes and Cancer, Helmholtz Zentrum Muenchen, Neuherberg, Germany; German Center for Diabetes Research (DZD), Neuherberg, Germany*

HELENA VIÑAS GAZA • *Hubrecht Institute-KNAW (Royal Netherlands Academy of Arts and Sciences), Oncode Institute, Utrecht, The Netherlands; University Medical Center Utrecht, Utrecht, The Netherlands*

BRUNO GEGENHUBER • *Cold Spring Harbor Laboratory, Cold Spring Harbor, NY, USA; School of Biological Sciences, Cold Spring Harbor Laboratory, Cold Spring Harbor, NY, USA; Department of Neurobiology, Harvard Medical School, Boston, MA, USA*

TAMAR GERA • *Department of Molecular Genetics, Weizmann Institute, Rehovot, Israel*

FRANZISKA GREULICH • *Metabolic Programming, TUM School of Life Sciences, Technical University of Munich, Freising, Germany*

MATTHIAS HEINIG • *Computational Health Center, Helmholtz Zentrum München Deutsches Forschungszentrum für Gesundheit und Umwelt (GmbH), Neuherberg, Germany; Department of Computer Science, TUM School of Computation, Information and Technology, Technical University of Munich, Garching, Germany; DZHK (German Centre for Cardiovascular Research), Partner Site Munich Heart Alliance, Munich, Germany*

FELIX JONAS • *School of Science, Constructor University, Bremen, Germany*

CÉLINE JOUFFE • *Institute for Diabetes and Cancer, Helmholtz Zentrum Muenchen, Neuherberg, Germany; Institute for Diabetes and Endocrinology, Helmholtz Zentrum Muenchen, Neuherberg, Germany*

DIVYA KRISHNA KUMAR • *Department of Molecular Genetics, Weizmann Institute, Rehovot, Israel*

YANG LI • *Institute of Systems Biomedicine, Department of Pathology, School of Basic Medical Sciences, Peking University Health Science Center, Beijing, China*

NICCOLÒ LIORNI • *Bioinformatics Unit, IRCCS Casa Sollievo della Sofferenza, San Giovanni Rotondo, Italy*
TOMMASO MAZZA • *Bioinformatics Unit, IRCCS Casa Sollievo della Sofferenza, San Giovanni Rotondo, Italy*
ANDREW D. MCCRAY • *Department of Molecular & Systems Biology, Dartmouth College, Lebanon, NH, USA; Dartmouth Cancer Center, Dartmouth College, Lebanon, NH, USA*
DENIS NALBANTOGLU • *Institute of Comparative Molecular Endocrinology, Ulm University, Ulm, Germany*
ALESSANDRO NAPOLI • *Bioinformatics Unit, IRCCS Casa Sollievo della Sofferenza, San Giovanni Rotondo, Italy*
SANDRA NITSCH • *Institute of Functional Epigenetics (IFE), Helmholtz Zentrum München, Neuherberg, Germany*
KYU-SEON OH • *Laboratory of Molecular Biology and Immunology, National Institute on Aging, National Institutes of Health, Baltimore, MD, USA*
XAVIER PASTOR • *Core Facility Genomics, Helmholtz Zentrum München Deutsches Forschungszentrum für Gesundheit und Umwelt (GmbH), Neuherberg, Germany*
JONATHAN M. PREUSS • *Institute of Comparative Molecular Endocrinology, Ulm University, Ulm, Germany*
TAMAS SCHAUER • *Institute of Epigenetics and Stem Cells, Helmholtz Munich, Munich, Germany*
JOSEPH D. SCHINDERLE • *Department of Pathology, University of Pittsburgh School of Medicine, Pittsburgh, PA, USA; Pittsburgh Liver Research Center, University of Pittsburgh School of Medicine, Pittsburgh, PA, USA*
ROBERT SCHNEIDER • *Institute of Functional Epigenetics (IFE), Helmholtz Zentrum München, Neuherberg, Germany; Faculty of Biology, Ludwigs-Maximilians Universität München, Planegg-Martinsried, Munich, Germany*
MYONG-HEE SUNG • *Laboratory of Molecular Biology and Immunology, National Institute on Aging, National Institutes of Health, Baltimore, MD, USA*
JESSICA TOLLKUHN • *Cold Spring Harbor Laboratory, Cold Spring Harbor, NY, USA*
NINA HENRIETTE UHLENHAUT • *Institute for Diabetes and Endocrinology (IDE), Helmholtz Zentrum Muenchen, Neuherberg, Germany; German Center for Diabetes Research (DZD), Neuherberg, Germany; Metabolic Programming, School of Life Sciences Weihenstephan, ZIEL – Institute for Food and Health, Technical University of Munich (TUM), Freising, Germany*
SABINE VETTORAZZI • *Institute of Comparative Molecular Endocrinology, Ulm University, Ulm, Germany*
MANLIO VINCIGUERRA • *Department of Translational Stem Cell Biology, Research Institute of the Medical University, Varna, Bulgaria; Faculty of Health, Liverpool John Moores University, Liverpool, UK*
THOMAS WALZTHOENI • *Core Facility Genomics, Helmholtz Zentrum München Deutsches Forschungszentrum für Gesundheit und Umwelt (GmbH), Neuherberg, Germany*
XIAOFENG WANG • *Department of Molecular & Systems Biology, Dartmouth College, Lebanon, NH, USA; Dartmouth Cancer Center, Dartmouth College, Lebanon, NH, USA*
YUCHENG WANG • *State Key Laboratory of Desert and Oasis Ecology, Key Laboratory of Ecological Safety and Sustainable Development in Arid Lands, Xinjiang Institute of Ecology and Geography, Chinese Academy of Sciences, Urumqi, China; Xinjiang Key Laboratory of Conservation and Utilization of Plant Gene Resources, Xinjiang Institute of*

Ecology and Geography, Chinese Academy of Sciences, Urumqi, China; College of Forestry, Shenyang Agricultural University, Shenyang, China

XUEJING WEN • *State Key Laboratory of Desert and Oasis Ecology, Key Laboratory of Ecological Safety and Sustainable Development in Arid Lands, Xinjiang Institute of Ecology and Geography, Chinese Academy of Sciences, Urumqi, China; Xinjiang Key Laboratory of Conservation and Utilization of Plant Gene Resources, Xinjiang Institute of Ecology and Geography, Chinese Academy of Sciences, Urumqi, China*

GILAD YAAKOV • *Department of Molecular Genetics, Weizmann Institute, Rehovot, Israel*

QI YIN • *Institute of Systems Biomedicine, Department of Pathology, School of Basic Medical Sciences, Peking University Health Science Center, Beijing, China; State Key Laboratory of Oral Diseases & National Center for Stomatology & National Clinical Research Center for Oral Diseases, West China Hospital of Stomatology, Sichuan University, Chengdu, Sichuan, China*

YUXIN YIN • *Institute of Systems Biomedicine, Department of Pathology, School of Basic Medical Sciences, Peking University Health Science Center, Beijing, China*

PETER ZELLER • *Hubrecht Institute-KNAW (Royal Netherlands Academy of Arts and Sciences), Oncode Institute, Utrecht, The Netherlands; University Medical Center Utrecht, Utrecht, The Netherlands*

Chapter 1

Native ChIP: Studying the Genome-Wide Distribution of Histone Modifications in Cells and Tissue

Sandra Nitsch and Robert Schneider

Abstract

For the genome-wide mapping of histone modifications, chromatin immunoprecipitation (ChIP) followed by high-throughput sequencing remains the benchmark method. While crosslinked ChIP can be used for all kinds of targets, native ChIP is predominantly used for strong and direct DNA interactors like histones and their modifications. Here we describe a native ChIP protocol that can be used for cells and tissue material.

Key words Chromatin immunoprecipitation (ChIP), ChIP-seq, Histone modifications, Native protein-DNA interactions

1 Introduction

Chromatin immunoprecipitation (ChIP) allows to quantitatively study the interaction between DNA-associated proteins and the genomic DNA. Both direct and indirect protein-DNA interactions as well as specific post-translational modifications of chromatin proteins can be investigated by ChIP [1, 2]. The origin of ChIP protocols dates back to 1984 when the Lis lab used UV irradiation to crosslink proteins to DNA in living bacterial cells and performed an IP of the bacterial RNA polymerase using specific antibodies [3]. Soon this crosslinked approach (X-ChIP) was advanced by Salomon et al. to study the distribution of histone H4 at heat shock genes using formaldehyde as a crosslinker, which is in contrast to UV light reversible [4]. In 1988, Hebbes et al. showed an association between histone acetylation and an active transcriptional state in vivo using a native ChIP (N-ChIP) approach without crosslinking [5]. While for X-ChIP chromatin is usually sheared by sonication or enzymatic digestion, during N-ChIP approaches chromatin is extracted in its native state and is fragmented enzymatically by micrococcal nuclease (MNase). Once the chromatin is

Franziska Greulich (ed.), *Chromatin Immunoprecipitation: Methods and Protocols*, Methods in Molecular Biology, vol. 2846, https://doi.org/10.1007/978-1-0716-4071-5_1,

fragmented, both approaches use immunoprecipitation with a specific antibody that recognizes the protein or modification of interest [6]. The use of native chromatin can improve epitope accessibility by avoiding, for example, masking of the epitope due to crosslinking and thus can increase IP efficiency. N-ChIP is particularly well suited for targets that are strong DNA interactors, such as histones because crosslinking is not required to stabilize their binding to DNA [7]. New approaches, such as CUT&RUN (Cleavage Under Targets and Release Using Nuclease) and CUT&Tag (Cleavage Under Targets and Tagmentation), have recently emerged as alternatives to ChIP-seq [8, 9]. These techniques require often lower input material and a reduced sequencing depth. However, more benchmarking and validation studies have to be performed for CUT&RUN and CUT&Tag, and the choice of method depends on the type of modification or protein studied [*see* 10 for a recent comparison]. Because of a broad range of validated antibodies and the availability of reference datasets, ChIP-seq remains currently the gold standard.

In this protocol, we describe an N-ChIP approach that is optimized for histone modifications but can be easily adapted to non-histone proteins like RNA polymerase II and transcription factors [11]. We discuss sample preparation using tissue culture cells as well as tissue as input material and provide an alternative procedure to extract pure mono-nucleosomes as input material that increases resolution and allows for Re-ChIP, a sequential ChIP approach using a second primary antibody. The purified DNA from this protocol can be quantified via quantitative PCR but can also be subjected to further processing to create libraries for ChIP sequencing to study genome-wide distribution patterns. Genome-wide analysis of targets of interest enables a closer examination of local regulation of transcription or can help to study the effect of environmental factors or diseases on chromatin [12–14].

2 Materials

2.1 Nuclei Preparation

1. 1 M KCl: Dissolve 74.55 g of Potassium chloride (KCl) in 900 mL of de-ionized H_2O. Adjust the volume to 1 L with H_2O and sterilize by autoclaving. The solution can be stored at 19–24 °C for years.
2. 5 M NaCl: Dissolve 292 g of Sodium chloride (NaCl) in 800 mL of de-ionized H_2O. Adjust the volume to 1 L with de-ionized H_2O and sterilize by autoclaving. Store the solution at 19–24 °C for years.
3. 1 M $MgCl_2$: Dissolve 203.3 g of Magnesium chloride ($MgCl_2 \cdot 6H_2O$) in 800 mL de-ionized H_2O. Adjust the volume to 1 L with de-ionized H_2O and sterilize by autoclaving. Store the solution at 19–24 °C for years.

4. 0.5 mM EGTA pH 8.0: Dissolve 19 g of Ethylene glycol-bis (2-aminoethylether)-N,N,N′,N′-tetraacetic acid (EGTA) in 90 mL H_2O. Adjust the pH with NaOH pellets (~4 g) to 8.0. Adjust the volume to 100 mL with de-ionized H_2O and sterilize by autoclaving. Store the 0.5 mM EGTA solution at 19–24 °C for years.
5. 0.5 mM EDTA pH 8.0: Dissolve 18.61 g Ethylenediaminetetraacetic acid (EDTA) disodium salt dehydrate in 80 mL de-ionized H_2O. Adjust the pH with NaOH pellets (~2 g) to 8.0. Adjust the volume to 100 mL with de-ionized H_2O and sterilize by autoclaving. Store the solution at 19–24 °C for years.
6. 1 M Tris–HCl pH 7.5: Dissolve 121.14 g Tris (hydroxymethyl)aminomethane (Tris base) in 700 mL de-ionized H_2O. Adjust pH to 7.5 with the appropriate volume of concentrated Hydrochloric acid (HCl) (~50 mL). Bring final volume to 1 L with de-ionized H_2O. Autoclave and store solution at 19–24 °C for years.
7. 1 M DTT: Dissolve 1.5 g of Dithiothreitol (DTT) in 8 mL of de-ionized H_2O. Adjust the total volume to 10 mL, dispense into 1 mL aliquots, and store them in the dark at −20 °C for up to 1 year. Do not autoclave DTT or solutions containing it. Only add to buffers just before use.
8. 100 mM AEBSF: Dissolve 0.9592 g of 4-(2-Aminoethyl)-benzenesulfonylfluoride hydrochloride (AEBSF·HCl) in 30 mL de-ionized H_2O. Fill to a final volume of 40 mL with de-ionized H_2O. Mix until completely dissolved. Use immediately or store in small aliquots at −20 °C for months.
9. 1 M NaBu: Dissolve 5.5 g Sodium butyrate (NaBu)in 40 mL de-ionized H_2O. Bring the final volume to 50 mL. Keep 1 mL aliquots at −20 °C for up to 1 year.
10. 10% (v/v) IGEPAL® CA-630 in de-ionized H_2O. Store at 4 °C for years.
11. 1 M Tris–HCl pH 8.0: Dissolve 121.14 g Tris (hydroxymethyl)aminomethane (Tris base) in 700 mL de-ionized H_2O. Adjust pH to 8.0 with the appropriate volume of concentrated Hydrochloric acid (HCl) (~42 mL). Bring final volume to 1 L with de-ionized H_2O. Autoclave and store solution at 19–24 °C for years.
12. 10% (v/v) Triton X-100in de-ionized H_2O. Store at 4 °C for years.
13. Micrococcal Nuclease Buffer: 50% (v/v) glycerol, 10 mM Tris–HCl pH 7.5, 50 mM NaCl.

14. Buffer I: 0.3 M sucrose in 60 mM KCl, 15 mM NaCl, 5 mM $MgCl_2$, 0.1 mM EGTA, 15 mM Tris–HCl pH 7.5. Add 0.5 mM DTT, 0.1 mM AEBSF, 5 mM NaBu, and 1× Protease inhibitor mix prior to starting (*see* **Note 1**).
15. Buffer II: 0.3 M sucrose in 60 mM KCl, 15 mM NaCl, 5 mM $MgCl_2$, 0.1 mM EGTA, 15 mM Tris–HCl pH 7.5. Add 0.4% (v/v) IGEPAL® CA-630, 0.5 mM DTT, 0.1 mM AEBSF, 5 mM NaBu, and 1× Protease inhibitor mix prior to starting.
16. Buffer III: 1.2 M sucrose in 60 mM KCl, 15 mM NaCl, 5 mM $MgCl_2$, 0.1 mM EGTA, 15 mM Tris–HCl pH 7.5. Add 0.5 mM DTT, 0.1 mM AEBSF, 5 mM NaBu, and 1× Protease inhibitor mix prior to starting.
17. MNase Digestion Buffer: 85 mM NaCl, 0.16 M sucrose, 10 mM Tris–HCl pH 7.5, 3 mM $MgCl_2$, 1.5 mM $CaCl_2$. Add 1 mM DTT, 0.1 mM AEBSF, and 5 mM NaBu prior to start.
18. Stop Buffer: 215 mM NaCl, 0.16 M sucrose, 10 mM Tris–HCl pH 8.0, 20 mM EDTA, 2% (v/v) Triton X-100. Add 0.5 mM DTT, 0.1 mM AEBSF, 5 mM NaBu, and 2× Protease inhibitor mix prior to starting.
19. Tubing Preparation Solution I: 2% (w/v) sodium bicarbonate ($NaHCO_3$), 1 mM EDTA pH 8.0.
20. Tubing Preparation Solution II: 1 mM EDTA pH 8.0.
21. Dialysis Buffer: 1 mM Tris–HCl pH 7.5, 0.2 mM EDTA. Add 0.1 mM AEBSF and 5 mM NaBu before starting the dialysis.
22. 1× PBS.
23. RNase A (DNase and protease-free, 10 mg/mL).
24. Micrococcal Nuclease (MNase) (*see* **Note 2**).
25. Centrifuge (cooled) for 1.5 mL tubes.
26. Centrifuge with swing-out rotor for 14 mL tubes (up to 10,000× *g*, cooled).
27. Dounce tissue homogenizer.
28. 14 mL polypropylene tubes (suitable for 10,000× *g* spins).
29. Dialysis tubing (3.5 kDa cutoff).
30. Magnetic stirrer and magnetic stirrer bars.
31. Thermo block for 1.5 mL tubes.
32. Nanodrop.
33. Water bath.
34. Parafilm.
35. Pasteur pipette.

2.2 Checking DNA Fragment Size and Chromatin Concentration

1. 1× TBE: 90 mM Tris base, 90 mM boric acid, 2 mM EDTA. Store at 19–24 °C. Prepare fresh once it starts to precipitate.
2. 6× DNA loading buffer Orange G: 0.4% (w/v), Orange G, 1.5% (w/v) Ficoll® 400. Stocks can be kept at −20 °C for years. The working aliquot can be stored at 19–24 °C.
3. Proteinase K (PCR grade, 20 mg/mL).
4. PCR purification kit (column range > 100 bp, for example, QIAquick PCR Purification Kit).
5. Phenol:Chloroform:Isoamyl Alcohol (25:24:1, v/v).
6. Bioanalyzer.
7. 2100 Bioanalyzer High Sensitivity (HS) DNA Kit.
8. Agarose gel electrophoresis systems.
9. Agarose (suitable for smaller DNA fragments).
10. Nucleic acid stain (e.g., ethidium bromide or SYBR® Safe).
11. UV lamp.
12. Qubit Fluorimeter.
13. Qubit dsDNA HS Assay Kit.
14. Nuclease-free H_2O (de-ionized as well as ultra-pure DNase/RNase Free).

2.3 Binding of Antibody to Agarose Beads

1. Blocking Buffer: 1× PBS, 0.5% (w/v) Bovine serum albumin (BSA).
2. Dynabeads Protein A or G.
3. Antibodies (*see* **Note 3**).
4. Magnetic separation rack for 0.2 and 1.5 mL tubes.
5. Rotating wheel.

2.4 Chromatin Immunoprecipitation

1. 20% (w/v) sodium dodecyl sulfate (SDS).
2. 10% (w/v) Sodium deoxycholate (DOC) in de-ionized H_2O. Must be protected from light and can be stored at 19–24 °C. Prepare a fresh stock solution once it starts to precipitate.
3. Wash Buffer: 150 mM NaCl, 10 mM Tris–HCl pH 8.0, 2 mM EDTA, 1% (v/v) IGEPAL® CA-630, 1% (v/v) Sodium deoxycholate. Add 0.5 mM DTT, 0.1 mM AEBSF, 5 mM NaBu, and 1× Protease inhibitor mix prior to starting.
4. 1× TE: 10 mM 1 M Tris–HCl pH 8.0, 1 mM EDTA. Store at 19–24 °C for years.
5. 0.5 M $NaHCO_3$: Dissolve 4.2 g $NaHCO_3$ in 90 mL de-ionized H_2O. Bring the final volume to 100 mL and store the solution at 19–24 °C for years.

6. Elution Buffer: 0.1 M Sodium hydrogen carbonate ($NaHCO_3$), 1% (w/v) SDS.
7. 1 M Tris–HCl pH 6.8: Dissolve 181.65 g Tris base in 600 mL de-ionized H_2O. Adjust pH to 6.8 with the appropriate volume of concentrated HCl (~120 mL). Bring final volume to 1 L with de-ionized H_2O. Autoclave and store solution at 19–24 °C for years.
8. 1.5/ 2 mL reaction tubes and DNA low-binding tubes
9. PCR machine with heated lid.

2.5 Quantification by qPCR

1. qPCR-related reagents (e.g., SYBR Green qPCR Master Mix) and primers (*see* **Note 4**).
2. qPCR machine.

2.6 Library Preparation for Sequencing

1. 0.1× TE: from 1× TE buffer in H_2O. Store at 19–24 °C for years.
2. 80% (v/v) ethanol in ultra-pure H_2O. Always prepare fresh.
3. 8-well 0.2 mL PCR tubes.
4. NEBNext® Ultra™ II DNA Library Prep Kit for Illumina®.
5. Absolute ethanol.
6. AMPure XP DNA purification beads.
7. Illumina sequencing platform.

3 Methods

The following steps need to be carried out on ice and with ice-cold buffers to which inhibitors should be added freshly if not otherwise indicated. Precool centrifuges and set the temperature on the heat blocks.

3.1 Nuclei Preparations from Cells

As a starting point, we recommend using one million cells per immunoprecipitation (IP). We successfully used this protocol for mouse embryonic stem cells and human HEK293T cells. Using different cell types and fewer cells is also possible but should be tested. Depending on the number of IPs, the number of cells, as well as the volume of buffers and units of enzymes, has to be scaled.

1. Harvest one million cells and pellet them at 500× *g* for 3 min at 4 °C.
2. Wash cells twice in 200 μL PBS and centrifuge each time at 500× *g* for 3 min at 4 °C.
3. After the second wash, resuspend cells in 100 μL Buffer I. Add 100 μL of Buffer II, mix by rotational tip movement, and incubate for 3 min on ice. Centrifuge for 3 min at 1000× *g*, 4 °C.

4. Resuspend nuclei in 200 μL MNase Digestion Buffer and centrifuge for 3 min at 2000× *g* at 4 °C. Resuspend nuclear pellet in 60 μL MNase Digestion Buffer.
5. Add 1 μL of 10 mg/mL RNase A and incubate for 5 min at 19–24 °C.
6. Add 4 U MNase and mix by pipetting. Incubate for 10 min at 20 °C in a heat block. Mix by pipetting after 5 min (*see* **Note 5**).
7. After 10 min, add 60 μL (equal volume) Stop Buffer. Adjust the volume to 300 μL by adding 180 μL of 1:1 MNase Digestion Buffer:Stop Buffer incubate samples 12–16 h on ice in a 4 °C room or a Thermo block set to 4 °C. No mixing is required.

3.2 Optional: Purification of Mono-nucleosomes from Cells

While Subheading 3.1 allows generating mono-, di-, and tri-nucleosomes as input for the IP, the only accurate method for purifying mono-nucleosomes is through a sucrose gradient. This increases the resolution of the ChIP assays and allows to study the co-occurrence of histone modifications on the same nucleosome (Re-ChIP or sequential ChIP).

1. Resuspend the cells in 4 mL of ice-cold Buffer I and divide them into two 14 mL polypropylene tubes.
2. Add 2 mL of ice-cold Buffer II in each tube, mix gently, and place on ice for 10 min.
3. Fill two new 14 mL polypropylene tubes with 8 mL of ice-cold Buffer III each. This buffer contains a higher sucrose concentration and acts as a sucrose cushion. Carefully layer 2 mL of each cell suspension on each sucrose cushion and cover the tubes with a piece of Parafilm.
4. Centrifuge in a prechilled swing-out rotor at 10,000× *g* for 20 min at 4 °C. During this centrifugation step, the nuclei will form a white pellet on the bottom of the tube, whereas the cytoplasmic components will remain in the top layer.
5. Carefully remove the supernatant with a Pasteur pipette in three steps and change the pipette for each step to avoid any contact between the solution on top and the pellet.
6. Resuspend the nuclei pellet into 1 mL of MNase Digestion Buffer and keep it on ice.
7. For the MNase fractionation, resuspend the purified nuclei in 1 mL of ice-cold MNase Digestion Buffer. Aliquot the suspension in two 1.5 mL tubes (500 μL of resuspended nuclei in each tube).
8. Add 1 μL of MNase (4 U/μL) to each tube and mix gently. Incubate the two tubes in a 37 °C water bath for appropriate times (*see* **Note 5**). Add EDTA pH 8.0 to a final concentration of 5 mM to stop the reaction.

9. Let the samples chill on ice and centrifuge for 10 min at 10,000× *g*, at 4 °C to pellet the nuclei.
10. Transfer the supernatant (soluble fraction of chromatin S1) into another 1.5 mL tube and store at 4 °C. Carefully resuspend the nuclei pellet in 500 μL of dialysis buffer to obtain the second chromatin fraction (S2).
11. Close one side of the washed dialysis tubing and transfer 500 μL of the resuspended nuclei pellet (S2) into the dialysis tube (*see* **Note 6**) and close the other side with a second clamp.
12. Submerge the tube for 12–16 h in 1 L of dialysis buffer. Dialysis is performed at 4 °C, in a beaker with constant mild stirring using a magnetic stirrer.
13. Transfer the dialyzed nuclei into a 1.5 mL tube and centrifuge for 10 min at 10,000× *g*, at 4 °C.
14. Transfer the supernatant to a new 1.5 mL tube and store at 4 ° C. This second soluble chromatin fraction S2 comprises larger fragments of chromatin (mainly between two and 6 nucleosomes) that were removed from the nuclei during the dialysis.
15. Resuspend the pellet, which is the chromatin fraction P, in 250 μL of dialysis buffer and store at 4 °C.
16. Take the Optical Density (OD) of each fraction at 260 nm using a Nanodrop.
17. Add 0.5 μg of the S1, S2, and P fraction to 10 μL 0.5% SDS and mix gently.
18. Follow Subheading 3.4, **step 4** in the protocol. The S1 fraction should consist of mainly mono-nucleosomes. The S2 fraction should contain larger fragments of chromatin (mostly between two and 6 nucleosomes), while chromatin fragments that are longer than 5–6 nucleosomes in length should be detected in the P fraction.

3.3 Nuclei Preparations from Tissue

We usually use around 50 mg of frozen mouse liver, which is sufficient for 6 IPs (*see* **Note 7**). Depending on the tissue type, this has to be optimized. It is possible to use a different amount of starting material for which the volume of buffers and enzymes has to be adjusted accordingly. All the following steps have to be performed on ice if not otherwise mentioned.

1. Add 4 mL of Buffer I to a pre-cooled 5 mL tissue homogenizer and add ~50 mg of frozen tissue.
2. Disrupt the tissue with ~15 strokes while keeping the homogenizer on ice. Transfer the cell suspension into two 2 mL reaction tubes and spin for 10 min at 6000× *g* at 4 °C.
3. Remove the supernatant and resuspend the cell pellet in 500 μL Buffer I.

4. Add 500 μL Buffer II and gently mix the cell suspension by pipetting. Incubate for 3 min on ice and centrifuge for 3 min at 3000× *g* at 4 °C.
5. Remove the supernatant and wash the pellet with 1 mL MNase Digestion Buffer. Centrifuge again for 3 min at 3000× *g* at 4 °C.
6. Remove the supernatant and resuspend the nuclei pellet in 400 μL MNase Digestion Buffer. Add 2 μL of 10 mg/mL RNase A, mix, and incubate for 5 min at 19–24 °C.
7. Add 20 U MNase and mix by rotational tip movement. Incubate for 15 min at 20 °C and mix again by rotational tip movement after 7 min incubation (*see* **Note 5**).
8. After 15 min, add 400 μL Stop Buffer. Adjust the volume to 1.8 mL by adding 1 mL of 1:1 MNase Digestion Buffer:Stop Buffer and incubate samples 12–16 h on ice in a 4 °C room or in a Thermo block set to 4 °C.

3.4 Checking DNA Fragment Size and Chromatin Concentration

1. The next day, centrifuge your sample at max speed for 10 min at 4 °C. The supernatant contains the released chromatin, while the pellet contains nuclear debris. Make sure to only transfer the supernatant (~90%) into a new precooled tube on ice and determine the exact volume to calculate the actual chromatin amount.
2. Transfer 20 μL released chromatin to a new tube and add 2 μL Proteinase K (stock at 20 mg/mL).
3. Incubate for 60 min at 55 °C and purify via PCR purification columns (range > 100 bp). Elute in 15 μL nuclease-free water and measure the DNA concentration using Qubit quantification or a Nanodrop to calculate the chromatin input per IP (*see* **Note 8**). Alternatively, purification via phenol/chloroform and precipitation can be used.
4. Add 3 μL 6× DNA loading buffer Orange G to the remaining eluate and run the samples on a 1.3% agarose gel in 1× TBE electrophoresis buffer. Stain the gel after the run in water containing 0.1 μg/mL ethidium bromide for 30 min. Wash with water for 15 min and visualize the chromatin fragments under a UV lamp. Alternatively, the gel can be prestained with other nucleic acid gel stains like SYBR Safe or 1 μL of the sample can also be directly loaded on a Bioanalyzer HS chip. A clear enrichment for mono-nucleosome over residual di- and tri-nucleosomes is optimal (Fig. 1a).
5. Per IP, use at least 5 μg of chromatin diluted in a minimum of 300 μL of 1:1 MNase Digestion Buffer:Stop Buffer. Depending on the number of IPs, add 1:1 MNase Digestion Buffer: Stop Buffer to the prepared chromatin. Keep 5% of chromatin as input on ice until Subheading 3.6, **step** 7.

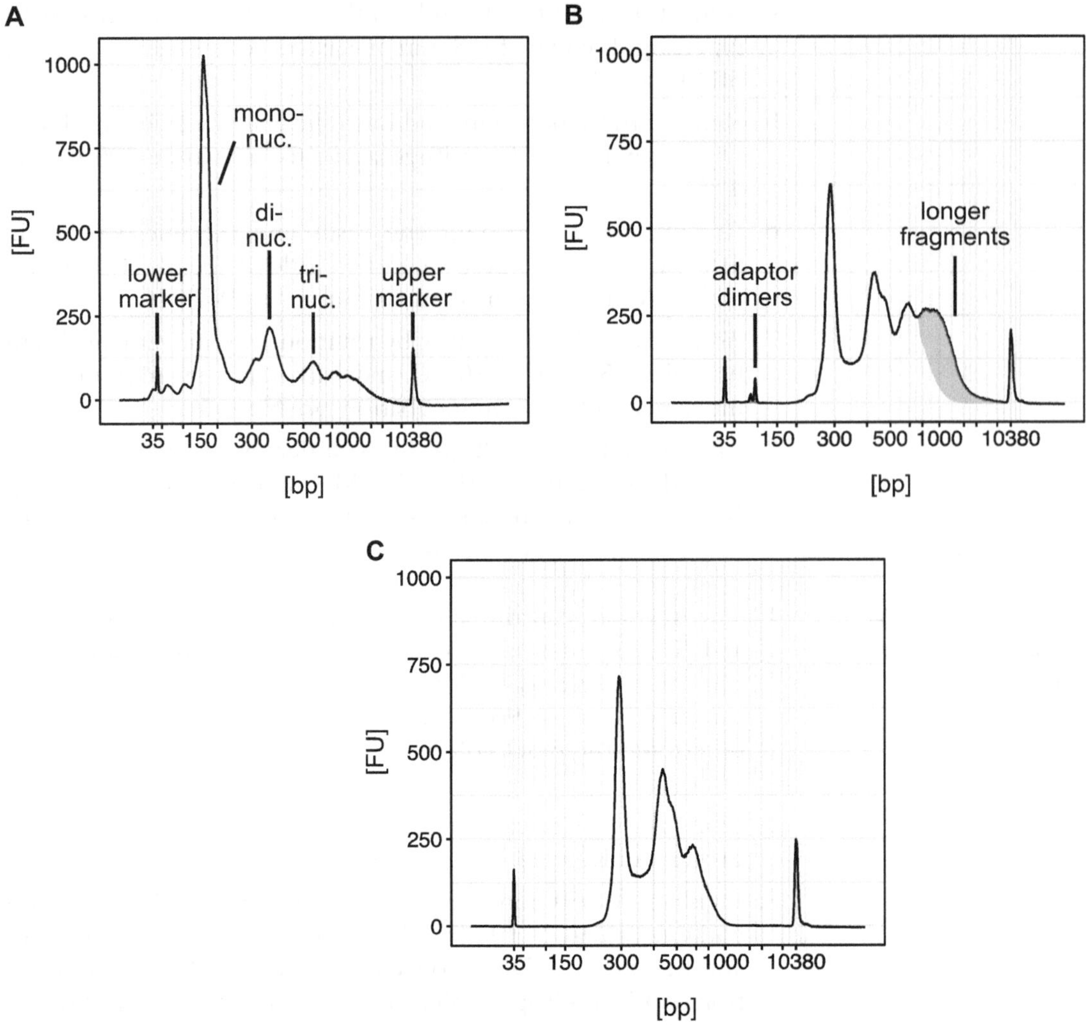

Fig. 1 N-ChIP input material and NGS library preparation. Shown are bioanalyzer electropherograms of MNase-digested chromatin derived from mouse liver before and after the library preparation using the same chromatin. Samples were analyzed on a bioanalyzer HS DNA chip. In all plots, the fluorescence unit (FU) is plotted on the y-axis, while the x-axis is the fragment length in base pairs (bp). (**a**) A successful MNase digestion results in a clear enrichment of mono-nucleosomes (~150 bp) over residual di-nucleosome (~300 bp) and tri-nucleosome (~500 bp) fractions. The lower marker (first peak at 35 bp) and the upper marker (last peak at 10,380 bp) are included in all runs as internal standards. (**b**) The digested chromatin was used for an IP with an antibody against H3K9ac (ab4441), and the recovered DNA was used for NGS library preparation. Primers and adaptors were ligated to the DNA fragments, therefore increasing their length (+120 bp). After the library preparation, adaptor dimers (~120 bp) and longer fragments (>900 bp) were detected. A double-size selection (0.9–0.54 (left-right) ratio) was performed to remove fragment sizes that are highlighted in grey (adaptor dimers and longer fragments). (**c**) After the double-size selection, no more adaptor dimers and longer fragments were detected, and the library was ready for sequencing

3.5 Binding of Antibody to Agarose Beads

1. Add 10 μL magnetic ProteinA/G beads (*see* **Note 9**) per IP to 500 μL Blocking Buffer. Mix via tube inversion and insert tubes after a quick spin into a magnetic rack. Remove the buffer without interfering with the beads and wash them with 500 μL Blocking Buffer.
2. Resuspend the beads in 600 μL of Blocking Buffer. Add an appropriate amount of antibody (*see* **Note 10**) and incubate for 2 h on a rotating wheel at 4 °C.
3. Wash antibody-bound beads once in 600 μL Blocking Buffer and resuspend them in 10 μL Blocking Buffer.

3.6 Chromatin Immunoprecipitation

1. To antibody-bound ProteinA/G beads in DNA low-binding tubes, add your chromatin solution and incubate for at least 3 h and up to 16 h on a rotating wheel at 4 °C.
2. Remove the supernatant and wash four times with 1 mL Wash Buffer (*see* **Note 11**). For the first wash, mix the beads with the Wash Buffer for 1 min and quickly spin down the tubes before placing them into the magnetic rack.
3. For the next 3 washes, rotate the slurry for 5 min at 4 °C. Quickly spin the tubes down, place them in a magnetic rack for 1 min, and withdraw the supernatant without taking any beads. After two washes, transfer the mixture to a new DNA low-binding tube.
4. Finally, add 1 mL of 1× TE at 19–24 °C to the beads and mix via tube inversions. Remove the TE buffer as above.
5. Add 100 μL of preheated (37 °C) Elution Buffer to the washed beads. Vortex briefly and incubate for 15 min at 37 °C shaking (1000 rpm).
6. Add 7 μL 1 M Tris–HCl pH 6.8 to the slurry to adjust the pH to approximately 8.0 and mix.
7. Add 20 μg of Proteinase K (stock at 20 mg/mL) to each sample and to the 5% input that has been kept on ice. Due to the small sample volume, transfer all samples into 8-well PCR strips and incubate for 12–16 h at 55 °C in a PCR machine with a heated lid.
8. The next day, quickly spin the samples down before placing them into a magnetic rack and transfer the IP elutants into new DNA low-binding tubes.
9. Purify inputs and IPs using PCR purification columns (range > 100 bp) according to the manufacturer's protocol and elute with 20 μL of nuclease-free water. Leave for 1 min and centrifuge for 1 min at max speed.
10. Measure the DNA concentration using Qubit quantification (*see* **Note 12**).
11. The DNA can be stored at −20 °C until further use or can be directly used for qPCR or library preparation.

3.7 Quantification of DNA by qPCR

For one PCR use 12.5 μL 2× SYBR Green qPCR Master Mix, 0.5 μL 10 μM forward and reverse primer, 0.5–1 μL template, and 10.5 μL nuclease-free water (*see* **Note 13**). We recommend at least three replicates for the inputs diluted 1:100 and three for the IP samples and to check genomic regions with expected enrichments (positive control) as well as regions with no expected enrichment (negative control). If the samples are used for next-generation sequencing, use as little material as possible or decrease the volume of the PCRs.

3.8 Library Preparation for Sequencing

We follow the protocol of NEBNext Ultra II DNA Library preparation for Illumina sequencing [15]. The manufacturer recommends a starting amount of 500 pg to 1 μg fragmented DNA in 0.1× TE. We usually prepare libraries starting with 5–10 ng DNA. Adjust the volume for all samples to 50 μL with 0.1× TE and then transfer them to 8-well PCR stripes before starting the library preparation. Depending on the input material, different adaptor dilutions for the adaptor ligation should be used. Since the input usually does not exceed 50 ng, the clean-up of adaptor-ligated DNA can be performed without size selection. For the PCR enrichment of adaptor-ligated DNA, the cycle number has to be adjusted for different input amounts (at least 3 cycles and up to 11 cycles). After finishing the library preparation, quantify the DNA using 1 μL via the Qubit dsDNA HS kit. Perform quality control by loading an appropriate amount of DNA into a High Sensitivity DNA chip and run the chip in a Bioanalyzer. The libraries should resemble the profile from your starting material but take the linker adaptor lengths plus primers into account (Fig. 1b). We usually perform a double-size selection with AMPure XP beads using a 0.9–0.54 (left-right) ratio which allows recovering DNA fragments between 175 bp and 900 bp (Fig. 1c and *see* **Note 14**).

1. Bring all samples to a total volume of 50 μL using 0.1× TE.
2. Extensively mix the AMPure beads before use and add 27 μL of slurry to your samples (0.9× ratio). Mix by pipetting at least 10 times and incubate for 1 min at 19–24 °C.
3. Quickly spin the tubes down, place them in a magnetic rack, and allow the beads to completely settle.
4. Transfer 74 μL clear supernatant to a new PCR tube and add 18.73 μL AMPure beads (0.54× ratio). Mix again by pipetting at least 10 times and incubate for 1 min at 19–24 °C.
5. Quickly spin the tubes down, place them in a magnetic rack, and allow the beads to completely settle.
6. Remove the supernatant and add 200 μL of freshly prepared 80% ethanol. Incubate for 30 s at 19–24 °C and remove the supernatant and all visible liquid.

7. Air dry the beads on the magnetic stand with the lid open for up to 5 min at 19–24 °C.
8. Remove the tubes from the magnetic stand and elute in 23 μL 0.1× TE. Mix by pipetting at least 10 times and incubate for 2 min at 19–24 °C.
9. Quickly spin the tubes down and place them in a magnetic rack. Allow the beads to completely settle to the magnet and transfer 20 μL of the clear supernatant to a new tube.
10. Load 1 μL of the purified library into a high sensitivity DNA chip and run the chip in a Bioanalyzer to check if the size selection worked (Fig. 1c).

4 Notes

1. Chemicals and enzymes should be DNase/RNase-free. All stock buffers should be filtered using a 0.2 μm pore size filter and can then be stored at 4 °C. The Elution Buffer should be stored at 19–24 °C. The Blocking Buffer should be stored at – 20 °C after filtering. DTT and inhibitors should only be added on the day of the ChIP, and all buffers should be kept on ice if not mentioned otherwise.
2. MNase activity can vary between different batches and different suppliers. We usually prepare 4 U/μL working stocks in Micrococcal Nuclease Buffer.
3. The efficiency and accuracy of any ChIP experiment are largely depending on the specificity of the antibody used. We highly recommend using ChIP-validated antibodies that are site- and modification-specific, for example, anti-H3K9ac antibody (ab4441, Abcam) or anti-H3K27ac antibody (ab4729, Abcam). We also recommend checking the information provided by the antibody distributing companies. Often the quality varies from batch to batch. As negative controls, rabbit IgG or mouse IgG antibodies can be used to measure nonspecific binding. For antibodies against histone modifications, peptide immunoblots using differently modified histone peptides (chemically synthesized) can be used to determine the site and modification specificity.
4. Determine the genomic regions in which your protein of interest or your preferred histone modification might be located via available databases (e.g., ENCODE) or published datasets. Select your genetic region and generate possible primers using primer designer tools like Primer-BLAST or Primer3. The PCR product size should be between 70 and 100 bp. As a negative control, choose genomic regions that are expected to be devoid of your protein of interest or your preferred histone modification.

5. Always prepare a test titration with every new MNase dilution prepared or when using a new batch. Test different units (e.g., 4, 8, 12, and 16 U) and various incubation times. During optimization, also the incubation temperature can be adjusted. For tissue material, starting units for MNase are higher. Therefore, test, for example, 15, 20, 25, and 30 U when using ~50 mg of tissue.
6. For the preparation of the dialysis tubing (3.5 kDa cutoff), cut the tubing into pieces of convenient length (10–20 cm). Boil the tubes for 10 min in 0.5 L of tubing preparation solution I and rinse the tubes twice in de-ionized H_2O. Boil the tubes for 10 min in 0.5 L of tubing preparation solution II. Allow the tubes to cool down. Tubes can be stored in tubing preparation solution II at 4 °C for years. Before use, wash the tubing twice inside and out with de-ionized H_2O.
7. For other tissues, for example, brain, heart, or kidney, we recommend verifying an efficient tissue homogenization by examining a small aliquot of cell suspension on a phase contrast microscope. We also recommend checking nuclei integrity after the nuclei preparation via DAPI staining. In case adipose tissue or muscle material is used, the starting material should be adjusted due to the lower cell density in these tissues. We usually prepare chromatin from snap-frozen mouse livers. We keep frozen livers at −80 °C for years.
8. Five to 50 μg chromatin should be used per IP. Depending on the target investigated and the quality of the antibodies used, a trial ChIP-qPCR should be performed to determine if the input material is sufficient.
9. Use Protein A beads for rabbit antibodies and Protein G beads for rat or mouse antibodies. Instead of magnetic beads, sepharose beads can also be used.
10. For ChIP-validated antibodies, use the recommended amount per IP. As controls, we use 1 μg of an anti-H3K9ac antibody (ab4441, Abcam) and 2 μg of an anti-H3K27ac antibody (ab4729, Abcam). If no recommendation for the antibody of choice is given, add between 1 and 5 μg for a trial ChIP and confirm the IP by qPCR.
11. If the background is too high, the NaCl concentration can be increased to 300 mM, SDS to up to 0.1%, NP-40 to 1%, and DOC to 0.5%. LiCl can also be used instead of NaCl. The use of various salts can help to improve the removal of nonspecific chromatin interactions with beads.
12. For the Qubit measurement, usually 1 μL of sample is enough. If the concentration is too low to be measured, the preparation of a library for NGS might still work. Use the maximum recommended number of PCR cycles during the PCR enrichment of adaptor-ligated DNA.

Table 1
qPCR program

Steps	Temperature °C	Duration	Number of cycle
Initial denaturation	95	10 min	1
Denaturation	95	15 s	45–50
Annealing	60	20 s	
Elongation	72	20 s	

13. The PCR program depicted in Table 1 may need to be adjusted depending on the qPCR master mix used and the PCR machine.

14. If adaptor dimers (~120 bp) or a significant amount of long fragments (>900 bp) are present in the profiles, a double-size selection is recommended. More information is given in the SPRIselect User Guide.

Acknowledgments

We would like to thank Pradeepa Madapura and Wendy Bickmore for sharing their protocols with us. We thank Lara Zorro Shahidian, Sylvain Daujat, and former and present Schneider lab members for discussions and their protocols. Work in the R.S. laboratory was supported by the DFG through SFB 1064 (Project-ID 213249687), SFB 1309 (Project-ID 325871075), and the Helmholtz Gemeinschaft.

References

1. Park PJ (2009) ChIP–seq: advantages and challenges of a maturing technology. Nat Rev Genet 10:669–680
2. Marinov GK (2018) A decade of ChIP-seq. Brief Funct Genomics 17:77–79
3. Gilmour DS, Lis JT (1984) Detecting protein-DNA interactions in vivo: distribution of RNA polymerase on specific bacterial genes. Proc Natl Acad Sci 81:4275–4279
4. Solomon MJ, Varshavsky A (1985) Formaldehyde-mediated DNA-protein cross-linking: a probe for in vivo chromatin structures. Proc Natl Acad Sci 82:6470–6474
5. Hebbes TR, Thorne AW, Crane-Robinson C (1988) A direct link between core histone acetylation and transcriptionally active chromatin. EMBO J 7:1395–1402
6. O'Neill L (2003) Immunoprecipitation of native chromatin: NChIP. Methods 31:76–82
7. Kasinathan S, Orsi GA, Zentner GE et al (2014) High-resolution mapping of transcription factor binding sites on native chromatin. Nat Methods 11:203–209
8. Skene PJ, Henikoff JG, Henikoff S (2018) Targeted in situ genome-wide profiling with high efficiency for low cell numbers. Nat Protoc 13:1006–1019
9. Kaya-Okur HS, Wu SJ, Codomo CA et al (2019) CUT&Tag for efficient epigenomic profiling of small samples and single cells. Nat Commun 10:1930
10. Hu D, Abbasova L, Schilder BM, et al (2023) CUT&Tag recovers up to half of ENCODE ChIP-seq peaks in modifications of H3K27. bioRxiv 2022.03.30.486382

11. Liu Z, Naler LB, Zhu Y et al (2022) nMOWChIP-seq: low-input genome-wide mapping of non-histone targets. NAR Genom Bioinform 4(2):lqac030
12. Barski A, Cuddapah S, Cui K et al (2007) High-resolution profiling of histone methylations in the human genome. Cell 129:823–837
13. Kidder BL, Hu G, Zhao K (2011) ChIP-Seq: technical considerations for obtaining high-quality data. Nat Immunol 12:918–922
14. Benabdallah NS, Williamson I, Illingworth RS et al (2019) Decreased enhancer-promoter proximity accompanying enhancer activation. Mol Cell 76:473–484.e7
15. Gautreau I (2018) Ultra II DNA Library Prep Kit for Illumina E7645/E7103 V.1. protocols.io. https://doi.org/10.17504/protocols.io.mc5c2y6

Chapter 2

Glucocorticoid Receptor-Dependent Binding Analysis Using Chromatin Immunoprecipitation and Quantitative Polymerase Chain Reaction

Denis Nalbantoglu, Jonathan M. Preuss, and Sabine Vettorazzi

Abstract

ChIP-qPCR offers the opportunity to identify interactions of DNA-binding proteins such as transcription factors and their respective DNA binding sites. Thereby, transcription factors can interfere with gene expression, resulting in up- or downregulation of their target genes. Utilizing ChIP, it is possible to identify specific DNA binding sites that are bound by the DNA-binding proteins in dependence on treatment or prevailing conditions. During ChIP, DNA-binding proteins are reversibly cross-linked to their DNA binding sites and the DNA itself is fragmented. Using bead-captured antibodies, the target proteins are isolated while still binding their respective DNA response element. Using quantitative PCR, these DNA fragments are amplified and quantified. In this protocol, DNA binding sites of the glucocorticoid receptor are identified by treatment with the synthetic glucocorticoid Dexamethasone in murine bone marrow-derived macrophages.

Key words Glucocorticoid receptor (GR), Chromatin immunoprecipitation (ChIP), Polymerase chain reaction (PCR), Dexamethasone, Co-factor, Protein–protein interaction, Macrophage, Immune system

1 Introduction

Chromatin immunoprecipitation (ChIP) provides a method to analyze specific interactions between DNA and DNA-binding proteins, such as transcription factors, nuclear receptors, cofactors, histones, and histone modifications. In a broader context, ChIP consists of five major steps: chromatin cross-linking, chromatin fragmentation, immunoprecipitation, DNA purification, and data analysis [1] (Fig. 1). During the process of cross-linking, DNA is reversibly connected to all proteins binding at that moment by using paraformaldehyde (PFA) as a cross-linking agent. For chromatin fragmentation, cells are first lysed and the cross-linked DNA-protein complexes are purified. To obtain soluble fragments,

Franziska Greulich (ed.), *Chromatin Immunoprecipitation: Methods and Protocols*, Methods in Molecular Biology, vol. 2846,
https://doi.org/10.1007/978-1-0716-4071-5_2,

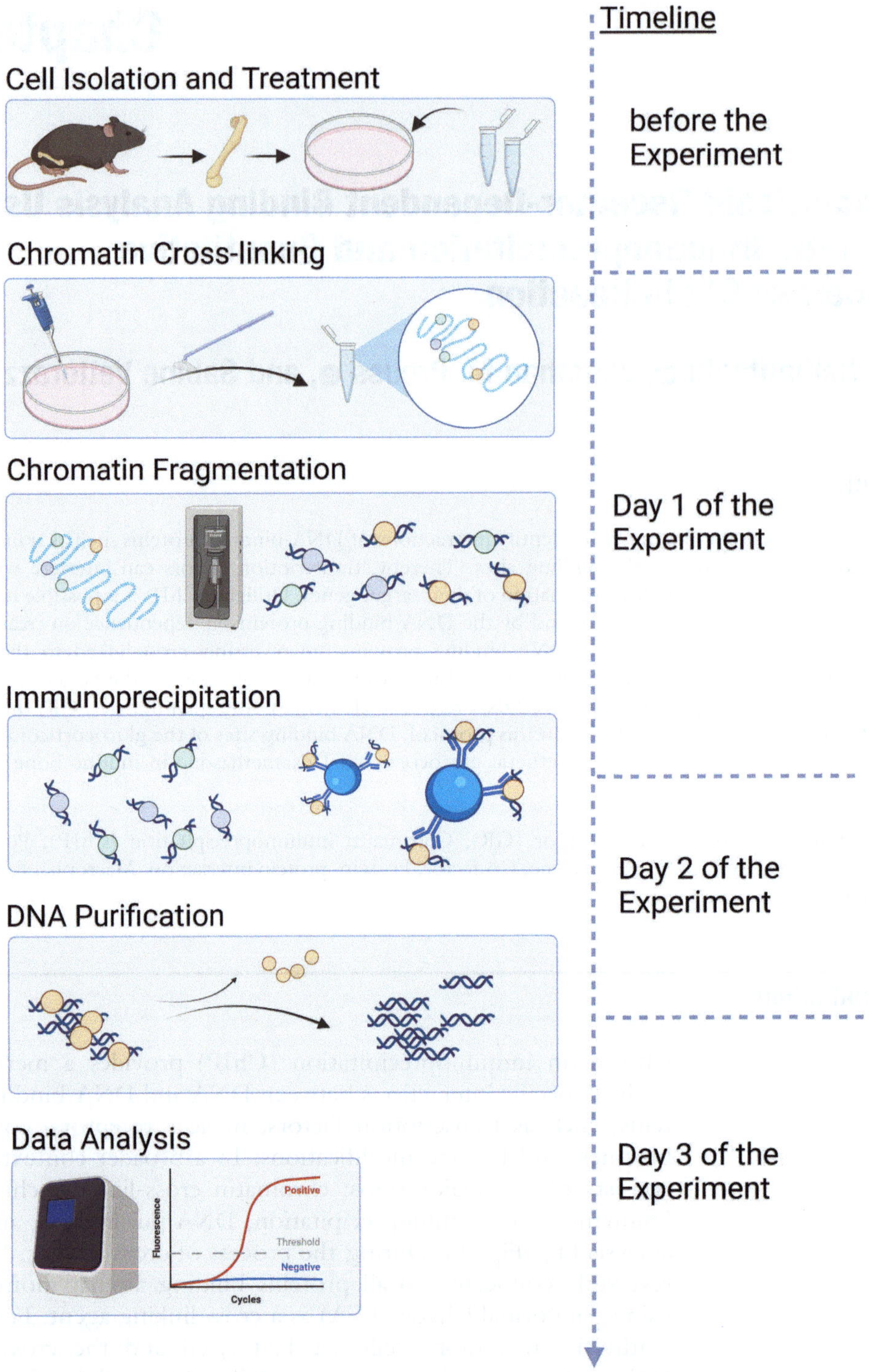

Fig. 1 Schematic overview of ChIP-qPCR. The major steps of the presented method consist of cell isolation and treatment, chromatin cross-linking, chromatin fragmentation, immunoprecipitation, DNA purification, and data analysis

the purified complexes are then sonicated, preferentially producing fragments ranging in length from 500 to 1000 base pairs (bp). During immunoprecipitation, fragments associated with the protein of interest are selected. The protein of interest is enriched by bead-immobilized antibodies. Subsequent washing removes the unbound fragments. After the elimination of unwanted fragments, the cross-linking between DNA and the protein of interest is reversed by heat incubation at a high salt concentration at 65 °C overnight (16–24 h) [1]. For downstream data analysis, this protocol utilizes quantitative polymerase chain reaction (qPCR), which facilitates the identification of genes that were previously bound to the targeted transcription factor.

In this protocol, ChIP-qPCR is used to analyze the binding of the glucocorticoid receptor (GR) in murine bone marrow-derived macrophages (mBMDM). The GR belongs to the superfamily of nuclear receptors [2, 3]. In its ligand-free state, the GR resides in the cytoplasm, bound to chaperon complexes, heat-shock proteins, and immunophilins in a sizable complex [4, 5]. Only upon binding to glucocorticoids (GCs), the GR performs conformational changes and translocates into the nucleus. There, it interacts both directly with the DNA and indirectly via other transcription factors to modulate gene transcription by transactivation or transrepression. As a ubiquitously expressed protein, the GR mediates a wide range of cellular and tissue-specific processes key to metabolism, development, and immune response [2]. Importantly, while endogenous GCs regulate the circadian rhythm and act as stress hormones, pharmacological GC-derivates like Dexamethasone (dex) are prominently used as anti-inflammatory drugs to combat chronic inflammatory disease. The method described here is tailored to identify DNA regions bound by the GR after dex treatment in mBMDMs. It can be surmised that this protocol also applies to other cell types or DNA-protein complexes; however, antibody concentration, incubation times, and sonication times may need to be adjusted.

2 Materials

2.1 Chromatin Cross-Linking

1. 1× PBS.
2. 16% (w/v) Paraformaldehyde (PFA).
3. 1 mM dexamethasone.
4. Fixation Buffer 1: 2 mM Disuccinimidyl glutarate (DSG). Dissolve 1 g DSG in 1.53 L of 1× PBS. Store at 4 °C for several weeks or at −20 °C for several months.
5. Fixation Buffer 2: 1% methanol-free Paraformaldehyde (PFA). Dilute 62.5 mL 16% (w/v) PFA in 937.5 mL PBS.

6. 1 M Glycine: Dissolve 7.51 g Glycine in 100 mL 1× PBS. To sterilize the solution, filter through a 0.22 μm filter. Store at 18–24 °C for a year.
7. Cooling centrifuge for 1.5/2 mL tubes.
8. Lab-grade scale.
9. Cell scraper.
10. Ø20 cm cell culture plates.
11. 1.5/2 mL tubes (*see* **Note 1**).

2.2 Chromatin Fragmentation

1. 10 mg/mL RNase A (DNase and protease-free).
2. 10 mg/mL Proteinase K: Dissolve 100 mg of Proteinase K (PCR-grade) in 100 mL TE-Buffer, aliquot, and store at −20 ° C Protease Inhibitor cocktail.
3. 0.5 M EDTA pH 7.5 or 8.0: Weigh in 18.61 g of Ethylenediaminetetraacetic acid (EDTA) disodium salt and dissolve it in 80 mL distilled H_2O. Adjust the pH with NaOH to pH 7.5 and pH 8.0, respectively. Fill up with distilled H_2O to a total volume of 100 mL and store at room temperature (18–22 °C).
4. 1 M Tris–HCl pH 7.5 or 8.0: Weigh in 12.11 g of Tris base and dissolve it in 80 mL distilled H_2O. Adjust the pH with HCl to pH 7.5 and pH 8.0, respectively. Fill up with distilled H_2O to a total volume of 100 mL and store at room temperature (18–22 °C).
5. 5 M NaCl: Dissolve 292 g of NaCl in distilled H_2O.
6. Fast IP: 167.5 mM NaCl, 5 mM EDTA pH 7.5, 50 mM Tris–HCl pH 7.5, 0.5% (v/v) IGEPAL® CA-630, 1% (v/v) Triton X-100. Add 33.3 mL of 5 M NaCl, 10 mL of 0.5 M EDTA pH 7.5, 50 mL of 1 M Tris–HCl pH 7.5, 5 mL of IGEPAL® CA-630, and 10 mL of Triton X-100. Fill to 1 L using desalt water. To sterilize the buffer, filter through a 0.22 μm filter. Store the buffer at 4 °C for up to 6 months.
7. Shearing Buffer: 1% (w/v) SDS, 10 mM EDTA pH 8.0, 50 mM Tris–HCl pH 8.0. Mix 50 mL of 20% (w/v) SDS, 20 mL of 0.5 M EDTA pH 8.0, and 50 mL of 1 M Tris–HCl pH 8.0. Add H_2O to an end volume of 1 L. To sterilize the buffer, filter through a 0.22 μm filter and store at room temperature (18–22 °C). The day the buffer is used, add a protease inhibitor tablet for every 10 mL of buffer needed.
8. Proteinase K Master mix 1: Add 1 μL of 10 mg/mL Proteinase K, 1 μL of 1 M Tris–HCl pH 8.0 and 0.5 μL of 0.5 M EDTA pH 8.0. Prepare the master mix fresh when needed.
9. TE Buffer: 1 mM EDTA pH 8.0, 10 mM Tris–HCl pH 8.0. Add 2 mL of 0.5 M EDTA pH 8.0 and 10 mL of 1 M Tris–HCl pH 8.0, and fill with H_2O to an end volume of 1 L.

10. 1 mL insulin syringes.
11. ChIP DNA Clean & Concentrator kit (e.g., from Zymo, *see* **Note 2**).
12. Agarose for DNA gels.
13. Agarose gel dye (e.g., SYBR Safe).
14. 100-bp DNA ladder.
15. Agarose-Gel Chamber.
16. Gel documentation system.
17. 86% (v/v) Glycerol as DNA loading dye.
18. 1.5 mL BioruptorPlus Diagenode tubes.
19. Bioruptor.
20. Heat block/shaker.

2.3 Immunoprecipitation (IP)

1. Dilution Buffer: 167 mM NaCl, 0.01% (w/v) SDS, 11% (v/v) Triton X-100, 1.2 mM EDTA pH 8.0, 16.7 mM Tris–HCl pH 8.0. Mix 33.4 mL of 5 M NaCl, 0.5 mL of 20% (w/v) SDS, 11 mL of Triton X-100, 2.4 mL of 0.5 M EDTA pH 8.0, and 16.7 mL of 1 M Tris–HCl pH 8.0. Add H_2O to an end volume of 1 L. To sterilize the buffer, filter through a 0.22 μm filter and store the buffer at 4 °C.
2. 5% (w/v) Bovine serum albumin (BSA).
3. Absolute Ethanol (Molecular Biology-grade).
4. Anti-GR antibody (Proteintech #24050-1-AP) (*see* **Note 3**).
5. Anti-IgG antibody (DA1E mAb IgG XP isotope control from Cell Signaling (#39000)) (*see* **Note 3**).
6. Protein A/G Magnetic Beads (*see* **Note 4**).
7. Invitrogen DynaBead Separator or comparable magnetic rack for 1.5 or 2 mL tubes.
8. Rotation wheel.
9. Suction Pump.

2.4 DNA Purification

1. Bead Elution Buffer: 1.05% (w/v) SDS, 0.88 M $NaHCO_3$. First, dilute 84 mg $NaHCO_3$ in 1 mL H_2O. Secondly, add 525 μL of this solution to 262.5 μL of 20% (w/v) SDS. Add H_2O to a total volume of 5 mL. It is important to prepare Bead Elution buffer fresh when you need it.
2. Proteinase K Master mix 2: Prepare 7 μL Master mix per sample. Mix 1 μL of 10 mg/mL Proteinase K, 4 μL of 1 M Tris–HCl pH 8.0, and 2 μL of 0.5 M EDTA pH 8.0. Prepare the master mix fresh when needed.

2.5 Quantitative PCR and Data Analysis

1. SYBR Green qPCR Master mix.
2. qPCR primers (*see* **Note 5**).
3. Nuclease-free water.
4. qPCR Thermocycler.

3 Methods

Before starting this protocol, note the following information:

1. All chemicals need to be handled carefully concerning potential toxicity and by trained professionals only. Both DSG and PFA are toxic and must be handled with appropriate caution. Subheading 3.1, **steps 1–4** should be performed under a fume hood, and respective aspirated liquids should be discarded in a special waste allocated for toxic substances.
2. The ChIP-qPCR protocol takes at least 3 days. An additional day might be needed in case of extensive data analysis. This does not include primer design, cell seeding, treatment, and harvest.
3. For this procedure, mBMDMs are isolated from murine bone marrow and differentiated for 5 days as described elsewhere [6]. After differentiation, mBMDMs are seeded in Ø20 cm plates. Seed one Ø20 cm plate with 20×10^6 cells for every biological replicate and every treatment planned for the experiment. For qPCR, at least three biological replicates are advised, yet we suggest using at least four biological replicates. For example, in an experiment that wants to compare two treatments (vehicle control and 0.1 μM dex), seed two Ø20 cm plates for each biological replicate. If cells from four different mice are used, 40×10^6 cells are required per biological replicate, resulting in a total of 160×10^6 seeded cells in eight Ø20 cm plates (*see* **Note 6**). The day before treatment, the mBMDM medium is changed to a charcoal-stripped mBMDM medium version to eliminate existing glucocorticoids present in the medium (*see* **Note 7**). To activate the GR and initiate translocation, mBMDMs are treated with 0.1 μM dexamethasone for 3 h.
4. To design primers suitable for the detection of GR target sequences, *see* **Note 5**.

3.1 Chromatin Cross-Linking

In this part of the protocol, DNA-binding proteins are reversibly cross-linked to the DNA, provided that the applied treatment conditions promote binding to their DNA-binding site. After cross-linking, the cells are harvested (Fig. 1).

Day 1 of the experiment

1. Aspirate medium from cells and wash once with 1× PBS.
2. Fix the cells by adding 10 mL Fixation Buffer 1 (*see* **Note 8**) to each plate and incubate them for 30 min at 4 °C (*see* **Note 9**). Aspirate Fixation Buffer 1 afterwards but refrain from washing the cells.
3. Start another fixation cycle by adding 10 mL Fixation Buffer 2 and incubate for 15 min at room temperature (18–22 °C) (*see* **Note 9**).
4. Proceed by adding 1.5 mL of 1 M Glycine and incubate for an additional 5 min at room temperature (18–22 °C) (*see* **Note 10**). To ensure proper PFA quenching, place the plates on a shaker (*see* **Note 11**).
5. Aspirate Fixation Buffer 2-Glycine and wash cells twice with ice-cold 1× PBS. Place the plates on ice.
6. On the ice, harvest the cells in the remaining PBS using a cell scraper. Transfer the content of each dish into an individual 2 mL tube (*see* **Note 1**) and centrifuge the cells for 5 min at 4 ° C and 400× *g* (*see* **Note 12**). Aspirate the supernatant and store pellets on ice (*see* **Note 13**).

3.2 Chromatin Fragmentation

As the next step, the harvested DNA is sonicated using a Bioruptor, breaking the chromatin into small fragments (Fig. 1). To verify the correct size of the resulting fragments, agarose gel electrophoresis is performed, ideally resulting in DNA bands of the size of 500–1000 bp.

1. Resuspend each pellet in 1 mL ice-cold Fast-IP buffer (*see* **Notes 14** and **15**) by pipetting up and down.
2. Incubate the resuspended pellets for 10 min on ice (*see* **Notes 16** and **17**).
3. During this incubation, pull the cells through a 1 mL insulin syringe once and centrifuge afterwards at 4 °C for 1 min at 12,000× *g*. Discard the supernatant and resuspend it in 1 mL ice-cold Fast-IP buffer. Repeat **steps 2** and **3** once.
4. After the second centrifugation, suspend the pellet in 1 mL Shearing buffer (*see* **Notes 16** and **17**).
5. Split the suspended chromatin for each treatment into two 1.5 mL BioruptorPlus Diagenode tubes, resulting in two technical replicates of each sample. Using Bioruptor, perform 12 shear cycles on "high" with 30 s of on/off periods. Store the samples at 4 °C in a fridge afterwards (*see* **Notes 17, 18,** and **19**).

Before proceeding with immunoprecipitation (IP), the base pair length of the sheared chromatin should be determined using gel electrophoresis. Ideally, the resulting length should be between 500 and 1000 base pairs.

6. Pool 10 μL of each technical replicate of a sample in a new tube (*see* **Note 1**) and add 80 μL TE Buffer and 5 μL 5 M NaCl. Put the samples on a heat block and incubate at 99 °C for 15 min shaking at 1000 rpm.
7. Cool the samples on ice until they reach room temperature (18–22 °C). When the samples have cooled down, add 1 μL of 10 mg/mL RNase. Incubate at 37 °C for 20 min at 300 rpm.
8. During incubation, prepare Proteinase K master mix 1 freshly. For each sample, 2.5 μL master mix is required.
9. Add 2.5 μL of the Proteinase K master mix 1 to every sample and incubate at 56 °C for 1 h at 300 rpm on a heat block.
10. Purify the chromatin using the ChIP DNA Clean & Concentrator kit according to the manufacturer protocol (*see* **Note 2**).
11. After purification, elute samples in 15 μL Elution Buffer (provided by the DNA extraction kit) and add 6 μL of 86% glycerol (*see* **Note 20**).
12. Run the samples on a 0.7% (w/v) agarose gel at 100 V (*see* **Note 21**). Use a 100 bp DNA ladder to cover the desired area of bp lengths.
13. To validate the quality of your purification, your bands should run at a length of 500–1000 bp (*see* **Note 22**) (Fig. 2).

3.3 Immunoprecipitation (IP)

Next, chromatin fragments that are bound to the DNA binding protein of interest are enriched by using bead-immobilized antibodies against the respective protein (Fig. 1). Therefore, antibodies are first linked to magnetic beads.

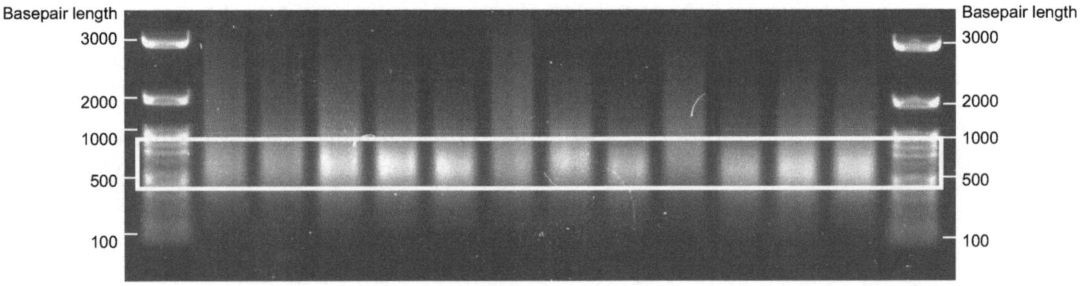

Fig. 2 Verification of successful chromatin fragmentation. Chromatin was sheared for 12 cycles on high settings with 30 s on/off periods on a Diagenode Bioruptor, reverse cross-linked, and loaded on a 0.7% agarose gel. An ideal fragment size is indicated by bands in the range of 500–1000 bp

1. Start the immunoprecipitation by centrifuging the samples from Subheading 3.2, **step 5** at 4 °C for 10 min at 14,000× *g*. This results in a clear supernatant, containing the sheared chromatin, and cell debris containing pellet.
2. Pool the supernatant containing the cleared chromatin of equal treatment in a fresh tube (*see* **Note 1**). To not disturb the pellet, transfer only 900 μL of the supernatant (90%).
3. Next, split the purified chromatin into two fractions. One fraction will be used as input and frozen away, and the other faction will be used as IP faction further.
4. To create the input fraction, transfer 50 μL of the clear chromatin in a fresh tube (*see* **Note 1**). This amounts to 5.5% of the total input.
5. Add 300 μL of absolute ethanol to each input and freeze them at −20 °C overnight (16–24 h).
6. For the IP fraction, dilute chromatin 1:10 with Dilution Buffer. Prepare two tubes per sample (*see* **Note 1**). One tube will be loaded with an antibody binding to the protein of interest (in this example, the GR), and the other tube will be loaded with an IgG control (*see* **Note 3**). Add 1350 μL Dilution Buffer to DNA low-binding tubes and add 1.5 μg of the respective antibody (*see* **Note 23**).
7. Add 150 μL of the clear chromatin per IP to the prepared tubes containing the Dilution Buffer-antibody mixture (*see* **Note 24**). Incubate on a rotation wheel for 12–16 h at 4 °C.
8. Aliquot protein A/G magnetic beads into fresh tubes. For every ChIP sample, 20 μL protein A/G magnetic beads are required in a separate tube (*see* **Notes 4** and **25**). The 20 μL of added magnetic beads results in 1.5% compared to the total volume of the added IP fraction later in **step 14**. In the example described here, a total of 320 μL protein A/G magnetic beads are used, divided into 16 tubes (4 biological replicates × 2 treatments × 2 antibodies (IgG and GR)).
9. Begin bead preparation by washing them twice with 1× PBS. To discard PBS after washing, place the tubes into a magnetic rack to concentrate beads on the tube wall. Remove liquid afterwards (*see* **Note 26**).
10. Dilute the 5% BSA stock in 1× PBS to a final concentration of 0.5% and wash the beads once. To discard BSA after washing, place the tubes into a magnetic rack to concentrate beads on the tube wall. Remove liquid afterwards (*see* **Note 26**).
11. After discarding the supernatant, add 1 mL of 0.5% BSA and block the beads for 12–16 h on a rotation wheel at 4 °C.

Day 2 of the Experiment

12. The next day, start with centrifuging the ChIP samples at 4 °C for 10 min at 14,000× *g*.
13. During that time, take the blocked beads and wash them with 1 mL Dilution Buffer once and suspend the beads in 1 mL Dilution Buffer (*see* **Note 15**). To discard PBS after washing, place the tubes into a magnetic rack to concentrate beads on the tube wall. Remove the liquid afterwards (*see* **Note 26**).
14. For every sample, pipet 20 μL of the beads into a new tube (*see* **Note 1**). At this point, the centrifugation of the ChIP samples should be done. Transfer 90% of the cleared ChIP samples to the beads (*see* **Note 27**) and incubate the samples at 4 °C for 3 h on a rotation wheel.
15. After incubation, collect the magnetic beads with a magnetic rack and aspirate the supernatant carefully, not disturbing the pellet (*see* **Note 26**).
16. Wash the beads by suspending them in 1 mL ice-cold Fast IP Buffer (*see* **Notes 15** and **26**). Pipet up and down carefully to not introduce air bubbles. Incubate the ChIP samples at 4 °C for 3 min on a rotation wheel.
17. Repeat the **steps 15** and **16** for a total of 5 wash steps. At the end, aspirate the supernatant and do not add more Fast IP Buffer.
18. During the wash steps, centrifuge the frozen input samples at 4 °C for 10 min at 21,000× *g*.
19. Aspirate the supernatant of the input samples and suspend the inputs in 300 μL of 70% ethanol. Centrifuge the samples at room temperature (18–22 °C) for 5 min at 21,000× *g*.
20. Aspirate the supernatant of the input samples and let the pellets dry in open tubes on a heat block at 56 °C for 5 min at 1000 rpm. During that time, freshly prepare the Bead Elution Buffer (*see* **Note 28**).

3.4 DNA Purification

To release the bead-bound chromatin fragments, samples are treated with a Bead Elution Buffer. Additionally, reverse cross-linking is performed to detach the DNA fragments from their associated DNA-binding proteins.

1. After the wash steps and drying of the input, add 50 μL Bead Elution Buffer to both the beads from Subheading 3.3, **step 17** and the dried input samples from Subheading 3.3, **step 20**. Vortex the tubes and incubate at 18°–24 °C for 15 min at 1000 rpm.

2. After incubation, transfer the supernatant into new DNA low-binding tubes (*see* **Note 26**). Repeat **steps 1** and **2** once to end up with a total of 100 μL eluate.
3. For reverse crosslinking, add 4 μL of 5 M NaCl per 100 μL eluate and vortex quickly. Incubate the samples overnight (16–24 h) at 65 °C without shaking.

Day 3 of the Experiment

4. The next day, cool the samples on ice and add 1 μL of 10 mg/mL RNase. Vortex and incubate at 37 °C for 30 min at 300 rpm on a heat block.
5. During incubation, prepare the Proteinase K master mix freshly. For each sample, 7 μL master mix is required.
6. Add 7 μL of the Proteinase K master mix 2 to every sample and incubate at 55 °C for 1.5 h at 300 rpm on a heat block.
7. Purify the DNA using the ChIP DNA Clean & Concentrator kit according to the manufacturer's protocol (*see* **Note 2**).
8. Elute the purified DNA in 25 μL.
9. The eluted ChIP DNA can be stored at −20 °C (*see* **Note 29**).

3.5 Quantitative PCR and Data Analysis

In the final step, qPCR of the enriched, fragmented DNA is performed to analyze whether the respective protein binds a specific DNA element or not.

1. For qPCR, add 50 μL nuclease-free H_2O to the eluted samples, diluting it 1:3.
2. Take 7 μL from the input samples and add it to 77 μL nuclease-free H_2O, diluting it 1:12.
3. Further, pool 2 μL from all input samples and dilute it 1:6 with nuclease-free H_2O.
4. Create five standards from the pooled input samples by 1:10 dilution (*see* Table 1). Use nuclease-free H_2O for the dilutions.

Table 1
Standard dilution series

Name	Dilution
Standard 1	1:2 from the pooled inputs
Standard 2	1:10 from Standard 1
Standard 3	1:10 from Standard 2
Standard 4	1:10 from Standard 3
Standard 5	1:10 from Standard 4

Table 2
qPCR Protocol

Phase	Temperature	Time	Cycles
Initial denaturation	95 °C	10 min	1×
Denaturation	95 °C	15 s	45×
Annealing	60 °C	1 min	

Table 3
Exemplary data of a ChIP-qPCR for a GR enhancer

Sample	C_t mean of triplicates	Adjusted input	%Input
Vehicle control GR antibody	29.39		0.02
Dex GR antibody	22.55		1.51
Vehicle control IgG antibody	30.38		0.01
Dex IgG antibody	29.98		0.01
Vehicle control GR antibody input	20.28	16.70	
Dex GR antibody input	20.08	16.50	
Vehicle control IgG antibody input	19.80	16.21	
Dex IgG antibody input	20.08	16.50	

5. Pipet 4.5 μL of the diluted ChIP samples, input, standards, and nuclease-free H_2O (as a negative control) into PCR tubes. Work in triplicates.
6. Add the qPCR master mix to each tube. The qPCR master mix is composed of 5 μL SYBR Green PCR master mix (2×), forward primer (10 μM), and reverse primer (10 μM) (*see* **Notes 6, 30,** and **31**).
7. Perform the qPCR according to the protocol in Table 2.
8. Analyze the standard curve of the qPCR and verify that the reaction efficiency lies between 90% and 100%.
9. Compare differential bound regions by comparing the normalized Cycles of Threshold (C_t) of all biological replicates to both vehicle control treated samples and IgG controls. In the example described in this protocol, cells are either treated with 0.1 μM dex or a vehicle control. ChIP is performed with an anti-GR antibody or an anti-IgG control antibody. Table 3 shows C_t values for an exemplary experiment of a "GR target gene."

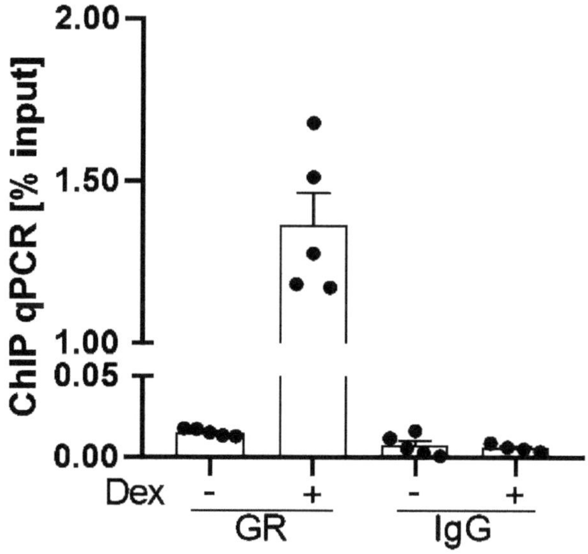

Fig. 3 ChIP qPCR for GR binding at its target gene enhancer region is shown as a percentage of input with/without 3 h 0.1 μM dex treatment for GR/IgG antibody used in ChIP

The adjusted input value is calculated using the following formula:

$$\text{adjusted input} = C_t[\text{input}] - log_2 df$$

df—dilution factor of the input.

Using these values, the percentage input (%Input) is determined according to this formula:

$$\%\text{Input} = 100 \times 2^{-(Ct \text{ mean of triplicates} - \text{adjusted input})}$$

10. In Fig. 3, these data points are blotted as an example. Normalized to 1% input, the amplification of an exemplary GR binding site is shown in the absence or presence of the GR-ligand dex. An anti-GR antibody or an anti-IgG antibody is used respectively.

4 Notes

1. Whenever possible, use DNA-low binding tubes to avoid chromatin loss.
2. Kits from other manufacturers work as well, as long as the size range covers 100–3000 bp for the purification.
3. Here, the anti-GR antibody from Proteintech (24050-1-AP) and the DA1E mAb IgG XP isotope control from Cell Signaling (#39000) were used. Generally, antibodies must be suited

for ChIP-qPCR. Here are some recommendations for finding a proper antibody:

(a) ChIP-validated antibodies can be bought from different manufacturers. However, it should not be assumed that these will work in 100% of cases.

(b) If the antibodies are also suitable for Western Blot, this method gives an option to verify epitope binding.

(c) Antibodies that are used for immunohistochemistry or immunofluorescence of PFA fixed samples over another pool of possible candidates to test.

(d) Polyclonal antibodies may have a higher chance of binding masked epitopes.

(e) Ensure antibody specificity by testing it in a knockout model, knockdown cell line or with blocking peptides.

4. In this protocol, magnetic protein A/G beads were used, yet other alternatives exist. Some of them are listed below:

   (a) Protein A conjugates: Protein A beads display the highest affinity against polyclonal antibodies of rabbit origin.

   (b) Protein G conjugates: In general, protein G beads exhibit affinity to a broader range of antibodies, especially most classes of mouse monoclonal IgGs.

   (c) Protein A/G conjugates: As a combination of the two previously mentioned conjugates, protein A/G beads provide high flexibility when it comes to choosing the used antibodies.

   (d) Secondary antibody conjugates: Instead of protein-binding conjugates, antibodies are bound to the beads. Anti-mouse beads that bind IgG1, IgG2a, and IgG2b regions and anti-rabbit beads binding all rabbit IgGs are available.

   (e) Magnetic beads: Magnetic beads offer easy handling as they are visible during workflow and show low unspecific binding. Special equipment in the form of a magnetic rack is required.

   (f) Agarose/sepharose beads: In contrast, agarose/sepharose beads are characterized by their high binding capacity of antibodies due to their porous design, which can also lead to increased nonspecific interactions. They come with an increased workload but don't require extra equipment. Instead of using a magnetic rack for washing steps, agarose/sepharose beads are washed by centrifuging them for 30 s at 300× *g*.

5. To design primers suitable for the detection of GR target sequences, we recommend using the UCSC genome browser (https://genome-euro.ucsc.edu). On the UCSC website, open

the genome browser and make sure to have JASPAR TFBS connect at My Data/Track hubs. Search for your gene of interest (e.g., *Tsc22d3* as a typical GR target gene) and open the JASPAR TFBS track on the lower part of the website. Here, filter for NR3C1 and hit "submit." On the genome browser website, the binding sites of NR3C1 should be displayed. If not, zoom out until binding sites are shown. Click on the NR3C1 binding site displayed above the DNA sequence and select "View DNA for this feature." Generally, choosing 300 base pairs up and downstream is recommended so no binding sites are missed. Hit "Get DNA" and copy the DNA sequence into the online primer design tool of NCBI (https://www.ncbi.nlm.nih.gov/tools/primer-blast/). Your target PCR product should have a length of 60–100 bp and a melting temperature between 50 and 65 °C. The ideal GC content lies between 50% and 60%. We recommend designing at least two primers per target. Choose "genomes for the selected organism" as the database and *Mus musculus* as the organism. If published primer sequences are shown, they are preferable.

6. For every biological replicate, at least 20×10^6 cells should be used. In case of multiple treatments per biological replicate, make sure to seed cells originating from the same stock. If primary cells or cells of limited quantity are used, one origin might not result in enough cells for all treatments. For example, one mouse usually results in roughly 20×10^6 cells, which corresponds to one treatment. To circumvent this bottleneck, pool mBMDMs from multiple mice before seeding to ensure a sufficient amount of cells. Make sure to mix the cells well before seeding to ensure an efficient and uniform dispersion.
7. If media with fetal calf or bovine serum (FCS, FBS) is used, it is important to deplete the cells from glucocorticoids present in the serum. This can be achieved by either starving the cells overnight (16–24 h) in BMDM Medium without any FCS/FBS or by using BMDM medium containing charcoal-stripped FCS/FBS. The latter is recommended, yet both versions avoid basal activation of the GR by FCS containing GCs, which would lead to a high background signal during expression analysis.
8. DSG functions as a protein-protein crosslinker that is required when DNA and proteins do not interact directly with each other. If only direct DNA-protein interactions are to be observed, start with Subheading 3.1, **step 3**.
9. The added volume may vary depending on the used plate size. It is advised to use enough buffer to cover the cells. Usually, half the medium volume that has been used to incubate the cells is sufficient.

10. Add 1.5 mL of 1 M Glycine per 10 mL PFA used.
11. Glycine is used to stop cross-linking. Prolonged crosslinking can reduce IP efficiency later, as antibody binding sites can be masked by excess PFA.
12. Store the tubes on ice until all cells are harvested.
13. The protocol can be paused at this point. Cell pellets can be stored at −80 °C for up to a year. If multiple experimental charges are combined, freeze all samples once at least overnight (16–24 h) to avoid variations.
14. Do not vortex your samples. The preferred method of mixing is pipetting.
15. Add protease inhibitors freshly to used buffers to avoid protein degradation. After adding protease inhibitors, the buffer may be used for a maximum of 1 week. If needed, store buffers at 4 °C.
16. Keep the Shearing Buffer cold, yet SDS must not precipitate. If the Shearing Buffer appears cloudy, warm the samples gently and vortex to solubilize.
17. Chromatin-containing samples should always be kept cold. Yet, it is preferred to use a 4 °C fridge rather than ice to avoid SDS precipitation.
18. The required cycle number can vary, mainly depending on the cell type and fixation. It is advised to perform a small-scale experiment beforehand using the desired cell type and apply different shear cycle numbers. To test proper conditions, a serial sonication procedure can be performed. Start the experiment as described in this protocol on a smaller scale until Subheading 3.2, **step 5**. Using the Bioruptor, test different shear conditions to find the optimal setup. For example, set the shear cycles on "high" and perform 2–4 cycles with 30 s of on/off periods. Aliquot 20 μL into a fresh 1.5 mL tube and repeat until you have performed a total of around 22 shear cycles. Perform agarose-gel electrophoresis with the collected samples as described here, starting at Subheading 3.2, **step 6**. Under optimal conditions, the agarose gel should show chromatin bands between 500–1000 base pairs (Fig. 2). If the chromatin is not sheared enough, major bands will appear above 1 kb, and "oversheared" DNA will be visible below 200 bp. If this does not result in satisfying results, the shear intensity and on/off periods might be altered as well.
19. To diminish the risk of technical variations later during data analysis, combine samples that are going to be compared in one sonication round. In this particular case, rather shear vehicle control and dex samples together instead of two sample sets of dex-treated samples.

20. Glycerol is very viscous. For better handling, use 200 μL filter tips instead of 10 μL filter tips. Additionally, cut off the end of the tips and draw glycerol up slowly. Prefer glycerol to DNA loading dye, as the contained dye might interfere with the fluorescence capacity of the DNA gel dye.
21. The timing strongly depends on the gel- and the resulting fragment size. It is advised to check the gel regularly.
22. Bands at a greater or lower length indicate insufficient or too vigorous shearing, respectively. The former can be solved by performing extra shear cycles, provided that samples are properly cooled. Yet, the risk remains that chromatin is lost at this point.
23. Depending on the antibody concentration, different volumes are needed. In the described case, 1.8 μL of the anti-GR antibody and 0.6 μL of the anti-IgG control were added per tube.
24. Up to six ChIP reactions (including IgG) control can be performed from 20 M cells.
25. 20 μL per sample is a suitable amount for the magnetic Protein A/G beads from Invitrogen. If other beads are used, the respective volume per sample may differ and has to be adjusted. Independent of the beads, it is recommended to add at least 10 μL extra to compensate for the pipetting-induced loss of beads.
26. If magnetic beads like magnetic Protein A/G beads from Invitrogen are used, washing steps are carried out using a magnetic rack. Proper adhesion of the beads to the magnetic rack takes time and is a crucial process to avoid bead loss. Therefore, only proceed with supernatant removal when the solution has become fully clear. For bead washing, remove the tubes from the magnetic rack and pipet up and down carefully and pay attention to not introduce air bubbles. Tubes with beads might be occasionally carefully vortexed in case some of the solution is stuck in the lid.
27. According to this protocol, you should have a total of 1500 μL ChIP sample, therefore, transfer 1350 μL.
28. Bead elution buffer has to be prepared freshly and cannot be stored.
29. Due to its low concentration, purified DNA is very unstable after ChIP, as it adsorbs stronger to the tube and fragments into smaller pieces [7, 8]. Therefore, the maximal advised storage time is 12–16 h at −20 °C.
30. These volumes account for only one tube. Multiply the numbers with the planned qPCRs and add 10 extra samples. These account for pipetting errors.

31. If it is planned to examine several target genes, additional master mixes have to be prepared, containing the respective primer pairs.

References

1. Greulich F, Mechtidou A, Horn T et al (2021) Protocol for using heterologous spike-ins to normalize for technical variation in chromatin immunoprecipitation. STAR Protoc 2(3): 100609
2. Vettorazzi S, Nalbantoglu D, Gebhardt JCM et al (2022) A guide to changing paradigms of glucocorticoid receptor function-a model system for genome regulation and physiology. FEBS J 289(19):5718–5743
3. Kumar R, Thompson EB (2005) Gene regulation by the glucocorticoid receptor: structure: function relationship. J Steroid Biochem Mol Biol 94(5):383–394
4. Pratt WB, Toft DO (2003) Regulation of signaling protein function and trafficking by the hsp90/hsp70-based chaperone machinery. Exp Biol Med (Maywood) 228(2):111–133
5. Escoter-Torres L, Caratti G, Mechtidou A et al (2019) Fighting the fire: mechanisms of inflammatory gene regulation by the glucocorticoid receptor. Front Immunol 10:1859
6. Wepler M, Preuss JM, Merz T et al (2019) Impaired glucocorticoid receptor dimerization aggravates LPS-induced circulatory and pulmonary dysfunction. Front Immunol 10:3152
7. Gaillard C, Strauss F (1998) Avoiding adsorption of DNA to polypropylene tubes and denaturation of short DNA fragments. Tech Tips Online 3(1):63–65
8. Zhong J, Ye Z, Lenz SW et al (2017) Purification of nanogram-range immunoprecipitated DNA in ChIP-seq application. BMC Genomics 18(1):985

Chapter 3

Chromatin Immunoprecipitation in Adipose Tissue and Adipocytes: How to Proceed and Optimize the Protocol for Transcription Factor DNA Binding

Céline Jouffe, Kenneth A. Dyar, and Nina Henriette Uhlenhaut

Abstract

Chromatin immunoprecipitation (ChIP) coupled to qPCR or sequencing is a crucial experiment to determine direct transcriptional regulation under the control of specific transcriptional factors or co-regulators at loci-specific or pan-genomic levels.

Here we provide a reliable method for processing ChIP from adipocytes or frozen adipose tissue collection, isolation of nuclei, cross-linking of protein-DNA complexes, chromatin shearing, immunoprecipitation, and DNA purification. We also discuss critical steps for optimizing the experiment to perform a successful ChIP in lipid-rich cells/tissues.

Key words ChIP, Adipocyte, Adipose tissue, Chromatin immunoprecipitation

1 Introduction

Transcription factors are proteins capable of direct binding to DNA. In association with co-regulators, they regulate the gene expression programs for an adaptive response to specific stimuli (hormones and metabolites). As it allows the identification of direct target genes, chromatin immunoprecipitation (ChIP) constitutes a key tool for studying the regulation mechanisms of specific transcription factors or co-regulators in the modulation of gene expression programs in physiological and pathological contexts.

ChIP protocols were primarily developed for cell culture and then optimized for frozen lean metabolic tissues [1]. However, performing ChIP in lipid-rich tissues, such as adipose tissues or adipocytes, remained challenging. We have established and optimized a reliable and reproducible ChIP method that is suitable for both cultured differentiated adipocytes [2] and snap-frozen adipose tissue biopsies [3]. We provide details about the steps to perform ChIP in adipose tissues or adipocytes (Fig. 1). Indeed,

Franziska Greulich (ed.), *Chromatin Immunoprecipitation: Methods and Protocols*, Methods in Molecular Biology, vol. 2846, https://doi.org/10.1007/978-1-0716-4071-5_3,

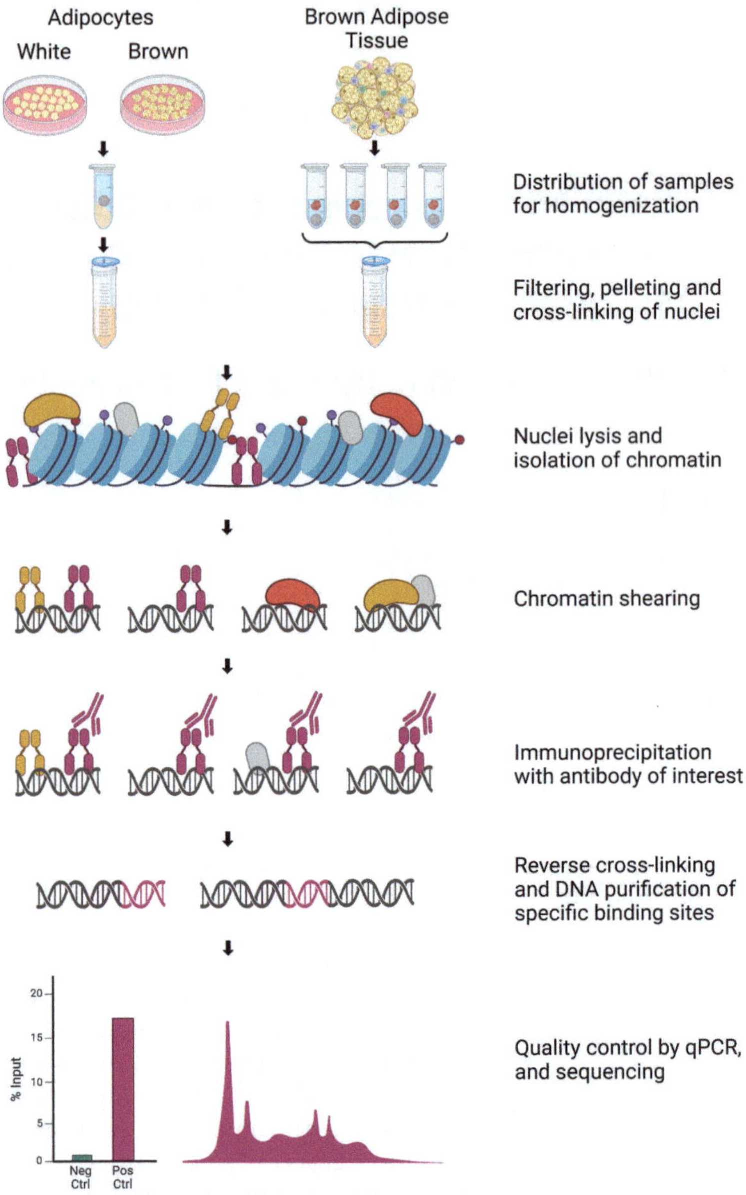

Fig. 1 Overview of ChIP protocol from nuclei isolation to quality control. Figure created with BioRender.com

after collection, the samples undergo nuclei isolation, cross-linking of protein-DNA complexes, chromatin shearing, and immunoprecipitation. The transcription factor-bound DNA (or "ChIPped" DNA) is purified, and quality control of enrichment via ChIP-qPCR is performed. While we do not discuss the specifics of library construction or data analysis, this ChIP method is suitable for further DNA sequencing [3].

2 Materials

2.1 Cells or Tissue Collection, Homogenization, Nuclear Isolation, and Cross-Linking

1. 1× Phosphate-buffered saline (PBS).
2. Homogenization buffer (prepare fresh and keep on ice): 10 mM HEPES-KOH pH 7.3, 10 mM KCl, 5 mM $MgCl_2$, 0.5 mM DTT, 1× Roche cOmplete™ Protease Inhibitor tablets (EDTA-free).
3. 1% (v/v) formaldehyde solution (prepare fresh): Dilute 16% (w/v) ultrapure formaldehyde (methanol-free) in 1× PBS.
4. 1 M glycine solution: Dissolve 75.07 g of glycine in 1 L of H_20. Sterilize through a 0.22 μM pore filter membrane and store at 22–23 °C.
5. 25 cm plate of differentiated adipocytes or frozen adipose tissue biopsies.
6. Cell scrapers.
7. Scalpels.
8. Forceps.
9. Liquid nitrogen and dry ice.
10. 5 mm stainless steel beads.
11. 70 μM cell strainers.
12. 1.5/2 mL DNase-free microcentrifuge tubes.
13. 15/50 mL DNase-free conical centrifuge tubes.
14. TissueLyser.
15. Cooling Microcentrifuge for 1.5 and 2 mL tubes.
16. Cooling centrifuge for 15 mL conical tubes.
17. Orbital shaker.

2.2 Nuclei Lysis and Chromatin Shearing

1. 20% (w/v) SDS solution.
2. Fast IP buffer: 150 mM NaCl, 50 mM EDTA pH 7.5, 5 mM Tris–HCl pH 7.5, 1% (v/v) Triton X-100, 0.5% (v/v) NP40. Once dissolved, sterilize through a 0.22 μM pore filter membrane and store at 4 °C. Before use, add fresh 1× Roche cOmplete™ Protease Inhibitor tablets (EDTA-free) and keep on ice.
3. Shearing buffer: 1% (w/v) SDS, 10 mM EDTA pH 8.0, 50 mM Tris–HCl pH 8.0. Once dissolved, filter sterilize and store at 22–23 °C. Before use, add fresh Roche cOmplete™ Protease Inhibitor tablets (EDTA-free) and keep on ice.
4. 1 mL syringes (24 G needles).
5. 15 mL Bioruptor tube.
6. Sonicator (Diagenode Bioruptor or similar).

2.3 DNA Fragments Profiling by De-Crosslinking

1. Proteinase K solution: combine freshly 1 μL of 10 mg/mL Proteinase K stock, 4 μL of 1 M Tris–HCl pH 7.5, 2 μL 0.5 M EDTA pH 8.0 per 50 μL of DNA sample.
2. 5 M NaCl.
3. 10 mg/mL RNase A.
4. 10 mg/mL Proteinase K.
5. PCR Purification Kit.
6. 3 M NaOAc pH 5.2.
7. Agarose, agarose gel running buffer and chambers, DNA loading dye, 100-bp DNA ladder, gel documentation system.
8. Thermomixer.

2.4 Chromatin Immunoprecipitation (ChIP)

1. Dilution buffer: 0.01% (w/v) SDS, 1.1% (v/v) Triton X-100, 1.2 mM EDTA pH 8.0, 16.7 mM Tris–HCl pH 8.0, 0.167 M NaCl. Once mixed, sterilize through a 0.22 μM pore filter membrane and store at 4 °C. Before use, add fresh 1× Roche cOmplete™ Protease Inhibitor tablets (EDTA-free) and keep on ice.
2. Beads blocking solution: Dissolve 0.5% (w/v) Bovine serum albumin (BSA) powder in 1× PBS. Once dissolved filter through a 0.22 μM pore filter membrane. Aliquots can be stored at −20 °C.
3. Tris-EDTA buffer.
4. Beads Elution Buffer (prepare fresh): Dissolve 84 mg $NaHCO_3$ in 1 mL H_2O. Combine 105 μL $NaHCO_3$ (1 M), 52.5 μL 20% (w/v) SDS, and 892.5 μL H_2O.
5. "ChIP grade" antibody raised against the protein of interest and species-specific control antibody.
6. Appropriate host-specific antibody-conjugated magnetic beads (depending on ChIP antibody source species).
7. 15 mL DNase-free conical centrifuge tubes.
8. 1.5 mL and 2 mL Low DNA Binding DNase-free microcentrifuge tubes.
9. Magnetic racks for microcentrifuge tubes.
10. Tube rotating wheel.

2.5 Quality Control and Library Preparation

1. Qubit and appropriate Quibit DNA quantification kit (e.g., dsDNA high sensitivity).
2. qPCR machine, master mix, primer for positive and negative control regions.

3 Methods

3.1 Cells or Tissue Collection, Homogenization, Nuclear Isolation, and Cross-Linking

1. Samples can be collected from cultured cells (*see* **Note 1**) or tissue. Collect the tissues and directly snap freeze in liquid nitrogen. Tissues can be stored at −80 °C until ready to proceed.
2. Using clean forceps and scalpels, cut frozen tissue on dry ice, and distribute 30 to 50 mg of frozen tissue into ice-cold 2 mL microcentrifuge tubes (*see* **Note 2**).
3. Add 1 mL of homogenization buffer and one 5 mm stainless steel bead to each tube containing frozen tissues or pelleted cells (*see* **Note 3**).
4. Homogenize the tissues or cells immediately using the Tissue-Lyser for 2 min at 30 Hz (*see* **Note 3**).
5. Filter the homogenate through a 70 μm cell strainer into a 50 mL conical centrifuge tube. Pool the lysates from the several 2 mL tubes at this step.
6. Transfer to a 15 mL conical centrifuge tube and centrifuge for 3 min at 2000× *g* at 22–23 °C (*see* **Note 4**).
7. Aspirate the supernatant (*see* **Note 5**), and resuspend the pellet completely in 10 mL of 1% (v/v) formaldehyde solution (*see* **Note 6**).
8. Incubate the solution for 15 min with gentle rotation at 22–23 °C.
9. Then add 1.5 mL of 1 M Glycine solution. Incubate with gentle rotation for 5 min at 22–23 °C.
10. Centrifuge the solution for 10 min at 2000× *g* at 22–23 °C.
11. Aspirate the supernatant.
12. Resuspend the pellet completely in ice-cold PBS, transfer it into a new ice-cold 2 mL microcentrifuge tube, and incubate it on ice for 10 min.
13. Centrifuge for 10 min at 2000× *g* at 4 °C and aspirate the supernatant (*see* **Note** 7). *See* Fig. 2a for an example of the nuclei pellet.

3.2 Nuclei Lysis and Chromatin Shearing

1. Add 1 mL cold Fast IP buffer to the pellet and resuspend the pellet completely.
2. Aspirate through a syringe (24G) and release into the tube. Incubate the tube for 10 min on ice.
3. Centrifuge 1 min at 13,800× *g* at 4 °C, and aspirate the supernatant delicately. Repeat the **steps 2** and **3** a second time. *See* Fig. 2b for an example picture of the chromatin pellet.

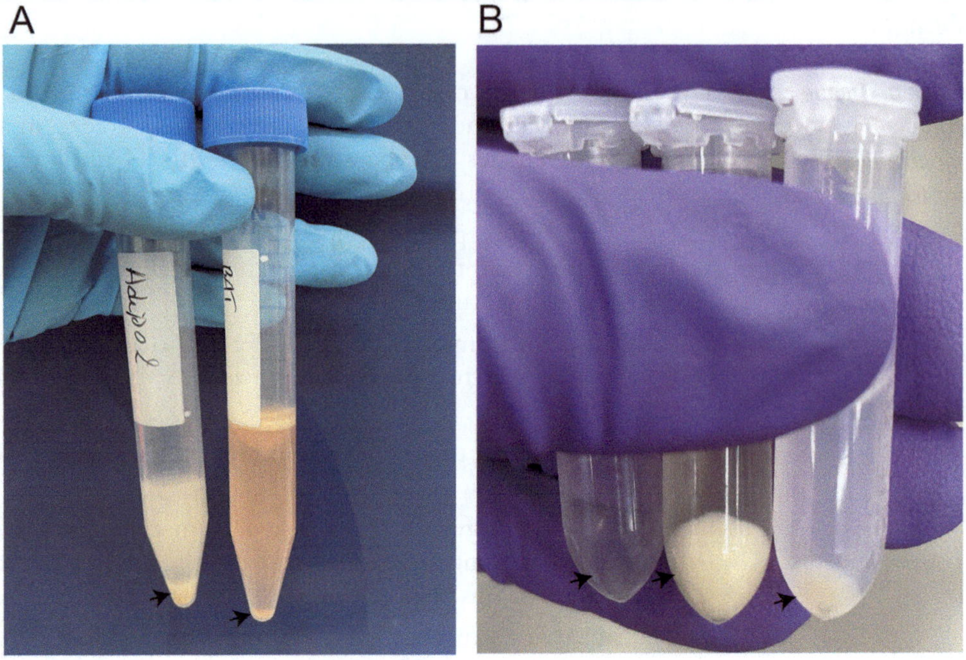

Fig. 2 Nuclei isolation from brown adipose tissue and adipocytes. (**a**) Nuclei (arrows) pelleted after tissue and cell homogenization. (**b**) Chromatin (arrows) pelleted after nuclei lysis

4. Resuspend the pellet completely in 1 mL ice-cold Shearing buffer, transfer it into a 15 mL Bioruptor tube, and proceed to the shearing (*see* **Note 8**).

3.3 DNA Fragments Profiling by De-Crosslinking

1. Take 50 μL of your sample and incubate overnight (15–16 h) at 65 °C with 2 μL of 5 M NaCl.
2. Then add 2 μL of 10 mg/mL RNase A and incubate for 30 min at 37 °C.
3. Add 7 μL of proteinase K solution and incubate at 56 °C for 2 h.
4. Purify the DNA using the PCR purification kit according to the manufacturer's manual.
5. Verify the DNA fragment size by running the samples on 0.7% (w/v) agarose gel.
6. The DNA profile should show a bulk between 300 bp and 1 kb (Fig. 3). Some additional shearing cycles may be added before proceeding to the immunoprecipitation. During these steps, the chromatin should be kept on ice at 4 °C.

3.4 Chromatin Immunoprecipitation (ChIP)

1. Centrifuge the sheared chromatin for 10 min at 13,800× *g* at 4 °C.
2. Pipet all the supernatant into a new 1.5 mL microcentrifuge tube.

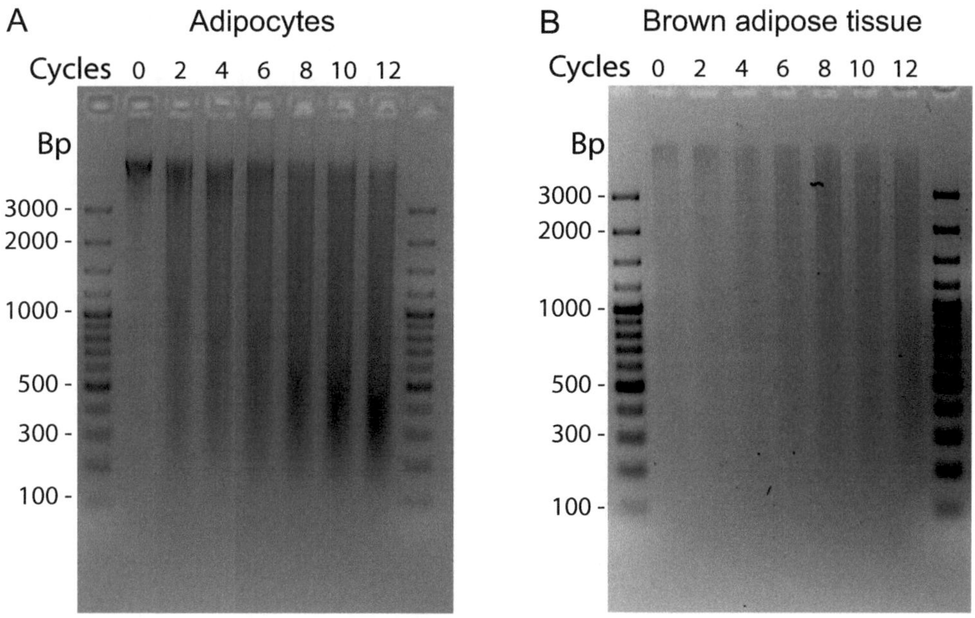

Fig. 3 Profiles of the sonicated DNA in adipocytes (**a**) and brown adipose tissue (**b**) after a time course shearing experiment

3. 72 μL of chromatin are used as 10% input and kept at −20 °C until needed.
4. For ChIP, transfer 800 μL of chromatin into a 15 mL conical tube and add 7.2 mL cold Dilution buffer. Add 8 μg of ChIP-grade antibody per IP (*see* **Notes 9, 10** and **11**). Incubate overnight (15–16 h) at 4 °C on a rotating wheel.
5. For each ChIP, prepare 100 μL of the appropriate host-specific antibody-coupled magnetic beads into a 2 mL microcentrifuge tube. Wash the beads twice with 1 mL ice-cold blocking solution (0.5% BSA solution) using a cooled adapted magnetic rack on ice. Add a third mL of cold-blocking solution and incubate overnight (15–16 h) at 4 °C on a rotating wheel.
6. After the incubation, clear the ChIP DNA by centrifugation at 3000× *g* at 4 °C for 20 min.
7. During the centrifugation, wash the magnetic beads by carefully aspirating the blocking solution using a cooled magnetic rack on ice. Add 1 mL of cold Dilution buffer, invert the tube gently, and put it back in the magnetic rack to aspirate the buffer. Add 100 μL of Dilution buffer per ChIP to the magnetic beads and resuspend the beads by inverting the tube. Distribute 20 μL of beads into 2 mL Low DNA Binding DNase-free microcentrifuge tubes (four tubes per ChIP). Keep the beads on ice.

8. After the centrifugation, use the top 90% of cleared chromatin, and distribute 1.8 mL into each 2 mL tube containing magnetic beads. Incubate the tubes for 6 h at 4 °C under gentle rotation.
9. After the incubation, proceed to the washes. Set up the tubes in the magnetic racks and aspirate the supernatant. Add 1 mL of cold Fast IP Buffer, gently invert the tubes, leave on ice for 1 min, and place them again in the magnetic racks.
10. Repeat **step 9** five times, for 6 washes in total.
11. During the washes, prepare fresh Beads Elution Buffer.
12. After the last wash of the beads, aspirate the supernatant, and add 100 μL of Tris-EDTA (TE) buffer in each tube. Combine all the beads from the same ChIP (four tubes) in one. Put the tubes back in the magnetic rack and aspirate the supernatant carefully.
13. Wash another time with 1 mL of TE buffer, and aspirate the supernatant carefully.
14. Add 100 μL of freshly made Beads Elution Buffer to the beads and place them into a thermomixer for 15 min at 22–23 °C at 1000 rpm.
15. Place tubes in a magnetic rack, and collect the supernatant into a clean 1.5 mL Low DNA Binding DNase-free microcentrifuge tubes.
16. Add another 100 μL of Beads Elution Buffer to the beads and repeat the incubation at 22–23 °C in the thermomixer at 1000 rpm for 15 min.
17. Place the tubes back in the magnetic rack and collect the additional 100 μL of supernatant, and combine with the first eluted 100 μL, yielding a total of 200 μL of "ChIPped" DNA.

3.5 DNA Purification

1. Add 8 μL of 5 M NaCl to the 200 uL "ChIPped" DNA from Subheading 3.4, **step 17**.
2. Add 28 μL of TE buffer to the 72 μL of input (from Subheading 3.4, **step 3**), and 4 μL of 5 M NaCl.
3. Incubate the ChIP and Input samples at 65 °C overnight (15–16 h).
4. Add 1 μL of 10 mg/mL RNase A to the ChIP and input samples and incubate for 30 min at 37 °C.
5. After the incubation, add 13 μL and 6.5 μL of proteinase K solution to the ChIP and input samples, respectively. Incubate at 56 °C for 2 h.
6. Purify the DNA from the ChIP and input samples using the PCR purification kit in 16 μL of Elution Buffer (EB) provided by the kit.

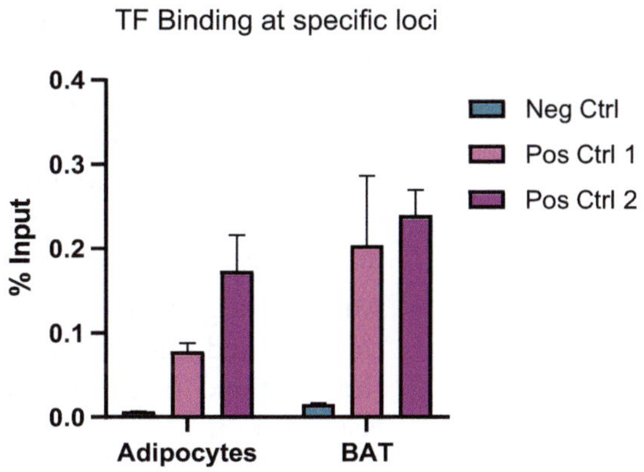

Fig. 4 Enrichment at specific loci in brown adipocytes and adipose tissue. The binding of the transcription factor of interest is found strongly enriched at positive control loci (pink and purple) compared with negative locus (green)

3.6 Quality Control and Library Preparation

1. Use 1 μL of each eluted sample and input to measure the DNA concentration using the Qubit and the appropriate Qubit kit (high sensitivity dsDNA) (*see* **Note 12**).
2. Dilute 1 μL of each sample and input in 59 μL of water, and perform quantitative PCR to verify the enrichment of the transcription factor at specific loci (Fig. 4, *see* **Note 13**).
3. Use the remaining 14 μL to prepare the libraries for next-generation sequencing according to the manufacturer's instructions.

4 Notes

1. From cell culture plates, aspirate the medium and wash the cells with ice-cold PBS. Aspirate the PBS and collect the cells in the 2 mL microcentrifuge tube using cell scrapers. Add 1 mL of cold PBS, centrifuge 0.9× *g* at 4 °C for 5 min, and aspirate the PBS. The pelleted cells can be snap-frozen and kept at −80 °C until ready to proceed (*see* **Note 2**).
2. The success of the ChIP depends on the presence of the transcription factor of interest in the nucleus. It is important to define the appropriate condition for the experiment (age, sex, diet, time of the day, absence/presence of ligand, etc.). The nuclei amount in the tissue is also crucial and should be empirically determined to define the quantity of tissue/cells to be used for the experiment. It is important to note that for specific tissues, adipose tissues in particular, several biological replicates could be pooled in 1 ChIP sample, 250–300 mg of tissue constituting a good starting point for ChIP in tissue.

3. The cells/tissue lysis constitutes an important step in isolating nuclei. To ensure good lysis, we empirically optimized the conditions using a ratio of one 25 cm plate of differentiated cells (around 20 million) or 30–50 mg of frozen tissue in 1 mL of lysis buffer, homogenized for 2 min at 30 Hz. However, depending on the cells/tissues, the conditions may be optimized (lysis buffer volume, time of homogenization).
4. After homogenization, it is important to check the integrity of the obtained nuclei. This can be done by staining an aliquot of the obtained nuclei pellet with a fluorescent DNA dye and visualizing it under a microscope.
5. After the centrifugation of the homogenate, the nuclei are in the pellet (Fig. 2a). For high lipid-content tissues or cells, a lipid layer is found on top of the homogenate. It is important to fully aspirate this layer carefully to "clean" the samples from any lipid content contamination.
6. Depending on the protein of interest (transcription factor or co-regulator), the fixation method can differ. For transcription factors, exhibiting direct protein-DNA interactions, fixation with formaldehyde (1%) should be sufficient. However, in the case of co-regulators, proteins without direct protein-DNA interactions, an additional cross-linker may be required (such as Disuccinimidyl glutarate, or DSG). In addition, the time and temperature of incubation with the cross-linker(s) must be optimized.
7. After the wash with PBS, the pellet is not always visible for Brown Adipose Tissue samples. To avoid losing material at this step, it is recommended to not fully aspirate the PBS.
8. The chromatin shearing must be optimized for each tissue/cell type. It is recommended to proceed to a time course of shearing before starting an experiment (Fig. 3).
9. The choice of the antibody is a crucial point for successful ChIP. ChIP-grade level can be guaranteed by manufacturers, but it is important to verify the specificity of the antibody by western blot or immuno-fluorescence using negative control samples (knock-out samples) and the subcellular localization of the protein of interest (in the nucleus).
10. A successful ChIP requires an optimal amount of antibody. The amount of antibody can be optimized by performing a titration experiment, including positive and negative loci controls. It is important to note that the suitability of an antibody can depend on batches.
11. When establishing the ChIP for a transcription factor for the first time, it is highly recommended to add a positive control antibody. ChIP-suitable RNA polymerase II or histone marks

constitute good positive controls to determine if the ChIP experiment worked.

12. The DNA concentration from ChIP and input samples depends on the material input and abundance of the transcription factor of interest. For transcription factors, "ChIPped" and Input DNA concentrations oscillate from 0.12 to 0.5 ng/μ L and from 25 to 80 ng/μL, respectively, to ensure further successful sequencing. If the DNA concentration is too low, quality control by qPCR of the ChIP is possible, allowing to determine the quality/specificity of the ChIP, but it is recommended to increase the material input before proceeding with sequencing.
13. The design of positive and negative loci is crucial to validate the enrichment of binding for a transcription factor. When no specific locus is known, it is possible to use available data to design primers for determining the enrichment by qPCR. Indeed, ChIP-seq data performed with RNA Polymerase II or histone mark (acetylation for example) will provide information on the potential sequences for binding of the transcription factor of interest. It is possible to combine the results with data from RNA-seq to focus on differentially expressed genes from the control condition versus knock-out or knock-down conditions, even from another tissue.

References

1. Mir AA, Dyar KA, Greulich F et al (2019) In vivo ChIP-Seq of nuclear receptors: a rough guide to transform frozen tissues into high-confidence genome-wide binding profiles. Methods Mol Biol 1966:39–70
2. Giroud M, Tsokanos FF, Caratti G et al (2021) HAND2 is a novel obesity-linked adipogenic transcription factor regulated by glucocorticoid signalling. Diabetologia 64(8):1850–1865
3. van Eenige R, In Het Panhuis W, Schonke M et al (2022) Angiopoietin-like 4 governs diurnal lipoprotein lipase activity in brown adipose tissue. Mol Metab 60:101497

Chapter 4

Bioinformatics Core Workflow for ChIP-Seq Data Analysis

Tamas Schauer

Abstract

Chromatin immunoprecipitation (ChIP) followed by next-generation sequencing (-seq) has been the most common genomics method for studying DNA–protein interactions in the last decade. ChIP-seq technology became standard both experimentally and computationally. This chapter presents a core workflow that covers data processing and initial analytical steps of ChIP-seq data. We provide a step-by-step protocol of the commands as well as a fully assembled Snakemake workflow. Along the protocol, we discuss key tool parameters, quality control, output reports, and preliminary results.

Key words ChIP-seq, Chromatin biology, Automated workflow, Step-by-step protocol, Quality control, Reproducibility

1 Introduction

ChIP-seq is the core technology of chromatin biology as it determines the genomic location of histone modifications and variants, binding sites of transcription factors, or chromatin remodeling enzymes. To date, there are more than nineteen thousand data series when searched for the keyword "ChIP-seq" in Gene Expression Omnibus. The richness of ChIP-seq datasets is accompanied by a great variety of data processing methods that might differ from entry to entry. Here, we cover the essential steps of ChIP-seq data processing and discuss the key parameters to be considered by the users. Such a core workflow includes analytical steps from the raw sequencing reads to preliminary results. In addition, we give insights into quality control and replicability and, to some extent, into the interpretation of results.

Manual installation of bioinformatics software might be tedious due to incompatible software, missing dependencies, and different platforms. Tracking software versions is also essential for reproducibility reasons. Therefore, we suggest using package man-

Franziska Greulich (ed.), *Chromatin Immunoprecipitation: Methods and Protocols*, Methods in Molecular Biology, vol. 2846, https://doi.org/10.1007/978-1-0716-4071-5_4,

agers, such as Conda, to automate installation and configure software environments. Conda environments can be saved, exported, and shared with other users.

Workflow management systems provide an infrastructure to create data processing workflows that can be automatically executed [1]. Tool versions, execution steps and parameters, as well as log, report, and result files can be tracked. Such workflows are not only easy to run but also easy to reuse, share, or modify by the users. Snakemake is one of the popular workflow managers used in bioinformatics [2]. Steps in a Snakemake workflow are defined in rules that execute commands to create output files from the input files. Dependencies between rules form a directed acyclic graph, which determines the order of executing jobs (Fig. 1). We provide an automatized, reproducible Snakemake workflow in our public GitHub repository, as well as detailed individual commands of the workflow in the methods section.

2 Materials

2.1 Data

1. Example datasets include ChIP-seq data of H3K4me3 [3], H3K9me3 [4] and H3K36me3 [5] in mouse embryonic stem cells. Accession numbers can be found in Table 1.
2. Genome annotation files are downloaded from the Ensembl FTP site: ftp://ftp.ensembl.org/pub/release-101
3. Blacklists [6] are a set of genomic regions with likely artifactual signals in ChIP-seq experiments, including regions with low-mappability or high signal in ChIP input controls.

   Source: https://github.com/Boyle-Lab/Blacklist/blob/master/lists/mm10-blacklist.v2.bed.gz

2.2 Hardware

Snakemake workflows can be easily adapted to cluster environments such as SLURM; however, for simplicity, we ran the workflow on an iMac desktop computer with an 8-Core Intel Core i7 processor, 64 GB RAM, and macOS version 12.6.3.

2.3 Software

1. Code, including instructions for setting up the Conda environment and running Snakemake, can be downloaded from GitHub: https://github.com/tschauer/chipseq_pipeline_2023
2. Conda is a package management system that allows the easy installation of software packages and their dependencies. In addition, Conda environments can be saved and shared with specific versions of software contributing to the reproducibility of results. Source: https://docs.conda.io/en/latest/
3. Snakemake [2] is a workflow management system that assembles reproducible data processing and analysis pipelines. Source: https://snakemake.readthedocs.io/en/stable/

Fig. 1 Overview of the Snakemake workflow as a rule graph

4. FastQC is a quality control tool that provides basic information about raw FASTQ files, such as base calling quality scores, sequence duplication levels, and adapter content. Source: https://www.bioinformatics.babraham.ac.uk/projects/fastqc/
5. Trim Galore is a wrapper script around Cutadapt [7] to automatically detect and remove commonly used adapter

Table 1
Accession numbers and basic information about the example dataset. SRA - sequence read archive (https:/www.ncbi.nlm.nih.gov/sra)

SRA accession	Stage	Assay	Source	Description	SampleName	Layout
SRR5479623	mESC	H3K9me3	Wang 2018 [4]	Mouse embryonic stem cells	GSM2588589	PAIRED
SRR5479624	mESC	H3K9me3	Wang 2018 [4]	Mouse embryonic stem cells	GSM2588590	PAIRED
SRR6956533	mESC	H3K36me3	Xu 2019 [5]	Mouse embryonic stem cells	GSM3084635	PAIRED
SRR6956534	mESC	H3K36me3	Xu 2019 [5]	Mouse embryonic stem cells	GSM3084636	PAIRED
SRR2147247	mESC	H3K4me3	Zhang 2016 [3]	Mouse embryonic stem cells	GSM1845286	PAIRED

sequences. Source: https://www.bioinformatics.babraham.ac.uk/projects/trim_galore/ and https://cutadapt.readthedocs.io/en/stable/

6. Bowtie2 [8] is a general-purpose short-read sequence aligner for mapping reads to the reference genome. Source: https://bowtie-bio.sourceforge.net/bowtie2/index.shtml
7. Samtools [9] is a software package that includes several tools for manipulating alignment files in SAM/BAM format. Source: https://www.htslib.org
8. Picard provides additional tools for working with SAM/BAM files, such as marking and removing read duplicates. Source: https://broadinstitute.github.io/picard/
9. Deeptools [10] is a software suite for processing, analyzing, and visualizing NGS data including ChIP-seq. Source: https://deeptools.readthedocs.io/en/develop/
10. Macs2 [11] is a general peak caller for ChIP-seq data allowing the detection of narrow or broad peak regions. Source: https://github.com/jsh58/MACS
11. Bedtools [12] is a collection of utilities for working with genomic intervals, such as merging or intersecting ChIP-seq peaks. Source: https://bedtools.readthedocs.io/en/latest/
12. MultiQC [13] aggregates quality control results of data processing steps and creates a visual overview of all the samples. Source: https://multiqc.info
13. IGV [14] is an interactive tool for browsing genomics datasets. Source: https://software.broadinstitute.org/software/igv/

3 Methods

This section describes the individual data processing steps of the example workflow. These commands can be executed manually from the command line. The same steps are incorporated in the Snakemake workflow available on GitHub. Note that the output directory structure is generated automatically when run by Snakemake; however, all files are in the current working directory when executed manually (not recommended).

3.1 Annotation Files

We recommend downloading the required genome annotation files at the beginning of each project. Then, genome annotation and sequencing data files are kept together in the same project folder, and their version numbers can be easily backtracked.

1. Get genome FASTA file (*see* **Note 1**):

```
wget -O genome.fa.gz ftp://ftp.ensembl.org/pub/release-101/
fasta/mus_musculus/dna/Mus_musculus.GRCm38.dna.primary_assem-
bly.fa.gz
gunzip genome.fa.gz
```

2. Get genome annotation GTF files:

```
wget -O genome.gtf.gz ftp://ftp.ensembl.org/pub/release-101/
gtf/mus_musculus/Mus_musculus.GRCm38.101.gtf.gz

gunzip genome.gtf.gz
```

3. Get blacklist BED file:

```
wget -O mm10-blacklist.v2.bed.gz https://github.com/Boyle-
Lab/Blacklist/raw/master/lists/mm10-blacklist.v2.bed.gz

gunzip mm10-blacklist.v2.bed.gz

cat mm10-blacklist.v2.bed | sed 's/^chr//' > GRCm38-black-
list.v2.bed
```

3.2 Read Processing

We recommend starting the data processing with quality control to see whether the number and quality of sequencing reads meet standard criteria (see Results for details).

1. Assess read quality using FastQC:

```
fastqc SRR2147247_1.fastq.gz -t 8 > SRR2147247_fastqc.out 2>
SRR2147247_fastqc.err
```

2. Trim adapters and low-quality read ends by Trim Galore (*see* **Note 2**):

```
trim_galore -j 8 --quality 28 --paired SRR2147247_1.fastq.gz
SRR2147247_2.fastq.gz > SRR2147247_trimgalore.out 2>
SRR2147247_trimgalore.err
```

3.3 Alignment

The alignment step determines the genomic positions of the sequencing reads. Although there are several fast short-read aligners available, we show the commonly used Bowtie2 tool here.

1. Create reference index from genome FASTA file using Bowtie2 (*see* **Note 3**):

```
bowtie2-build --threads 8 genome.fa genome
```

2. Map reads to the reference using Bowtie2 and create a BAM file using Samtools (*see* **Note 4**):

```
bowtie2 -x genome --threads 8 --local --very-sensitive-local
--no-unal --no-mixed --no-discordant --dovetail -I 10 -X
700 -1 SRR2147247_1_val_1.fq.gz -2 SRR2147247_2_val_2.fq.gz
2> SRR2147247_bowtie2.err | samtools view -Sbh -o SRR2147247.
mapped.bam
```

3.4 Filtering and Quality Control

Aligned reads go through several quality filters, such as mapping quality, the removal of potential PCR duplicates, and inspection of fragment size by paired-end reads.

1. Filter, sort, and index alignment using Samtools (*see* **Note 5**):

```
samtools view -bS -@ 8 -q 12 SRR2147247.mapped.bam | samtools
sort -@ 8 - | tee SRR2147247.filtered.bam | samtools index -
SRR2147247.filtered.bam.bai > SRR2147247_samtools.out
```

2. Remove PCR duplicates using Picard (*see* **Note 6**):

```
picard MarkDuplicates -I SRR2147247.filtered.bam -O
SRR2147247.removeddup.bam -CREATE_INDEX TRUE -REMOVE_DUPLI-
CATES TRUE -METRICS_FILE SRR2147247_picard.out 2>
SRR2147247_picard.err

samtools index SRR2147247.removeddup.bam
```

3. Create a histogram of fragment size distribution using Deeptools:

```
bamPEFragmentSize -hist SRR2147247.fragmentsize.png -p 8 -T
"Fragment size" --maxFragmentLength 800 --samplesLabel
SRR2147247 -b SRR2147247.removeddup.bam > SRR2147247_frags.out
```

3.5 Genome Coverage

Coverage tracks are generated from the aligned reads by counting the number of reads per consecutive genomic windows (bins) or base pair position.

1. Create binned and smoothed genome coverage using Deeptools (*see* **Note 7**):

```
bamCoverage --bam SRR2147247.removeddup.bam -o SRR2147247.
coverage.bw --blackListFileName GRCm38-blacklist.v2.bed --
binSize 20 --smoothLength 60 --extendReads --maxFragmentLength
700 --minMappingQuality 12 --normalizeUsing "CPM" --numberOf-
Processors 8 > SRR2147247_bamcov.out 2> SRR2147247_bamcov.err
```

3.6 Peak Calling

Genomic regions enriched for aligned reads are peaks that peak finder tools can call. Macs2 provides several options for identifying peak regions.

1. Call peaks using Macs2 (*see* **Note 8**):

```
macs2 callpeak --treatment SRR2147247.removeddup.bam --name
SRR2147247 --gsize 1.87e9 --format BAMPE --nomodel --nolambda
--cutoff-analysis --broad --broad-cutoff 0.001 --pvalue 0.001
--outdir . > SRR2147247_macs2.out 2> SRR2147247_macs2.err
```

2. Remove blacklist region from peaks:

```
bedtools intersect -v -a SRR2147247_peaks.broadPeak -b GRCm38-
blacklist.v2.bed | cut -f 1,2,3 > SRR2147247_filtered.bed
```

3.7 Metaplots

Average ChIP-seq enrichment is commonly visualized surrounding annotated genomic features such as genes, transcription start sites (TSS), or transcription termination sites (TTS or TES).

1. Create TSS Aligned or Gene Body Scaled Matrix (*see* **Note 9**)

```
computeMatrix reference-point --numberOfProcessors 8 -S
SRR2147247.coverage.bw -R genome.gtf --referencePoint 'TSS'
--binSize 20 --skipZeros --downstream 2000 --upstream 2000 -o
SRR2147247.TSS_matrix.gz > SRR2147247_TSSmat.out 2>
SRR2147247_TSSmat.err
```

2. Plot Metagene Profile

```
plotProfile -m SRR2147247.TSS_matrix.gz --plotHeight 10 --
plotWidth 12 -o SRR2147247.TSS_plot.png
```

3.8 Correlation Analysis

We calculate pairwise correlation coefficients to investigate the relationship between samples and visualize them as a heatmap. Overlaps between peak regions are judged by the Jaccard similarity index.

1. Generate multi bigwig summary (*see* **Note 10**):

```
multiBigwigSummary bins --numberOfProcessors 8 --bwfiles *.bw
--blackListFileName GRCm38-blacklist.v2.bed --smartLabels --
outFileName scores_per_bin.npz --outRawCounts scores_per_bin.
tab
```

2. Plot correlation matrix (*see* **Note 11**):

```
plotCorrelation --corData scores_per_bin.npz --corMethod
spearman --skipZeros --plotTitle "Spearman Correlation of Read
Counts" --whatToPlot heatmap --colorMap RdBu_r --plotNumbers
--plotHeight 14.25 --plotWidth 16.5 -o heatmap_SpearmanCorr_-
readCounts.png --outFileCorMatrix SpearmanCorr_readCounts.tab
```

3. Calculate Jaccard similarity (*see* **Note 12**):

```
Rscript --vanilla jaccard.R *filtered.bed heatmap_Jaccard_-
peaks.pdf
```

3.9 Quality Summary

We create a helpful interactive report of the processing steps by MultiQC.

1. Run MultiQC in your working directory:

```
multiqc . --config multiqc_config.yaml
```

3.10 Genome Browser

Finally, we interactively visualize the genome coverage (bw, BigWig files) and peak files (BED files) in the IGV browser. Sessions can be saved in XML files, whereas snapshots are saved as PNG images.

4 Results

4.1 Quality Control

To judge the quality of the sequencing data, we ran FastQC on the raw FASTQ files (See Wikipedia about the file format: https://en.wikipedia.org/wiki/FASTQ_format). FastQC extracts and visualizes the distribution of the base calling quality (Phred score, Fig. 2a). A Phred score of 30 indicates a base call accuracy of 99.9%. The score usually decreases towards the end of the read. Another critical piece of information in the FastQC report is the sequencing adapter content (*see* **Note 2**, Fig. 2b). These are unwanted sequences derived from the library preparation. Adapter content typically increases towards the end of the read when the reads are longer than the actual DNA fragments.

The number of aligned reads provides further insights into data quality. Overall alignment rates reported by Bowtie2 are typically greater than 80% (Fig. 3). Lower alignment rates could arise from library contamination by other species. The number of reads

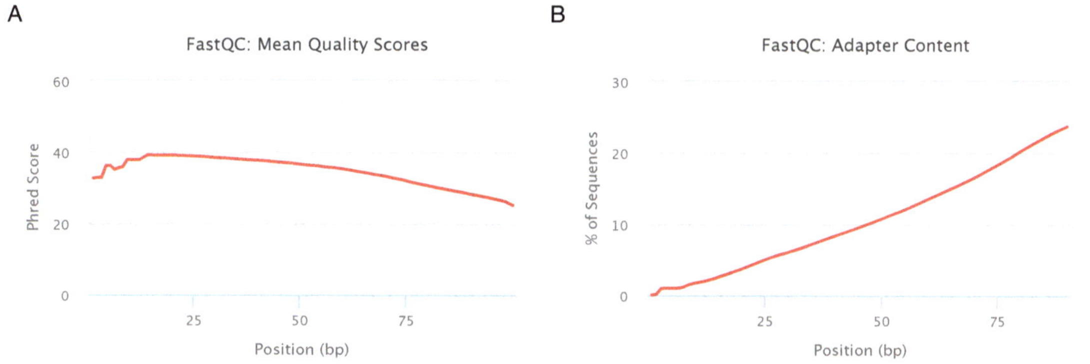

Fig. 2 Sequencing read quality of the sample SRR2147247. (**a**) Distribution of base calling quality scores along the sequencing reads by FastQC. Quality usually decreases towards the end of the reads. (**b**) Sequencing adapter content along the sequencing reads by FastQC. Illumina universal adapters show an increasing percentage towards the end of the reads

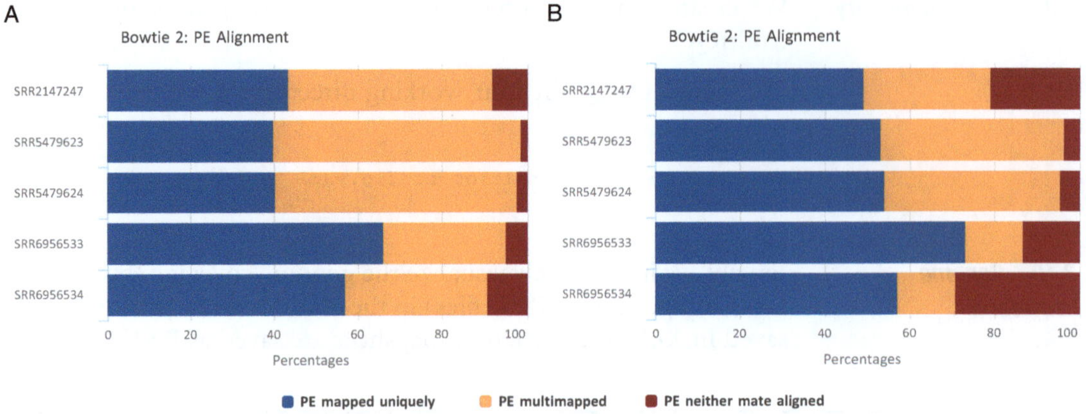

Fig. 3 Alignment summary. (**a**, **b**) Percentage of uniquely mapped, multimapped, and not aligned reads in local (**a**) or in end-to-end (**b**) alignment mode by Bowtie2 and plotted by MultiQC. Local mode leads to higher overall alignment rates, but also a higher number of multimappers

aligned to multiple positions in the genome (multimappers) varies from library to library between 10% and 50%. However, this depends on the aligner parameters (*see* **Note 4**, Fig. 3). For example, higher multimapper rates might occur for chromatin marks enriched at repetitive elements.

Aligned reads in SAM/BAM file format are filtered by several quality criteria (see Wikipedia: https://en.wikipedia.org/wiki/SAM_(file_format)). The mapping quality score (MAPQ) is calculated from the probability that the alignment is not the true position of the read. MAPQ ranges between 0 and 42 (for Bowtie2); the higher the score, the better the alignment. We set the threshold in between, so multimappers are likely excluded, but we allow for some mismatched bases (*see* **Note 5**). Next, duplicated reads are removed. These probably originate from the same DNA molecule (*see* **Note 6**, Fig. 4a). A high number of duplicates suggest potential PCR artifacts in the sequencing library. Finally, it is worth looking at the chromosomal distribution of the reads, as this could also suggest anomalies in the data (Fig. 4b). For example, a high number of mitochondrial reads might be an issue for ATAC-seq libraries [15], but it is less likely in ChIP-seq data (Fig. 4b).

Careful inspection of the chromatin fragment length is important quality control (Fig. 4c, d), especially in the case of MNase-based ChIP protocols or related techniques such as Cut&Run [16]. Over-digestion leads to a higher proportion of subnucleosomal fragments (<120 bp), which might not be desired when histone marks are the target of interest. Optimal MNase-ChIP-seq samples are highly enriched for mono-nucleosomal fragments (120–220 bp) and contain some proportion of longer fragments. Ideally, samples with similar fragment length distribution should be directly compared to each other.

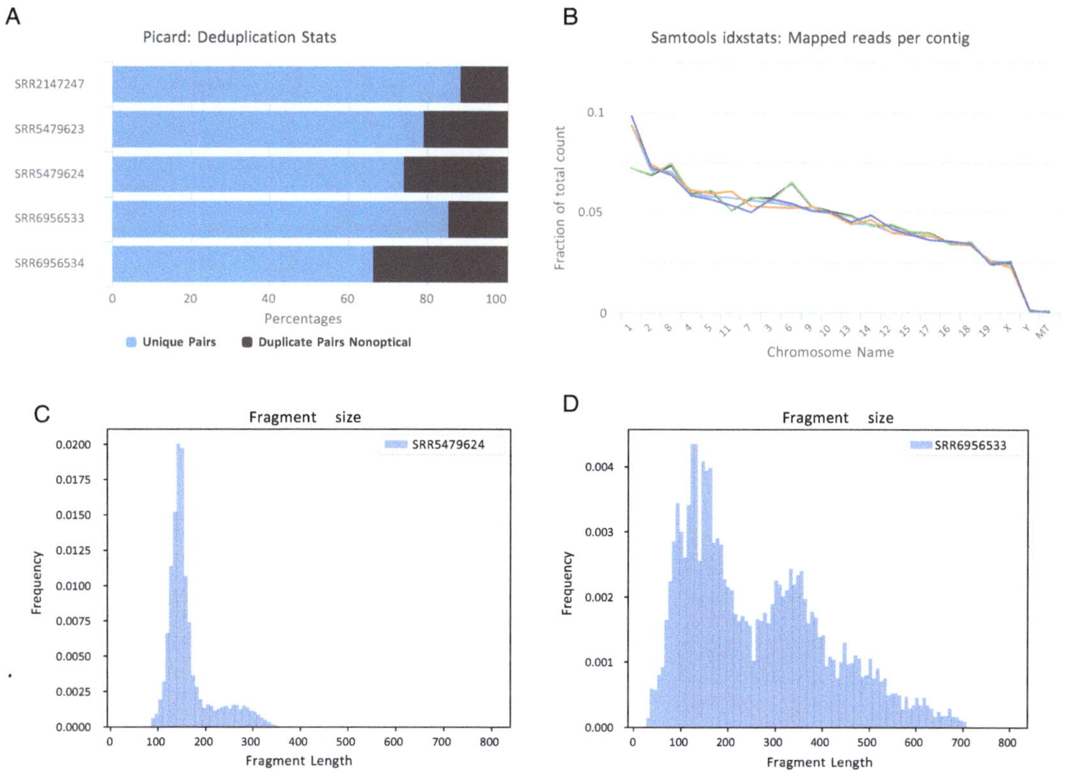

Fig. 4 Quality metrics and filtering of aligned reads. (**a**) Percentage of unique and duplicated read pairs calculated by Picard and plotted by MultiQC. A high percentage of duplicates suggests potential PCR artifacts. (**b**) Chromosomal distribution of reads calculated by Samtools and plotted by MultiQC. The percentage of mitochondrial contamination is very low in these ChIP-seq libraries. (**c**, **d**) Fragment length distribution of read pairs obtained by Deeptools. Most of the fragments are between 100-200 bp in sample SRR5479624 (**c**), whereas there is a higher proportion of longer fragments in sample SRR6956533 (**d**)

4.2 Genome Coverage and Peaks

Visual inspection of genome coverages (BigWig file format: https://genome.ucsc.edu/goldenPath/help/bigWig.html) and peaks (BED file format: https://en.wikipedia.org/wiki/BED_(file_format)) in the genome browser is essential. ChIP enrichment, replicate similarity, and peak calling can be immediately judged in the IGV browser (Fig. 5a). In the example datasets, as expected, we see high, narrower peaks of H3K4me3 close to transcription start sites (TSS), broader regions over gene bodies marked by H3K36me3 and H3K9me3 signal that seems to be anti-correlated with the other two histone marks. There is a visible difference in ChIP enrichment/efficiency between the H3K36me3 replicates as the signal over gene bodies is greater in the first replicate (Fig. 5a, yellow arrows). Browsing the genome can help to fine-tune peak-finding settings (*see* **Note 8**). Depending on the focus of the study, only the largest peaks should be considered (high specificity; Fig. 5b, arrows and *see* **Note 8**), or all potential regions of interest should be captured (high sensitivity).

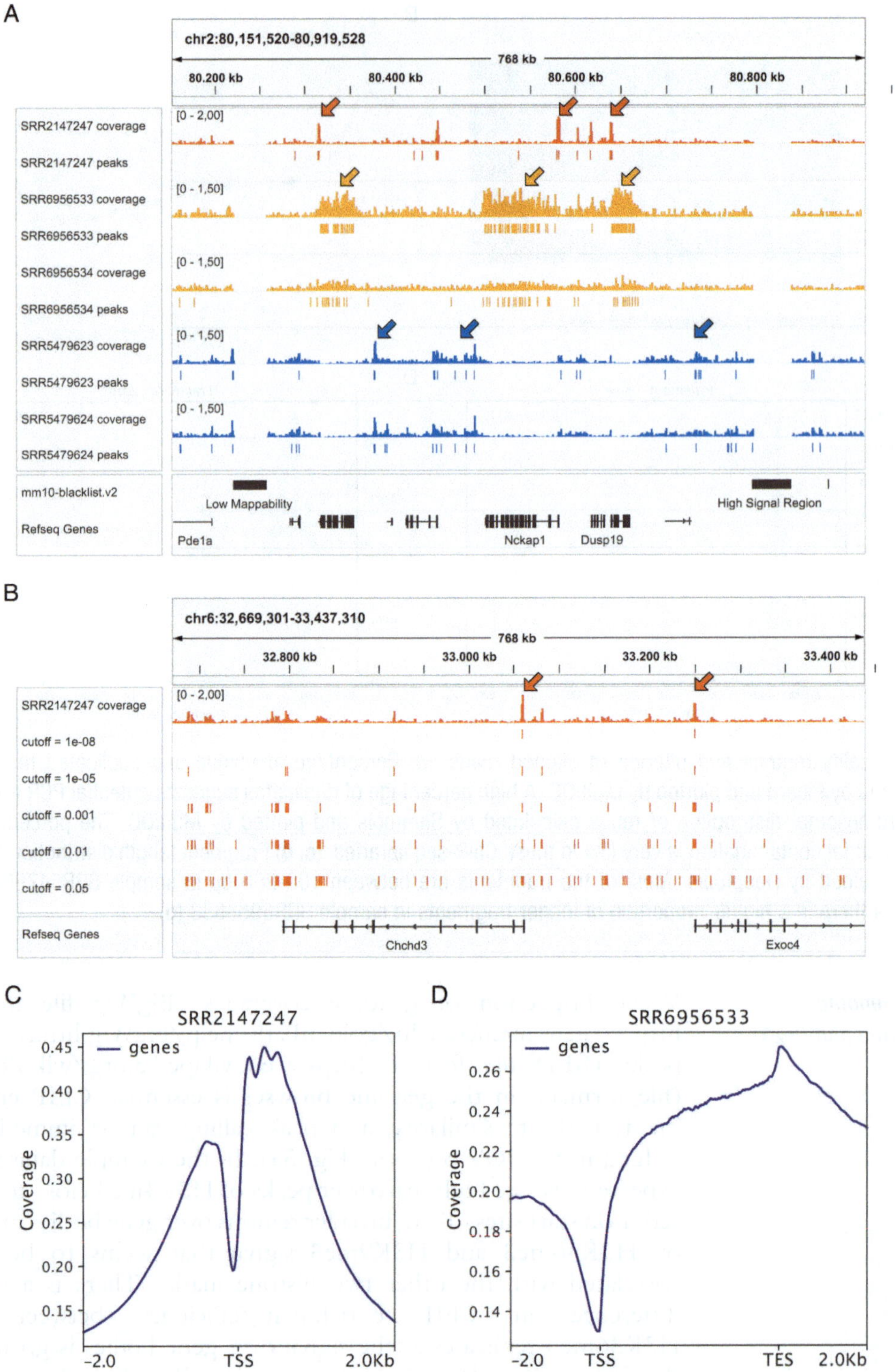

Fig. 5 Genome coverage and peak regions. (**a**) Genome coverage and corresponding peak regions of H3K4me3 (SRR2147247), H3K36me3 (SRR6956533 and SRR6956534), as well as H3K9me3 (SRR5479623 and SRR5479624) in the IGV browser. Arrows indicate narrow TSS peaks of H3K4me3 (dark orange), gene body signal of H3K36me3 (yellow), and repressive regions of H3K9me3 (blue), respectively. Blacklist regions

Nonetheless, browser snapshots don't necessarily provide a genome-wide overview but might help to generate hypotheses for downstream analyses. To summarize the data systematically, profile plots surrounding the TSS for H3K4me4 (*see* **Note 9**, Fig. 5c) or scaled over the gene body for H3K36me3 (Fig. 5d) are generated. Thus, the observation in the genome browser about the enrichments of these histone marks is confirmed on the genome-wide average.

4.3 Sample Similarity

The workflow provides two metrics to assess the similarity between the samples. First, the correlation coefficient on the normalized read counts measures the relationship of the ChIP-seq signal between two samples (*see* **Note 11**). As expected, the replicates of the same histone mark cluster, whereas active (H3K4me4 and H3K36me3) and repressive mark (H3K9me3) separate on the correlation heatmap (Fig. 6a). Secondly, the Jaccard index measures the similarity between two sets of peaks (*see* **Note 12**). Clustering results of Jaccard similarity are comparable to the correlation coefficients; however, it is lower between the active histone marks (Fig. 6b). Jaccard similarity might be sensitive to the number of called peaks, which could explain the relatively low values between the H3K36me3 replicates (Figs. 5a and 6b).

5 Notes

1. We downloaded the genome annotation files from the Ensembl database. Alternatively, the UCSC annotation can be used. The genome file and its annotation must have the same source and version. Note that chromosome names have a "chr" prefix in the UCSC, for example, "chr1", "chr2", etc., whereas they are numbered in the Ensembl annotation, for example, "1", "2", etc. For a detailed comparison, see the link:
   https://genome.ucsc.edu/FAQ/FAQgenes.html#ens
2. Trim Galore automatically detects the adaptor sequence, that is, Illumina, Nextera, or small RNA; therefore, there is no need to specify which adaptor was used.
3. Genome indexing by Bowtie2 should be executed once, and index files can be re-used for subsequent alignment runs. The advantage of workflow managers is that they enable re-entry

Fig. 5 (continued) and reference gene annotation are shown at the bottom. (**b**) Genome coverage and peak regions of H3K4me3 (SRR2147247) using various p-value cutoffs for peak calling. Arrows indicate the largest peaks that are called even with the most stringent cutoff. (**c**, **d**) Metaplots centered at the TSS for H3K4me3 (SRR2147247) and scaled between TSS and TES for H3K36me3 (SRR6956533)

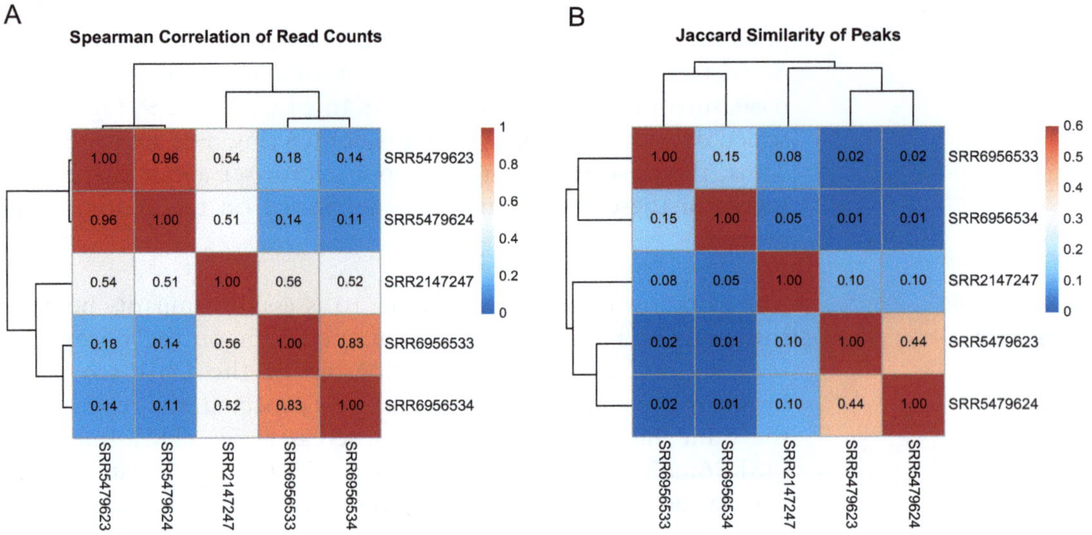

Fig. 6 Sample similarity metrics. (**a**) Spearman's correlation heatmap of read counts. (**b**) Jaccard similarity heatmap of called peaks. Replicates of H3K36me3 (SRR6956533 and SRR6956534) cluster with the other active mark H3K4me3 (SRR2147247) and separates from the repressive H3K9me3 (SRR5479623 and SRR5479624)

from the last successful step, thus making sure that steps such as genome indexing are carried out only when the workflow is run the first time.

4. Bowtie2 parameters to be considered are whether end-to-end or local alignment should be performed. Local mode omits read ends to find the best alignment. This usually leads to better overall alignment rates; however, it might result in more multimapper reads due to shorter read length. Paired-end reads carry information about the chromatin fragment length. The default upper limit of Bowtie2 is 500 bp (-X parameter), which might need to be longer if all fragments should be considered. Alternatively, lower and upper limits can be set to obtain sub- or mono-nucleosomal fragments depending on the application (-I sets the lower limit). However, these limits can be set in downstream processing steps too.

5. Mapping quality is an important parameter for filtering reads. Using higher thresholds, reads that are more "uniquely" aligned and carry fewer mismatches are kept.

6. In the current workflow, we remove duplicated reads likely derived from the same DNA fragment. These can be PCR duplicates (i.e., PCR amplification artifacts) or optical duplicates (i.e., sensor artifacts of the sequencer). However, the same sequence can originate from independent DNA

molecules, in which case duplicates should not be removed. Unique molecular identifiers (UMI) have been developed to reliably distinguish between PCR duplicates and distinct molecules [17]; however, they are not commonly used for ChIP-seq libraries.

7. Deeptools provides various settings to create genome coverage files. Reads are extended to fragments defined by the read mates for paired-end data. Binning and smoothing parameters might be changed by the user. Deeptools can filter read pairs by fragment length (if not done on the alignment level). Finally, we apply the simplest normalization method: read counts per million (CPM).
8. Peak finding parameters are set to broad as the example data are histone modifications. The user might want to pay attention to the --nomodel or --nolambda parameters (see details in Macs2 documentation). The latter could be considered for very broad regions (e.g., H3K36me3), as Macs2 will use a genome-wide background lambda instead of a local lambda parameter. In addition, Macs2 provides the option for cutoff analysis that helps to find the desired p-value cutoff. Visual inspection of peaks in the genome browser is highly recommended. We provide an example region in Fig. 5b showing various cutoffs of the --broad-cutoff and --pvalue parameters (the values were set the same for both parameters).
9. Profile plots relative to a reference point are recommended when the shape of the signal surrounding the reference should be maintained (e.g., ±1 nucleosome at TSS). In addition, the scale-regions mode is helpful for chromatin marks on the gene body (e.g., H3K36me3). Alternatively, other genomic regions (BED files) can be provided to visualize the average ChIP-seq signal, for example, at peaks or at a pool of peaks.
10. The default multibigwig summary bin size is 10 kb. A smaller bin size gives higher resolution but potentially more noisy results.
11. Spearman's correlation measures monotonic relationships. It is based on the rank of the data and therefore more robust to outliers. Alternatively, Pearson's correlation can be used, which measures linear relationships.
12. The custom Rscript for Jaccard similarity can be found in the GitHub repository. The Jaccard similarity is the ratio of the intersection length (in bp) to the union length of two peak sets.

References

1. Wratten L, Wilm A, Goke J (2021) Reproducible, scalable, and shareable analysis pipelines with bioinformatics workflow managers. Nat Methods 18(10):1161–1168
2. Molder F, Jablonski KP, Letcher B et al (2021) Sustainable data analysis with Snakemake. F1000Res 10:33
3. Zhang B, Zheng H, Huang B et al (2016) Allelic reprogramming of the histone modification H3K4me3 in early mammalian development. Nature 537(7621):553–557
4. Wang C, Liu X, Gao Y et al (2018) Reprogramming of H3K9me3-dependent heterochromatin during mammalian embryo development. Nat Cell Biol 20(5):620–631
5. Xu Q, Xiang Y, Wang Q et al (2019) SETD2 regulates the maternal epigenome, genomic imprinting and embryonic development. Nat Genet 51(5):844–856
6. Amemiya HM, Kundaje A, Boyle AP (2019) The ENCODE blacklist: identification of problematic regions of the genome. Sci Rep 9(1): 9354
7. Martin M (2011) Cutadapt removes adapter sequences from high-throughput sequencing reads. 17(1):3
8. Langmead B, Salzberg SL (2012) Fast gapped-read alignment with Bowtie 2. Nat Methods 9(4):357–359
9. Danecek P, Bonfield JK, Liddle J et al (2021) Twelve years of SAMtools and BCFtools. Gigascience 10(2)
10. Ramirez F, Ryan DP, Gruning B et al (2016) deepTools2: a next generation web server for deep-sequencing data analysis. Nucleic Acids Res 44(W1):W160–W165
11. Gaspar JM (2018) Improved peak-calling with MACS2. bioRxiv:496521
12. Quinlan AR, Hall IM (2010) BEDTools: a flexible suite of utilities for comparing genomic features. Bioinformatics 26(6):841–842. https://doi.org/10.1093/bioinformatics/btq033
13. Ewels P, Magnusson M, Lundin S et al (2016) MultiQC: summarize analysis results for multiple tools and samples in a single report. Bioinformatics 32(19):3047–3048
14. Robinson JT, Thorvaldsdottir H, Winckler W et al (2011) Integrative genomics viewer. Nat Biotechnol 29(1):24–26
15. Corces MR, Trevino AE, Hamilton EG et al (2017) An improved ATAC-seq protocol reduces background and enables interrogation of frozen tissues. Nat Methods 14(10): 959–962
16. Meers MP, Bryson TD, Henikoff JG et al (2019) Improved CUT&RUN chromatin profiling tools. Elife 8
17. Smith T, Heger A, Sudbery I (2017) UMI-tools: modeling sequencing errors in Unique Molecular Identifiers to improve quantification accuracy. Genome Res 27(3): 491–499

Chapter 5

Differential Analysis of Protein-DNA Binding Using ChIP-Seq Data

Christiane Boeckel, Xavier Pastor, Matthias Heinig, and Thomas Walzthoeni

Abstract

Chromatin immunoprecipitation in combination with next-generation sequencing (ChIP-Seq) allows probing of protein-DNA binding in a rapid and genome-wide fashion. Herein we describe the required steps to preprocess ChIP-Seq data and to analyze the differential binding of proteins to DNA for perturbation experiments. In these experiments, different conditions are compared to find the underlying biological mechanisms caused by the stimulus or treatment. In addition, we provide a sample analysis using the steps outlined in the chapter.

Key words ChIP-Seq, DiffBind, Differential ChIP-Seq analysis

1 Introduction

The phenotype of organisms is determined by the expression of an organism's genetic code (genotype) and by environmental factors. The DNA sequence directly impacts the protein composition, while the environment affects the epigenome. The epigenome describes (reversible) molecular changes on the chromatin without altering the genomic sequence, but the genomic activity. These alterations in gene regulation can be observed, for example, in the methylation patterns of the DNA or molecular modifications of histones, changing the accessibility of DNA regions and thus the ability of regulatory proteins, such as transcription factors [1], to interact with the DNA.

To study the interaction of proteins with DNA, antibody-based chromatin immunoprecipitation (ChIP) techniques have been developed and, with the advent of high throughput methods, enable rapid and genome-wide analyses. This includes microarray-

Franziska Greulich (ed.), *Chromatin Immunoprecipitation: Methods and Protocols*, Methods in Molecular Biology, vol. 2846,
https://doi.org/10.1007/978-1-0716-4071-5_5,

based protocols (ChIP-chip) [2] and the now standard technique that is based on next-generation sequencing (NGS) known as ChIP-Seq [3, 4].

In ChIP techniques, the samples are chemically cross-linked, thus generating stable (and reversible) bonds between DNA and the bound proteins. The chromatin is then sheared (mechanically or enzymatically), followed by immunoprecipitation (IP) using a specific antibody for the protein of interest. The IP step separates unbound DNA fragments from protein-bound DNA fragments, which are then purified and used to create a DNA library that is subsequently sequenced by NGS [5]. Aligning the sequenced fragments to a reference genome allows us to identify the origin of the sequences. Genomic regions with an enrichment in fragments can then be identified and assigned as putative protein binding sites [3].

Identification of specific protein binding sites by ChIP-Seq can be impaired by nonspecific binding. This can occur for several reasons: DNA shearing is not entirely random, and open chromatin regions are favored compared to other regions and yield more protein–DNA complexes. In addition, regions with abundant protein binding, such as promoters, can give rise to false positive signals due to the unspecific binding of the antibody [6, 7]. To correct these artifacts, technical control samples are required that are treated equally but without the IP step (input controls), or a mock IP with an unspecific antibody is carried out. The selection of either method depends on the samples and other factors [6].

Also, repetitive regions of the genome can produce artificial signals due to their under-representation in the reference genome. To deal with repetitive regions and other regions that lead to nonspecific signals, the most used method is to use blacklists of regions known to be noisy and exclude them from the analysis. The ENCODE consortium compiled one of the most comprehensive sources for blacklists [8].

Besides identifying the binding sites of proteins, researchers are often interested in the underlying mechanisms that a particular factor or histone modification exerts when the biological system is perturbed. Perturbations challenge the system with a drug or any other factor thought to induce a change in the biological system. Also of interest are comparisons between healthy and diseased states to find potential molecular sites for interventions.

In this chapter, we will explain the required preprocessing steps and how to use DiffBind [9], an R package available from Bioconductor [10], which allows computing differentially bound sites from ChIP-Seq experiments. DiffBind covers the required steps and provides regions with significant changes between different conditions. These are the basis of further downstream analyses such as motif analyses, pathways enrichment analysis, and genome browser visualizations.

Although the protocols described in this chapter focus on the analysis of ChIP-Seq data, the procedures are generally applicable to peak-based high throughput omics technologies (e.g., ATAC-Seq), with some tuning of the parameters necessary to fit their specificities.

Our goal is to provide a general guideline for experimentalists and bioinformaticians to enable ChIP-Seq data analysis with two-group experimental designs.

2 Methods

2.1 Data Preprocessing

Bioinformatic processing of ChIP-Seq, Cut&Run, or Cut&Tag data starts with short nucleotide sequences (reads) stored in FASTQ files [11, 12]. These sequences result from the DNA fragments bound to the immunoprecipitated protein and from random fragments, which are artifacts inherent to the sample preparation and sequencing (background signals) [6].

The preprocessing steps are required to perform quality control (QC) steps, to the alignment of the reads to the genome, and to identify the regions with significant enrichments of reads beyond the background signals.

2.1.1 Standard Workflow

A standard workflow (Fig. 1) of this preprocessing consists of five steps: quality control, read trimming, alignment, filtering and deduplication of aligned reads, and eventually, calling peaks [13].

Checking the quality of the reads is an important first step of the workflow that can eliminate potential biases or quality issues of the data. Some very common quality issues are sequencing adapter contamination and low-quality bases that occur due to the nature of the sequencing process. Therefore, these can impact the alignment success and are removed with read trimming tools. To locate the origin of the reads within the genome, the trimmed sequences are mapped to a reference genome. Due to sequencing errors, artifacts, repetitive regions, and natural sequence variation of the tested samples (genotypes), sequenced reads can map to the genome with varying quality. Reads, which map equally well to multiple locations or map poorly to any of the sites, are being filtered out from the alignment. In the case of paired-end data where two reads are generated from the 5′ and 3′ end of the DNA fragment, only reads where both segments are correctly aligned should be kept. Furthermore, exact duplicate reads or fragments, which can result from optical duplicates during sequencing or the over-amplification during the library preparation (PCR artifacts), are being removed because it is unlikely to find identical reads for this type of data. Up until this step, the preprocessing is similar for different types of NGS data. The peak calling, however, is specific to

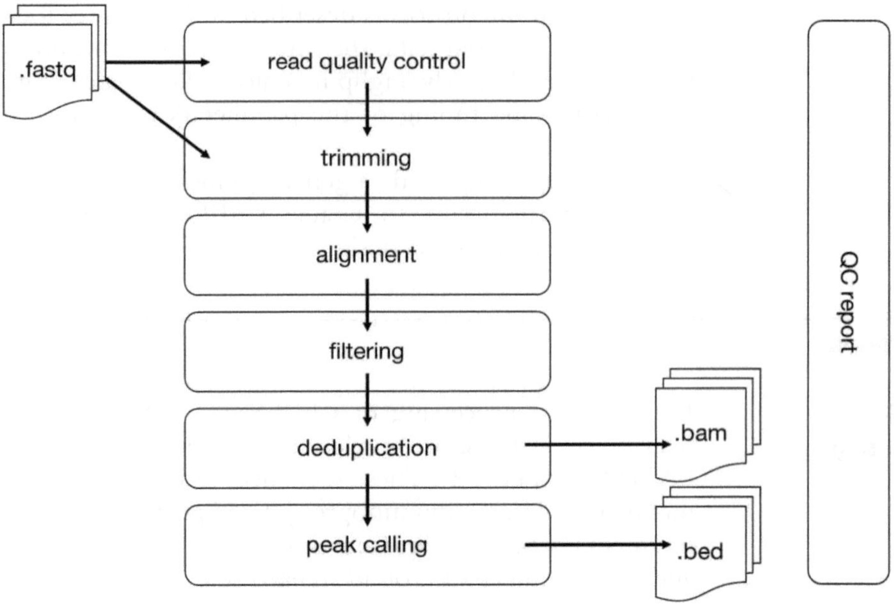

Fig. 1 Standard ChIP-Seq preprocessing workflow. The preprocessing of ChIP-Seq data starts from FASTQ files containing sequencing reads. They are checked for their quality and trimmed from adapters and low-quality bases. The trimmed reads are then aligned to a reference genome and filtered from low-quality alignments or duplicate reads. Peaks are then called to identify genomic regions with enriched binding, which are stored in BED files. Eventually, a quality report will be generated that includes statistics of each workflow process

ChIP-Seq-like data and will be explained in more detail. QC metrics of all the individual steps are collected in a combined QC report. For a detailed discussion of the QC metrics, *see* **Note 1**.

2.1.2 Peak Calling

The goal of ChIP-Seq analysis is to identify genomic regions that are bound to a protein of interest. The immunoprecipitation step during the sample preparation enriches DNA fragments, which are linked to this protein. However, this method is not powerful enough to avoid sequencing other random DNA fragments, which are considered noise or background. Therefore, peak callers have been developed to identify the regions that are significantly enriched beyond the background or noise. Peak callers are software tools that statistically test whether the observed number of aligned reads in a region exceeds the random expectation based on the background or control model. Peak calling usually consists of two steps: First, windows of potential peaks are defined based on the ChIP signal. Then, the ChIP signal is compared to a modeled or measured control signal to test for significant enrichment against the background within these peaks [14] (Fig. 2a). A popular peak caller is MACS (and its versions MACS2 and MACS3), which models local background signals assuming a Poisson distribution [15]. All peak callers produce files containing the locations of the

A

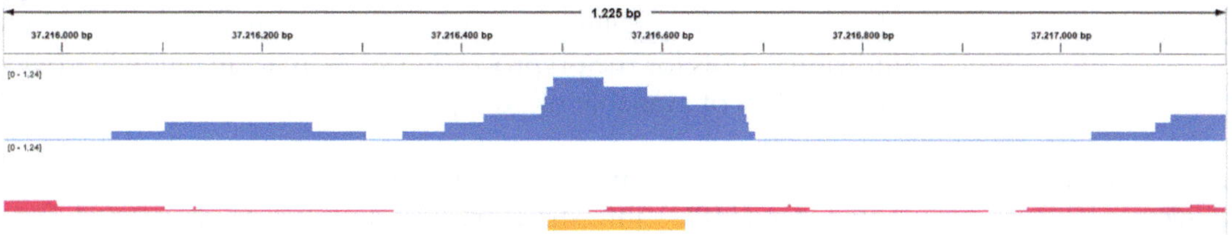

B

chromosome	start	end	name	score	strand
chr1	191398	191686	peak_1	33	.
chr1	270689	271540	peak_2	25	.
chr2	777656	780041	peak_3	266	.

Fig. 2 (**a**) Signal tracks of a called peak. Peak calling involves statistical testing if the observed ChIP-Seq signal exceeds the random expectation based on the background or control model. This genome visualization shows the ChIP-Seq signal (blue track) and its technical control (red track). At the position where the ChIP-Seq signal is significantly enriched over the background model, which was computed based on the control sample, a peak was called (yellow bar). (**b**) Browser Extensible Data (BED) file format. BED files are tab-delimited files which contain genomic locations. The first three columns are mandatory and indicate the chromosome (column 1), its start (column 2), and end position (column 3). In addition to the basic three columns, the BED6 format includes three more columns. These additional columns specify the name of the feature (column 4), a score (column 5), and the strand (either "+" or "−") on which the genomic feature is located (column 6). Please note that ChIP-Seq peaks do not contain any strand-specific information; therefore, a "." is used in the strand column

peaks, often along with a score that indicates the significance of the enrichment. Genomic windows like peaks are usually stored in the Browser Extensible Data (BED) format (Fig. 2b) that follows a standardized format and can be used as input for differential peak analysis.

2.1.3 Workflow Resources

A workflow consists of using many individual tools, and data flow between the tools is required. To that end, workflow languages such as Nextflow [16], Snakemake [17], and CWL [18] have been developed that facilitate the interconnection of individual processes and automation of the execution. Furthermore, these domain-specific languages allow parallel processing of the data and offer executors for various computational infrastructures. Coordinated efforts of the bioinformatics community have led to a collection of standardized high-quality and publicly available Nextflow pipelines for various use cases from the nf-core website [19]. The nf-core community has developed pipelines for ChIP-Seq and Cut&Run or Cut&Tag datasets (see https://nf-co.re/pipelines). Using these pipelines has the benefit that analyses are reproducible and comparable across research groups. Additionally, a vast number of adjustable parameters allow for customizations. Due to the availability of these pipelines, we will not explain the usage of the individual tools for the preprocessing steps in detail but will provide an overview of available nf-core pipelines in the following section.

2.1.4 Run nf-core Pipelines

nf-core pipelines can be run locally or on a high-performance computer (HPC) infrastructure. All software tools needed for a specific Nextflow pipeline are automatically installed using the package manager Conda or run with containers such as Docker or Singularity.

To run the ChIP-Seq workflow available from nf-core, the general steps are as follows:

1. Install Nextflow using conda (and Docker or Singularity).
2. Configure Nextflow to run on your system. For detailed information, please refer to https://nf-co.re/docs/usage/configuration.
3. To test the pipeline, run it in test mode with the following command in a terminal (the following lines starting with $ indicate command line prompts):

```
$ nextflow run nf-core/chipseq -r 2.0.0 -profile test,docker
--outdir
```

   Where the -r parameter specifies a particular version of the pipeline.

4. To run your samples with the pipeline, create a comma-separated sample sheet with the following header line followed by the samples (*see* **Note 2**):

   The sample table (Table 1) needs to be saved and provided as a comma-separated (csv) file. Make sure that the column separator is ",".

5. Run the pipeline with the following command:

```
$ nextflow run nf-core/chipseq --input <path/to/samplesheet.
csv> --outdir <path/to/outdirectory> -profile singularity
```

Table 1
Example sample sheet for a ChIP-seq workflow

sample	fastq_1	fastq_2	antibody	control
<samplename>	<path/to/first/fastq/file>	<path/to/second/fastq/file>	<antibody>	<samplenameofcontrol>
...				

where the paths indicated with < > must be specified according to your operating system (e.g., /analysis/output).

For more specifications and recommendations regarding the selection of parameters, please *see* **Note 3**.

The ChIP-Seq pipeline will create an output directory with a specified structure that contains all the output files. To run DiffBind, one needs to specify the BAM files containing the alignment position of all reads and the corresponding BED files containing the peaks. The BAM files can be found at results/bwa/mergedLibrary/ with the file extension *dedup.sorted.bam. In addition to the BAM files, DiffBind will need the corresponding index files (*.bam.bai) to be present in the same directory as the BAM files. The BED files are located at results/bwa/mergedLibrary/macs/<peak-type>/. MACS can call narrow or broad peaks and generates *.narrowPeak and *.broadPeak files, respectively.

Cut&RUN and Cut&TAG data can be preprocessed with the cutandrun pipeline. For more information on how to run this pipeline, *see* **Note 4**. Moreover, the type of immunoprecipitated protein and its expected chromatin profile dictates the selection of the peak calling parameters. *See* **Note 5** for more recommendations in that regard.

2.2 Differential Peak Analysis with DiffBind

A typical perturbation experiment often aims to compare a treated condition to a mock-treated or nontreated control. In the case of peak-based data, one seeks to find peaks where binding strength differs significantly between conditions. Then, binding strength is quantified using the ChIP-Seq signal within these peaks, specifically by the number of reads aligned to those regions. These comparisons are similar to differential gene expression analyses, which are based on read count data with R packages such as DESeq2 [20] or edgeR [21].

DiffBind was developed to specifically target differential analysis of ChIP-Seq peaks [9]. In comparison to other available differential ChIP-Seq tools, DiffBind performed well in terms of precision and recall for various use cases [22]. Hence, how to perform the differential analysis with DiffBind will be explained in this chapter and is supplemented with an exemplary analysis of a ChIP-Seq dataset that is available from https://github.com/Bioinformatics-Munich/diffchipseq and Zenodo [23].

2.2.1 DiffBind Workflow Introduction

DiffBind can be used to analyze peak data (occupancy) and peak count data (affinity). The dba() function of the package creates dba objects from a provided sample sheet. The main data is organized in data frames with the genomic locations of the peaks. These peaks are filtered to remove blacklisted regions before the remaining peaks are aggregated into consensus peaks. The number of reads within these consensus peaks is then added to the dba object using dba.count(). At this point of the analysis, it is advised to perform quality checks to estimate the biological variation across the samples based on the read counts. This is necessary because the outcome may influence the following normalization step and the design formula for the differential analysis [8].

2.2.2 Grey- and Blacklists

Accumulation of ChIP-Seq data over time revealed the existence of problematic genomic regions, which often show high ChIP-Seq signals regardless of tissue or experimental setup [7, 8]. As a result, the ENCODE consortium made the effort to collect them in exclusion lists, known as blacklists [8].

Moreover, there are experiment-specific anomalous regions, which depend on the type of tissue, the antibody, and other technical biases introduced during the experimental procedure. These are referred to in DiffBind as greylists and can be constructed from mock or input controls (*see* **Note 6**).

Peaks within these regions should be removed from the analysis because they can impact the normalization, thresholds, or other factors influencing the final results. This step can be performed with the dba.blacklist() function.

2.2.3 Consensus Peak Calling with DiffBind

One important feature of DiffBind is the consensus peak calling with dba.count(). Due to biological variation, DiffBind focuses on consensus peaks rather than individual peaks. Therefore, DiffBind considers all peaks overlapping in at least two samples as consensus peaks. The summits and borders of these consensus peaks are being recalculated based on the reads of all samples within these windows to combine the information of all samples despite the biological variation (shown in Fig. 3). Hence, all the reads within this window

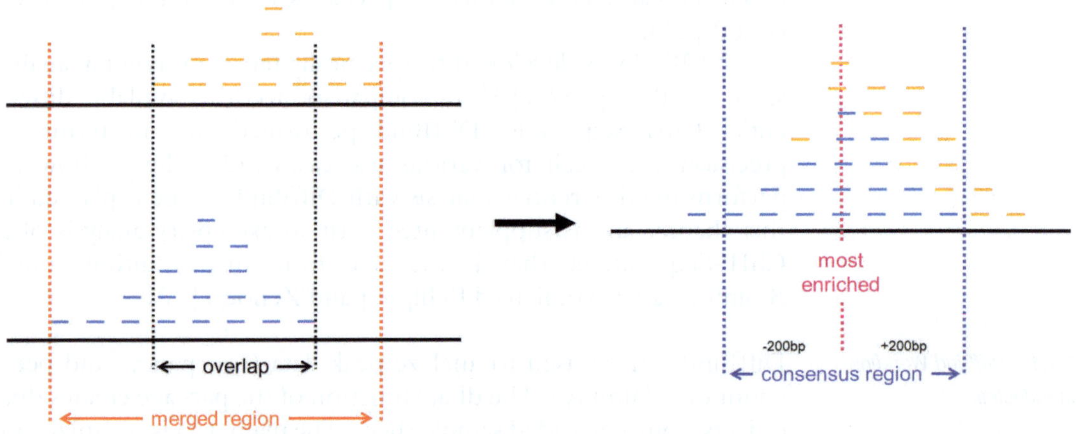

Fig. 3 Generation of consensus peaks. Peaks covering the same region do not exactly overlap. First, overlapping peaks are merged, and then the peak summit is defined as the position with the maximum number of reads. To standardize the consensus peaks, the borders of the peaks are set to a fixed distance (here 200 bp) upstream and downstream of the summit

are being piled up and the position with the maximum number of reads will be defined as the summit. Moreover, the summit will be extended by a defined number of bases (*see* **Note 5** for recommendations) in both directions to get peaks of a fixed width.

2.2.4 Quality Checks and Sample Visualization

Ideally, quality control of the data should be performed before the differential analysis. This includes investigating the samples with regards to their similarity, in particular the similarity between replicates of the same conditions. The ENCODE consortium has established guidelines for quality controls specific to transcription factor and histone ChIP-Seq data. These guidelines are described in **Note 1**. Since these guidelines have not been developed for comparative ChIP-Seq analyses, additional quality checks need to be considered for differential analysis.

The assessment of similarity among samples can be employed by techniques such as principal component analysis (PCA) and correlation heatmaps. These methods provide a graphical representation of the data that helps in visually identifying any unusual patterns or outliers among replicates.

Replicates are required by DiffBind to model the inherent biological variability, since it is expected to observe biological variation between samples, even under identical conditions.

There are no universal recommendations on the number of replicates that should be used for quantitative ChIP-Seq analyses. More replicates increase the robustness and sensitivity of the analysis as it has been shown in a comprehensive study for differential analyses of RNA-Seq data [24, 25]. Furthermore, one should also consider the potential loss of replicates during the experimental procedure.

1. Principal component analysis (PCA)

ChIP-Seq samples are usually compared based on the number of reads per feature (consensus peaks). The samples are represented by an $n \times p$ data matrix X of consensus peaks and the corresponding counts of the samples, which cannot be visualized directly in a low-dimensional space. PCA allows to summarize the data using a smaller number of representative variables that explain most of the variability in the original dataset. A PCA calculation and visualization can be performed with the dba.plotPCA() function of DiffBind. This function first performs the PCA in which principal components (PC) scores are calculated to visualize the data in a low dimensional space. The principal components are, by definition, ordered descending by the proportion of variance explained as shown in Fig. 4a, where the first two PCs are used to visualize the samples in a two-dimensional scatter plot. Samples that are further away from each other are more dissimilar than samples in close proximity. Ideally, in the PCA plot, replicates of the same condition

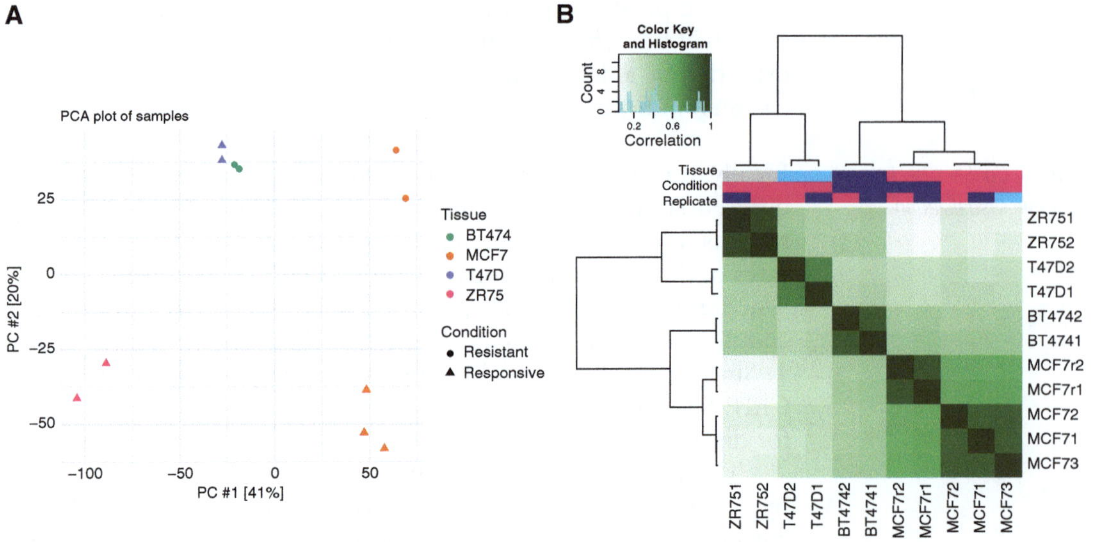

Fig. 4 Sample visualization. (**a**) Exemplary PCA plot. PCA plots are a low-dimensional representation of high-dimensional data. Each sample is shown as a point in a PCA space of the principal components (PCs) 1 and 2. The closer the data points, the more similar the respective samples are. The first PC, which explains 41% of the variance, separates between the different tissues (depicted by color). However, neither PC1 nor PC2 can resolve the different conditions (depicted by shape). (**b**) Exemplary correlation heatmap. The heatmap visualizes the pairwise correlation between samples. A darker color indicates higher similarity, whereas a lighter color intensity represents a weak correlation. The samples are ordered by their hierarchical clustering, which is also depicted on both axes as dendrograms. Additionally, sample descriptions, such as tissue type, condition, and replicate number, are color-coded above the heatmap. Here, all replicates of the same condition and tissue group together. Furthermore, the samples cluster according to the cell line (tissue) but not the tested condition

should form clusters and should be separated from other conditions. If a separation of conditions is not apparent using the first 1–3 PCs, this may indicate (unknown) confounding effects or high similarity of samples across conditions that will likely influence the results. Therefore, the analysis results should be interpreted with caution. PCA plots of high-quality data will cluster replicates of the same condition together. PCA plots are also useful to identify outliers, which are single replicates that are fundamentally dissimilar to the rest of the corresponding replicates.

2. Correlation Heatmaps

In addition to PCA, correlation matrices generated from the pairwise correlation of the samples can be used to visualize the samples' similarity. There are several ways to compute correlation coefficients. For count data, the Pearson coefficient is an appropriate measure. It can range from −1 to 1, with 1 representing perfect and −1 perfect negative correlation. Correlation coefficients of 0 indicate that the samples do not correlate. Depending on the investigated conditions and general biological variation, samples may show a similarly high correlation. When plotted in a heatmap, the calculation of the correlation coefficients is usually complemented with hierarchical clustering to show which samples group based on the ChIP-Seq signal similarity. It is not possible to specify cutoffs for correlation coefficients of high- or low-quality datasets, respectively, as the similarity highly depends on the experimental setup. However, replicates of the same condition should form clusters, which should be visible by the branching in the dendrogram.

Using the in-built R plotting function plot() on a dba object automatically creates a correlation heatmap with an additional hierarchical clustering dendrogram (Fig. 4b).

2.2.5 Normalization Methods in DiffBind

Normalization of ChIP-Seq affinity data is pivotal because it contains noise and technical biases in addition to the biological signal, which restrains direct comparisons across samples or peaks. Selecting a proper normalization strategy is not trivial, as it requires knowledge about the expected outcome and can alter or remove actual biological effects.

DiffBind offers three main normalization methods. Two of them are based on algorithms developed for differential expression analysis of RNA-Seq data, namely, DESeq2 [20] and edgeR [21]. The inherent normalization of DESeq2 is performed by calculating the geometric mean for each peak across samples. edgeR, on the other hand, computes a trimmed mean of M-values, meaning it normalizes excluding a fixed percentage of features with extreme fold changes. Both of these strategies assume that the majority of the features (e.g., genes or peaks) are nondifferential.

This normalization strategy might be applicable for experiments where no global changes are expected (e.g., targeting histone modifications), but it might not be appropriate for experiments that induce many changes in the binding profile.

The design of ChIP-Seq is inherently aimed at identifying regions of genomic enrichment that are likely different between the studied conditions. It is therefore possible that a significant proportion of these enriched regions may indeed be differential, particularly in experiments focusing on transcription factor binding or specific histone modifications that are expected to change under the experimental conditions.

This characteristic of ChIP-Seq experiments is what distinguishes them from gene expression studies, where a considerable proportion of genes are expected to show basal expression levels across different cell types or states. As such, the normalization strategies that work well for RNA-Seq data may not always be the most appropriate for ChIP-Seq data.

Therefore, a simpler normalization method was implemented in DiffBind, which only accounts for differences in sequencing depth. The default for the dba.normalize() method is a normalization by the total library size. Furthermore, any of these three methods may be applied on the reads in peaks (RiP) but also only on the background reads. The number of background reads is estimated by binning the genome into windows of 10 kbp or greater, which are much larger than the (expected) peaks. Any differences between samples within these windows likely result from bias rather than actual biological signals.

DiffBind also provides an option for spike-in normalization, a strategy that can help account for technical variations in ChIP-seq data. This method involves adding a known quantity of chromatin, typically from a different species to the experimental chromatin before immunoprecipitation [26]. Recent genome profiling techniques such as Cut&RUN and Cut&TAG simplify the use of spike-ins using specific reagents [27].

In general, enrichment in histone modification ChIP-Seq samples may be more balanced across conditions than transcription factor ChIP-Seq samples. Therefore, a normalization native to DESeq2 or edgeR on RiP is most likely inappropriate for ChIP-Seq data that is enriched for TF binding events.

Whatever normalization strategy may be chosen, DESeq2 or edgeR native methods on RiP harbor the risk of removing true differential peaks. On the other hand, normalizing only by library size makes no assumption about the balance of differential binding, but it increases the false positive error. It is therefore pivotal to carefully consider which normalization strategy should be applied. Due to the focus on neutral reads, a DESeq2 or edgeR normalization on the background reads harbors less risk of removing true differential peaks while reducing the number of false positive peaks. Although giving universal recommendations is impossible, background normalization may be a good compromise for many use cases.

2.2.6 Contrasts and Differential Analysis

To run the differential analysis in DiffBind, it is necessary to specify the experimental design and the desired comparisons with the dba.contrast() function. Due to the complexity of various experimental setups, we will not explain how to select appropriate design formulas that represent the underlying experiment. Instead, we refer to the DESeq2 vignette [28] and the guide to creating design matrices by Law et al. [29]. Although the guide to creating design matrices is targeted towards the linear model R package limma [30], the presented design formulas are also applicable for DiffBind.

Table 2
Example sample sheet for a differential binding analysis using DiffBind

SampleID	Tissue	Factor	Condition	Replicate	bamReads	Peaks	PeakCaller
<sampleid>	<tissue>	<antibody>	<condition>	<replicate>	<path/to/ bam/file>	<path/to/ peak/file>	<fileformat>
...							

2.2.7 Exemplary Differential Peak Analysis

To understand the individual steps of the differential analysis workflow, we explain the workflow with the example dataset of the DiffBind vignette [31]. In this experiment, cancer cell lines are treated with the anti-cancer drug tamoxifen. A comparison between estrogen receptor alpha (ERa) binding is made of cell lines, which are resistant or responsive to the treatment. Although it is possible to run the complete analysis with one piece of piped code, we recommend applying each DiffBind function separately. This allows to investigate and understand the intermediate output and adjust individual functions step by step.

1. Install R (>= version 4.0).
2. Install the BiocManager and DiffBind by running the following commands in R.

```
if (!require("BiocManager", quietly = TRUE))
 install.packages("BiocManager")
 BiocManager::install("DiffBind")
 # Load libraries
 library(DiffBind)
 # run install.packages('dplyr') if this package is not
installed
 library(dplyr)
```

3. Create a sample sheet similar to the one in Table 2.

   DiffBind allows for different file formats but requires specific naming conventions for columns. Please refer to the exemplary analysis [23] for details.

4. Create the dba object based on the sample sheet.

```
 tamoxifen <- dba(sampleSheet = "location/of/the/sample/
sheet")
```

5. Apply the appropriate blacklist. The reads were aligned to the hg38 genome; hence, we selected the DBA_BLACKLIST_HG38 blacklist. Moreover, we set the greylist option to

TRUE to invoke a computation of the greylist based on the provided control samples. Note that this calculation of greylists is computationally expensive and may take a while.

```
tamoxifen <- dba.blacklist(tamoxifen, blacklist = DBA_BLACK-
LIST_HG38, greylist = TRUE)
```

6. Add the count data to the dba object and compute the consensus peaks. This is a computationally intensive step, which may exhaust the computational resources (*see* **Note 7**).

```
tamoxifen <- dba.count(tamoxifen, score = DBA_SCORE_NORMA-
LIZED)
```

7. Check the sample quality and correlation with a PCA plot and a correlation heatmap. The PCA plot can be generated with the DiffBind function dba.plotPCA( ). This can result in overlapping sample labels. To resolve this issue, please see the code in our supplied RMarkdown.

```
dba.plotPCA(tamoxifen)
 plot(tamoxifen)
```

8. Normalize the raw counts. Because we do not want to make assumptions about the data distribution, we perform a simple normalization by the library size.

```
tamoxifen <- dba.normalize(tamoxifen, background = 10000,
normalize = DBA_NORM_LIB)
```

9. Specify the desired contrast and underlying experimental design. The experimental design considers only two different conditions, "resistant" and "responsive," and aims to find differences between them. Hence, we chose a simple design formula taking into account only "Condition" as a variable. Furthermore, we reorder the list of conditions, starting with the control group "responsive," because we want to ensure that a positive $\log_2$ fold change will correspond to increased binding in the "resistant" group.

```
tamoxifen <- dba.contrast(tamoxifen, design = "~ Condition",
reorderMeta = list(Condition = "Responsive")
```

10. Run the differential peak analysis.

```
tamoxifen <- dba.analyze(tamoxifen, method = DBA_DESEQ2,
bBlacklist = DBA_BLACKLIST_GRCH38, bGreylist = TRUE)
```

11. Extract the analysis results. By default, the dba.report() function will apply a threshold on the significance value (FDR) and filter the results. Additionally, we recommend applying a threshold on the $\log_2$ fold change (fold = 1). To get a full list of peaks, including nondifferential ones, set the threshold option to 1 (th = 1) without setting the fold option.

```
tamoxifen.DB <- dba.report(tamoxifen, fold = 1)
 write.csv(tamoxifen.DB, 'differential_binding.csv', row.names
= FALSE, quote = FALSE)
 write.csv(dba.report(tamoxifen, th = 1), 'differential_bin-
ding_all.csv', row.names = FALSE, quote = FALSE)
```

2.3 Visualization of Results

Visualizing analysis results is key to conveying the findings to fellow scientists in an easily interpretable manner. Differential analysis results are often depicted with volcano plots or MA plots, but also PCA plots and correlation heatmaps can be used to highlight the findings. In addition to their use as quality control plots, PCA plots and correlation heatmaps are helpful for hypothesis generation because they can illuminate biological patterns, especially for complex experimental designs with more than two conditions.

2.3.1 Volcano Plots

One very popular visualization strategy for differential analysis results is volcano plots. They show the statistical significance of the data against the magnitude of the differences between the compared conditions, the $\log_2$ fold change. Each point in the plot represents a feature, for example, a peak. Significant peaks are usually highlighted with color, and therefore, also the chosen thresholds become apparent from plots. Peaks with decreased differential binding can be seen on the left side, while peaks with increased binding can be found on the right side of the plot. Moreover, highly differential peaks will appear as outliers (Fig. 5a). DiffBind offers the dba.plotVolcano() function for this purpose. Please see Boeckel et al. [23] for exemplary code.

2.3.2 MA Plots

MA plots were initially created for microarrays to show possible biases in log2-fold change due to the expression levels of the genes. However, they are also helpful for the visualization of differential ChIP-Seq analysis results. Here, the $\log_2$ fold change (M-value) is plotted against the mean normalized read count over all samples (log concentration; A-value) for each peak. Figure 5b depicts an

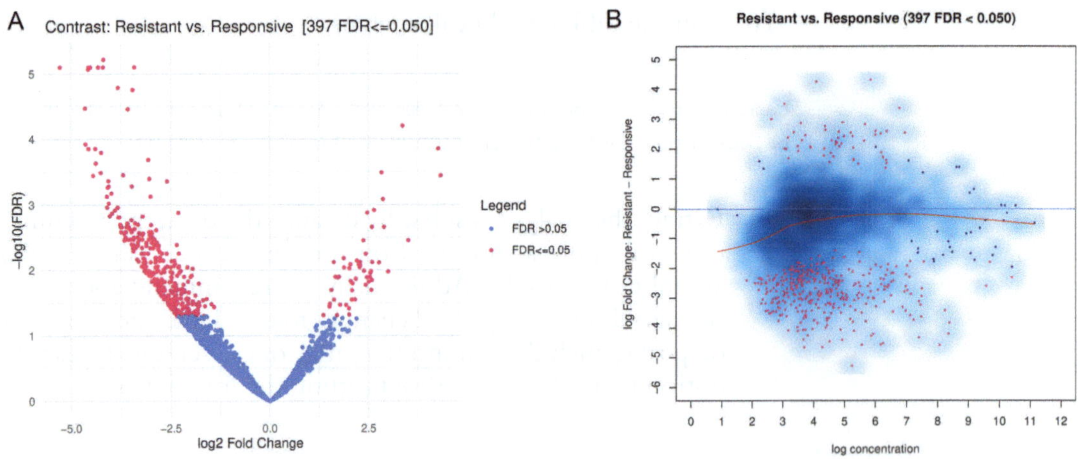

Fig. 5 (**a**) Exemplary volcano plot. Volcano plots show the negative decadent logarithm of the significance value (here FDR) against the $\log_2$ fold change of each peak. All peaks in pink represent significantly differential peaks. The population on the left side of the plot represents differential peaks with decreased binding in the resistant group, whereas the peaks on the right-hand side have increased binding in the resistant samples compared to the responsive ones. (**b**) Exemplary MA plot. MA plots are scatter plots, which show the $\log_2$ fold change (M-value) against the mean normalized read count over all samples (log concentration; A-value). All nondifferential peaks are smoothened to a density cloud (dark to light blue background). All outliers and differential peaks are depicted as individual points. Significant differential peaks are highlighted in pink. The blue line divides the plot into peaks with increased and decreased binding, respectively. All data points on the blue line are nondifferential. Additionally, the red line represents a fitted Loess curve, which depicts local biases within the data. This plot is an example of an unbalanced dataset with decreased transcription factor binding in the resistant group compared to the responsive samples; hence, the red curve is consistently below the blue line

exemplary MA plot. The plot was generated using DiffBinds dba. plotMA() function. Please see the exemplary analysis [23] for details and example code. The blue line separates the plot into peaks with increased and decreased binding at a $\log_2$ fold change of zero with no difference between the compared conditions. DiffBind smoothes the data points into a density cloud (dark blue) per default and only shows the highly differential peaks as individual points. All significantly differential peaks are highlighted in pink. The shape and position of the cloud indicate intensity-dependent biases in the data. This is supported by the red line, which is a Loess fit curve that shows local trends among the data. If the cloud and the red line are located above or below the blue line at a $\log_2$ fold change of zero, the binding events are unbalanced between the conditions. As discussed in Subheading 2.2.5, this may be an actual biological difference due to the enrichment of certain events. However, suppose a balanced distribution for the peaks is expected, then a native DESeq2 or edgeR normalization can correct for this and normalize the data in a way that the red Loess fit curve approaches the blue line.

As for the volcano plot, the MA plot can highlight highly differential peaks.

2.4 Potential Downstream Analyses

Different downstream analyses are possible depending on the type of protein that was immunoprecipitated. Certain histone modifications are associated with the presence of *cis*-regulatory elements, which regulate gene expression.

2.4.1 Peak Annotation: Linking Peaks to Genes

One approach to put the peaks in a functional context is associating the peaks with genes.

This annotation might be trivial when the peaks are expected to be observed for promoters and can be found in the vicinity of the transcriptional start site (TSS). Overlaps of peaks with regions around the TSS indicate their influence on the gene's expression. However, the annotation of enhancers is more complex, as they can be located distant from the regulated gene's TSS and can be found in intergenic regions as well as introns and even exons of neighboring genes.

The web-based tool GREAT (Genomic Regions Enrichment of Annotations Tool) aims to not only perform these annotations but also provides testing for statistical enrichment of ontologies and pathways with the annotated genes [32]. GREAT offers three different algorithms to determine all genes, which are potentially regulated by *cis*-regulatory elements at the position of the peaks. Per default, the basal plus extension approach is applied, which determines regulatory domains for each gene by taking basal windows 5 kb upstream and 1 kb downstream around the TSS and extending them to the next gene, which is less than 1000 kbs away. Each peak that falls into such a regulatory domain will be associated with the corresponding genes within that window. Due to the lack of an output that includes the annotations for each peak, we recommend running the GREAT algorithm in R using rGREAT [33] and to perform the pathway enrichment analysis based on the list of annotated gene symbols.

1. Install R (>= version 4.0).
2. Install the required R packages by running the following lines of code in R:

```
if (!require("BiocManager", quietly = TRUE))
 install.packages("BiocManager")
 BiocManager::install("rGREAT")
```

3. Extract the position of the transcriptional start sites (TSS) of your genome of interest in the form of a GRanges object [34]. We will use the human genome hg38 in accordance with the dataset used for DiffBind.

```
install.packages("TxDb.Hsapiens.UCSC.hg38.knownGene")
library(TxDb.Hsapiens.UCSC.hg38.knownGene)
txdb <- TxDb.Hsapiens.UCSC.hg38.knownGene
genes_gr <- genes(txdb)
tss_gr <- promoters(genes_gr, 0,1)
```

4. Define regulatory domains around the TSS of the gene using the basal plus extension algorithm by GREAT.

```
etss <- extendTSS(tss_gr, genome = "hg38", gene_id_type =
"ENTREZ", basal_upstream = 5000, basal_downstream = 1000,
extension = 1000000)
```

5. Determine overlaps between peaks of interest with the defined regulatory domains.

```
basal_ext <- findOverlaps(tamoxifen.DB, etss, type='any',
ignore.strand=TRUE)
basal_ext_gr <- tamoxifen.DB[queryHits(basal_ext)]
basal_ext_gr$gene_id <- etss[subjectHits(basal_ext)]$gene_id
```

6. Convert the Entrez gene IDs to gene symbols and Ensembl gene IDs using biomaRt [35].

```
mart <- useEnsembl(biomart = "genes", dataset = "hsapiens_gen-
e_ensembl")
res_basal <- getBM(attributes = c('entrezgene_id', 'ensembl_-
gene_id', 'external_gene_name'),
filters = 'entrezgene_id',
values = basal_ext_gr$gene_id,
mart = mart)
```

7. Merge the gene info with the differential gene list.

```
basal_ext_res <- merge(as.data.frame(basal_ext_gr, row.names
= NULL), res_basal, by.x="gene_id", by.y="entrezgene_id", all.
x=T, all.y=T)
basal_ext_res <- unique(basal_ext_res)
```

8. Extract the list of identified genes and perform a pathway enrichment analysis of your choice.

```
write.csv(basal_ext_res, "annotated_peaks.csv", row.names = F,
quote = F)
```

2.4.2 Motif Discovery

Transcription factors are proteins that influence gene expression by binding to DNA segments in *cis*-regulatory regions. These DNA regions, also known as transcription factor binding motifs (TFBMs), follow a specific pattern, which allows the binding of the respective transcription factor. ChIP-Seq of immunoprecipitation of transcription factors often tries to find novel TFBMs. The website meme-suite.org offers multiple web-based tools related to motif discovery, enrichment analysis, scanning, or comparison.

De-novo motif discovery based on ChIP-Seq data may be performed with MEME-ChIP [36]. Its primary purpose is the identification of novel and known motifs by focusing on the center of the provided sequences. Additionally, it compares the identified motifs to a selected database of known motifs.

A comparison of motifs within different subsets of sequences can be done with AME [37]. For example, one could compare the sequences from differential peaks with increased binding to differential peaks with decreased binding. In this case, one can provide the interesting set of sequences as the primary sequences and compare it to the other set of sequences by using the "User-provided control sequences" option.

1. Install R (<= version 4.0).
2. Install BSgenome [38] to extract the sequences of differential peaks.

```
if (!require("BiocManager", quietly = TRUE))
 install.packages("BiocManager")
 BiocManager::install("BSgenome")
 library(BSgenome)
```

3. Load the required genome.

```
bs_genome <- getBSgenome("BSgenome.Hsapiens.UCSC.hg38")
```

4. Extract the sequences of the peaks of interest and save them in a fasta file.

```
seq <- getSeq(bs_genome, tamoxifen.DB)
 names(seq) <- as.character(tamoxifen.DB)
 writeXStringSet(seq, "peak_sequences.fa", format="fasta")
```

We recommend performing downstream analyses such as pathway enrichment and motif analyses on separated subsets of peaks with increased and decreased binding, respectively. Because these peaks represent contrary conditions, investigating them as a combined set will lead to indistinct results.

2.4.3 Visualization of Results in a Genome Browser

The results can be visualized in a genome browser. Very popular are the UCSC Genome Browser [39] and the Integrative Genomics Viewer (IGV) [40]. Both offer a web application. While both browsers are very versatile and provide many options to visualize the data, only IGV provides an app that can be installed locally on a PC. It allows visualizing different types of genomic data, such as BAM, BED, BedGraph, or bigwig files.

Typical ChIP-Seq or peak-based data visualization includes the position of the peaks (provided from a BED file), which can be displayed as blocks in the genome along with the gene annotation. Furthermore, wig or bigwig files are used to show (normalized) read signals across the genome (*see* **Note 8**). For differential peaks, differences between conditions can be highlighted by showing their signal intensity within these windows (Fig. 6).

3 Notes

1. Besides the standard QC metrics for sequencing data that can be provided by tools like FastQC and the aligners, there are other specifics to be considered for ChIP-Seq data:
   - A bias in the GC content is common in ChIP-Seq, Cut&Run, or Cut&Tag data because of preferential binding of the protein of interest and is not a quality issue in this case.
   - For chromatin profiling that involves the use of transposases such as for Cut&Tag or ATAC-Seq, biases in the per base sequence composition at the beginning of the sequences are expected and negligible.
   - Fraction of Reads in Peaks (FRiP), between 0 and 1, tends to have small values for transcription factor analysis and larger ones for histone modifications. In a guideline published in 2012, the ENCODE consortium suggested FRiP scores of at least 0.01 [13].
   - Strand cross-correlation plots, where in general two peaks are to be observed, a smaller one for the read length ("phantom" peak) and a higher one for the average fragment length of the library ("ChIP" peak). A small ChIP peak or the absence of it may indicate a failed experiment.
   - Normalized Strand Coefficient (NSC) is the ratio between the fragment length cross-correlation peak and the background cross-correlation, with values between 1 and infinite. Experiments with very small values (below 1.05) are usually very poor in performance [13], although cases where the transcription factor is known to bind only to a low number of regions may also show extremely low values.

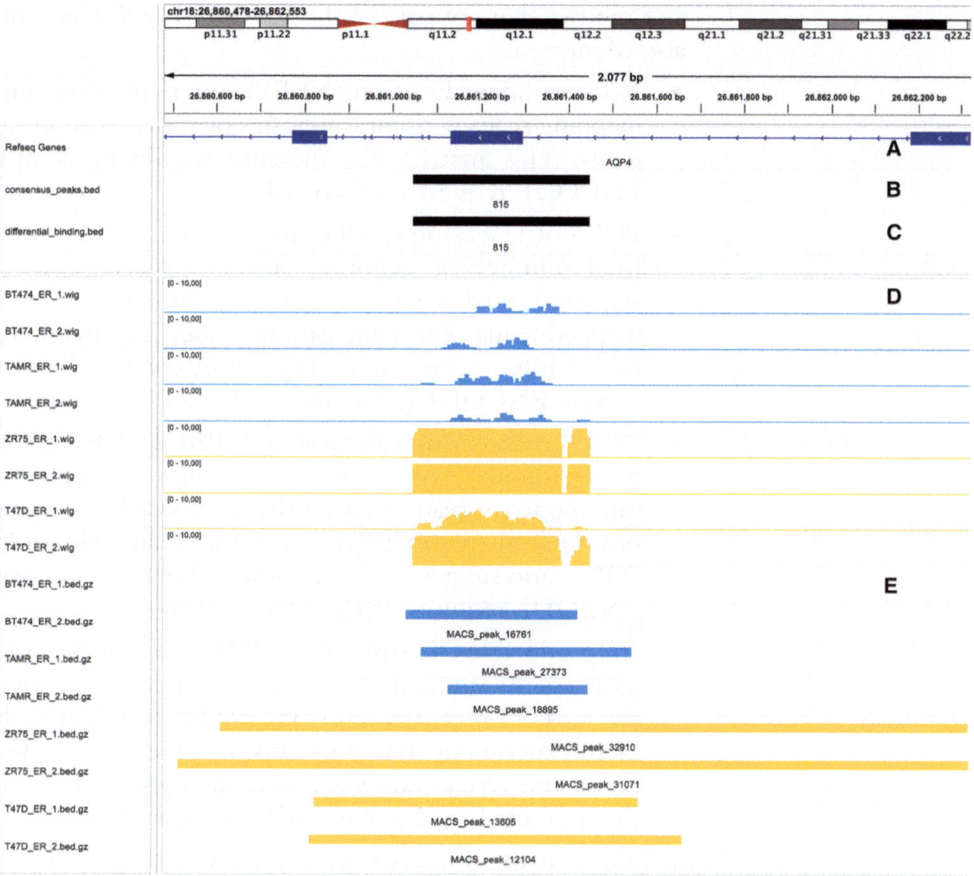

Fig. 6 Visualization of differential ChIP-Seq data in IGV. Various file types of ChIP-Seq data can be loaded and depicted in genome viewer like IGV. Each line represents a track. At the top, a gene annotation is shown (**a**). Peaks from BED files are represented as blocks (**b**, **c**, **e**). The depicted genomic region includes a consensus peak (**b**). All individual samples, except for BT474_ER_1, have a peak within this window. However, their exact width or position differs between the samples (**e**). DiffBind also identified this peak as differential (**c**) between the conditions resistant (blue) and responsive (orange). The signal within these peaks can be visualized with signal tracks in a wig format (**d**). Scaling them all to the same intensity window (0–10.00) allows comparison between the signal tracks and highlights the difference in signal intensity between the conditions

- Relative Strand Correlation (RSC) is the ratio between the peaks of the strand cross-correlation, with values between 0 and infinite. Experiments with very small values (below 0.8) are usually very poor in performance [13]. However, cases where the transcription factor is known to bind only to a low number of regions (below 200) may also show extremely low values.

Further metrics that are part of the ENCODE guidelines may be also of interest:

- Nonredundant Fraction (NRF) is the number of uniquely mapping reads without duplicates over the total number of reads. This metric is a measure of library complexity. ENCODE request values of >0.9
- PCR Bottlenecking Coefficients 1 (PBC1) is the ratio of the total number of genomic positions where exactly one sequence read starts to the total number of uniquely mapped reads. The value can range between 0 and 1, with higher PBC1 ratios suggesting a more complex library. The current ENCODE guidelines suggest a PBC1 value of $\geq$0.9
- PCR Bottlenecking Coefficients 2 (PBC2) is the ratio of the total number of genomic positions where exactly one-sequence read starts to the total number of genomic positions where exactly two-sequence reads start. A higher PBC2 ratio suggests a more complex library. Submission to ENCODE requires PBC2 $\geq$10.
- Irreproducible Discovery Rate (IDR) is a statistical method to assess the similarity between two replicates. It compares the ranks of measurements observed in different replicates of an experiment. Based on this, so-called self-consistency ratios and rescue ratios are calculated and at least one of them should be $\leq$2 to have sufficient quality for ENCODE.

Please note that these metrics are currently not generated by the nf-core ChIP-Seq ($\leq$v2.0.0) or cutandrun ($\leq$v3.1), but it is planned to be implemented in the future.

For further information, the following references may be of interest:

- Output of nf-core ChIP-Seq workflow explained:

  https://nf-co.re/chipseq/2.0.0/output
- ENCODE Quality metrics explained:

  https://genome.ucsc.edu/ENCODE/qualityMetrics.html

  https://www.encodeproject.org/data-standards/

2. Controls are listed as samples in the sample sheet. For these lines, antibody and control are not specified. Furthermore, if you are working with single-end data, only indicate the path for fastq_1. In any case, the sample sheet should contain all five columns listed.

3. To run the ChIP-Seq workflow, nextflow (https://www.nextflow.io) and Java must be installed. We recommend installing them with conda (https://docs.conda.io/en/latest/).

   To check if everything is properly set up, a test run of the ChIP-Seq workflow can be done:

```
$ nextflow run nf-core/chipseq -profile test,YOURPROFILE --
outdir <OUTDIR>
```

   The following parameters are mandatory to run the workflow:

--input	Path to comma-separated file containing information about the samples in the experiment.
--outdir	The output directory where the results will be saved. You have to use absolute paths to store on Cloud infrastructure.
--fasta	Path to FASTA genome file. If you don't have the appropriate alignment index available, this will be generated for you automatically. Combine with --save_reference to save the alignment index for future runs.
--gtf	Path to GTF annotation file.

   The following parameters, although optional, are strongly recommended:

--blacklist	If provided, alignments that overlap with the regions in this file will be filtered out (see https://sites.google.com/site/anshulkundaje/projects/blacklists). The file should be in BED format.

   The following parameters may be of interest:

--narrow_peak	MACS2 is run by default with the --broad flag. Specify this flag to call peaks in narrowPeak mode.
--save_reference	If the pipeline generates the BWA index, use this parameter to save it to your results folder. These can be used for future pipeline runs with the --bwa_index parameter, reducing processing times.
--deseq2_vst	Use vst transformation instead of rlog with DESeq2, which is much faster to compute.

   For more information on how to run the ChIP-Seq workflow, you can check the documentation at https://nf-co.re/chipseq.

4. For Cut&Run and Cut&Tag experiments, we recommend using the nf-core/cutandrun pipeline, due to their low background signal in comparison to ChIP-Seq data.

   To use this pipeline, you first have to create a sample sheet with the following header (Table 3).

Table 3
Example sample sheet for a Cut&Run workflow

group	replicate	fastq_1	fastq_2	control
<groupname>	<replicate>	<path/to/first/fastq/file>	<path/to/second/fastq/file>	<groupnameofcontrol>
...				

The group name hereby refers to the antibody.

Execute the following command to run the pipeline.

```
$ nextflow run nf-core/cutandrun -r 3.0 -profile test,docker
--outdir
```

For the cutandrun pipeline, the BAM files can be found at results/02_alignment/bowtie2/target as *.target.dedup.sorted.bam. Compared to the chipseq pipeline, peaks are called with SEACR for Cut&Run or Cut&Tag data, because these methods produce little background reads.

These files can be found at results/03_peak_calling/04_called_peaks/seacr with the .bed file extension.

5. When analyzing ChIP-Seq data, one must remember the expected chromatin profile. While transcription factors produce distinct, sharp peaks, histone modifications usually have a wider range. Some histone modifications may produce peaks in the range of hundreds of bp, whereas others, such as H3K27me3 profiles, can cover multiple hundred kilobases. These differences in peak width and shape should affect tool and parameter choices.

   One crucial aspect is the peak calling process. The popular peak caller MACS—and its versions MACS2 and MACS3—offers two modes for peak calling, narrow or broad [15]. Narrow peak calling is appropriate for ChIP-Seq targeting TFs and specific histone modifications. Broad peak calling also includes narrow peak calling, followed by the merging of adjacent narrow peaks. MACS2 might underperform for broad and diffuse peaks. In this case, peak callers developed for broad peaks like SICER [41] or BroadPeak [42] may be a better choice. In the differential peak analysis with DiffBind, the expected chromatin profile affects the selection of an appropriate extension width during the consensus peak calling in dba.count(). For TF ChIP-Seq peaks, an extension of a few hundred base pairs in each direction is feasible. This value should be increased for mid-size and broad peaks. Eventually, the type of immunoprecipitated protein can also dictate the necessary normalization strategy to remove biases. For an incomplete classification of histone marks into narrow and broad peaks, please refer to the Encode list of ChIP-Seq standards (https://www.encodeproject.org/chip-seq/histone/).

6. DiffBind provides ENCODE exclusion lists, which are limited to human, mouse, worm, and fly genomes [8]. Blacklists for other genomes can be obtained from other sources specified and constructed for use with DiffBind.

   Greylists, as well as blacklists, can be computed with the GreyListChIP package [43].

   Including the input control bam files in the sample sheet and setting the option greylist = T in the dba.blacklist function to TRUE will automatically invoke the calculation of greylists with the GreyListChIP.

   Both customized blacklists and greylists can be applied with DiffBind by executing the following steps.

   1. Load blacklist as a data frame into R.
   2. Create a GRanges object from the data frame.

```
$ makeGRangesFromDataFrame(df, keep.extra.columns = TRUE,
ignore.strand = FALSE, seqinfo = NULL)
```

   3. Provide the GRanges object as an argument for the blacklist or greylist option of the dba.blacklist() function.

7. Per default, the dba.count() function will run in parallel on all available cores to speed up the computation. However, the execution may throw an error if memory resources are limited. To circumvent this issue, run dba.count() with the bParallel option set to FALSE or set bLowMem to TRUE (https://support.bioconductor.org/p/52689/).

8. Bigwig files of the individual samples can be retrieved from the nf-core pipeline output. In the case of the ChIP-Seq pipeline, they are normalized to one million mapped reads to make them comparable across samples. These files can be found in the results directory under bwa/bigwig/*.BigWig. Furthermore, the nf-core pipelines generate IGV sessions, including all files of interest that were generated during preprocessing. These can be found at igv/<peaktype>. The respective files for the Cut&RUN pipeline are located at 03_peak_calling/03_bed_to_bigwig/*.BigWig and 04_reporting/igv.

References

1. Allis CD, Jenuwein T (2016) The molecular hallmarks of epigenetic control. Nat Rev Genet 17(8):487–500
2. Buck MJ, Lieb JD (2004) ChIP-chip: considerations for the design, analysis, and application of genome-wide chromatin immunoprecipitation experiments. Genomics 83(3):349–360
3. Pellegrini M, Ferrari R (2012) Epigenetic analysis: ChIP-chip and ChIP-seq. Methods Mol Biol 802:377–387

4. Park PJ (2009) ChIP-seq: advantages and challenges of a maturing technology. Nat Rev Genet 10(10):669–680
5. Furey TS (2012) ChIP-seq and beyond: new and improved methodologies to detect and characterize protein-DNA interactions. Nat Rev Genet 13(12):840–852
6. Xu J, Kudron MM, Victorsen A et al (2021) To mock or not: a comprehensive comparison of mock IP and DNA input for ChIP-seq. Nucleic Acids Res 49(3):e17
7. Jain D, Baldi S, Zabel A et al (2015) Active promoters give rise to false positive 'Phantom Peaks' in ChIP-seq experiments. Nucleic Acids Res 43(14):6959–6968
8. Amemiya HM, Kundaje A, Boyle AP (2019) The ENCODE blacklist: identification of problematic regions of the genome. Sci Rep 9(1): 9354
9. Stark R, Brown G (2011) DiffBind: differential binding analysis of ChIP- Seq peak data. https://bioconductor.org/packages/devel/bioc/vignettes/DiffBind/inst/doc/DiffBind.pdf. Accessed 20 March 2023
10. Gentleman RC, Carey VJ, Bates DM et al (2004) Bioconductor: open software development for computational biology and bioinformatics. Genome Biol 5(10):R80
11. Schmidt D, Wilson MD, Spyrou C et al (2009) ChIP-seq: using high-throughput sequencing to discover protein-DNA interactions. Methods 48(3):240–248
12. Kaya-Okur HS, Wu SJ, Codomo CA et al (2019) CUT&Tag for efficient epigenomic profiling of small samples and single cells. Nat Commun 10(1):1930
13. Landt SG, Marinov GK, Kundaje A et al (2012) ChIP-seq guidelines and practices of the ENCODE and modENCODE consortia. Genome Res 22(9):1813–1831
14. Thomas R, Thomas S, Holloway AK et al (2017) Features that define the best ChIP-seq peak calling algorithms. Brief Bioinform 18(3): 441–450
15. Zhang Y, Liu T, Meyer CA et al (2008) Model-based analysis of ChIP-Seq (MACS). Genome Biol 9(9):R137
16. Di Tommaso P, Chatzou M, Floden EW et al (2017) Nextflow enables reproducible computational workflows. Nat Biotechnol 35(4): 316–319
17. Molder F, Jablonski KP, Letcher B et al (2021) Sustainable data analysis with Snakemake. F1000Res 10:33
18. Crusoe MR, Abeln S, Iosup A et al (2022) Methods included: standardizing computational reuse and portability with the common workflow language. Commun ACM 65(6): 54–63
19. Ewels PA, Peltzer A, Fillinger S et al (2020) The nf-core framework for community-curated bioinformatics pipelines. Nat Biotechnol 38(3):276–278
20. Love MI, Huber W, Anders S (2014) Moderated estimation of fold change and dispersion for RNA-seq data with DESeq2. Genome Biol 15(12):550
21. Robinson MD, McCarthy DJ, Smyth GK (2010) edgeR: a Bioconductor package for differential expression analysis of digital gene expression data. Bioinformatics 26(1): 139–140
22. Eder T, Grebien F (2022) Comprehensive assessment of differential ChIP-seq tools guides optimal algorithm selection. Genome Biol 23(1):119
23. Boeckel C, Pastor X, Heinig M et al (2023) Tutorial on differential ChIP-Seq analysis using the DiffBind package. Zenodo. https://doi.org/10.5281/zenodo.7764086
24. Schurch NJ, Schofield P, Gierlinski M et al (2016) How many biological replicates are needed in an RNA-seq experiment and which differential expression tool should you use? RNA 22(6):839–851
25. Stark R, Hadfield J (2016) Characterization of DNA-Protein Interactions: design and analysis of ChIP-Seq experiments
26. Chen K, Hu Z, Xia Z et al (2015) The overlooked fact: fundamental need for spike-in control for virtually all genome-wide analyses. Mol Cell Biol 36(5):662–667
27. Meers MP, Bryson TD, Henikoff JG et al (2019) Improved CUT&RUN chromatin profiling tools. elife 8:e46314
28. Love MI, Anders S, Huber W (2023) Analyzing RNA-seq data with DESeq2. http://bioconductor.org/packages/release/bioc/vignettes/DESeq2/inst/doc/DESeq2.html. Accessed 20 Mar 2023
29. Law CW, Zeglinski K, Dong X et al (2020) A guide to creating design matrices for gene expression experiments. F1000Res 9:1444
30. Ritchie ME, Phipson B, Wu D et al (2015) limma powers differential expression analyses for RNA-sequencing and microarray studies. Nucleic Acids Res 43(7):e47
31. Ross-Innes CS, Stark R, Teschendorff AE et al (2012) Differential oestrogen receptor binding is associated with clinical outcome in breast cancer. Nature 481(7381):389–393
32. McLean CY, Bristor D, Hiller M et al (2010) GREAT improves functional interpretation of

cis-regulatory regions. Nat Biotechnol 28(5): 495–501

33. Gu Z, Hubschmann D (2023) rGREAT: an R/bioconductor package for functional enrichment on genomic regions. Bioinformatics 39(1):btac745
34. Lawrence M, Huber W, Pages H et al (2013) Software for computing and annotating genomic ranges. PLoS Comput Biol 9(8):e1003118
35. Durinck S, Spellman PT, Birney E et al (2009) Mapping identifiers for the integration of genomic datasets with the R/Bioconductor package biomaRt. Nat Protoc 4(8):1184–1191
36. Machanick P, Bailey TL (2011) MEME-ChIP: motif analysis of large DNA datasets. Bioinformatics 27(12):1696–1697
37. McLeay RC, Bailey TL (2010) Motif Enrichment Analysis: a unified framework and an evaluation on ChIP data. BMC Bioinform 11:165
38. Pages H (2023) BSgenome: software infrastructure for efficient representation of full genomes and their SNPs. https://bioconductor.org/packages/BSgenome
39. Nassar LR, Barber GP, Benet-Pages A et al (2023) The UCSC Genome Browser database: 2023 update. Nucleic Acids Res 51(D1): D1188–D1195
40. Thorvaldsdottir H, Robinson JT, Mesirov JP (2013) Integrative Genomics Viewer (IGV): high-performance genomics data visualization and exploration. Brief Bioinform 14(2): 178–192
41. Xu S, Grullon S, Ge K et al (2014) Spatial clustering for identification of ChIP-enriched regions (SICER) to map regions of histone methylation patterns in embryonic stem cells. Methods Mol Biol 1150:97–111
42. Wang J, Lunyak VV, Jordan IK (2013) BroadPeak: a novel algorithm for identifying broad peaks in diffuse ChIP-seq datasets. Bioinformatics 29(4):492–493
43. Brown G (2022) GreyListChIP: Grey lists – mask Artefact regions based on ChIP inputs

Chapter 6

An Optimized High-Resolution Mapping Method for Glucocorticoid Receptor-DNA Binding in Mouse Primary Macrophages

Suhail A. Ansari and Nina Henriette Uhlenhaut

Abstract

ChIP-exo is a powerful tool for achieving enhanced sensitivity and single-base-pair resolution of transcription factor (TF) binding, which utilizes a combination of chromatin immunoprecipitation (ChIP) and lambda exonuclease digestion (exo) followed by high-throughput sequencing. ChIP-nexus (chromatin immunoprecipitation experiments with nucleotide resolution through exonuclease, unique barcode, and single ligation) is an updated and simplified version of the original ChIP-exo method, which has reported an efficient adapter ligation through the DNA circularization step. Building upon an established method, we present a protocol for generating NGS (next-generation sequencing) ready and high-quality ChIP-nexus library for glucocorticoid receptor (GR). This method is specifically optimized for bone marrow–derived macrophage (BMDM) cells. The protocol is initiated by the formation of DNA-protein cross-links in intact cells. This is followed by chromatin shearing, chromatin immunoprecipitation, ligation of sequencing adapters, digestion of adapter-ligated DNA using lambda exonuclease, and purification of single-stranded DNA for circularization and library amplification.

Key words ChIP-exo, ChIP-nexus, Glucocorticoid receptor, Bone marrow–derived macrophages, Exonuclease digestion

1 Introduction

The glucocorticoid receptor (GR) obtains its regulatory specificity by recognizing specific genomic glucocorticoid response elements (GREs). Chromatin immunoprecipitation followed by sequencing (ChIP-seq) has been widely adopted to determine GR binding to genomic loci [1–6]. However, it has inherent high background noise and a resolution of several hundred base pairs. ChIP-exo circumvents this limitation by allowing to determine sequence-specific footprints of TF binding [7, 8]. In this method, the lambda exonuclease treatment trims one strand of the double-stranded DNA in a 5′-to-3′ direction and stops when it reaches near

Franziska Greulich (ed.), *Chromatin Immunoprecipitation: Methods and Protocols*, Methods in Molecular Biology, vol. 2846,
https://doi.org/10.1007/978-1-0716-4071-5_6,

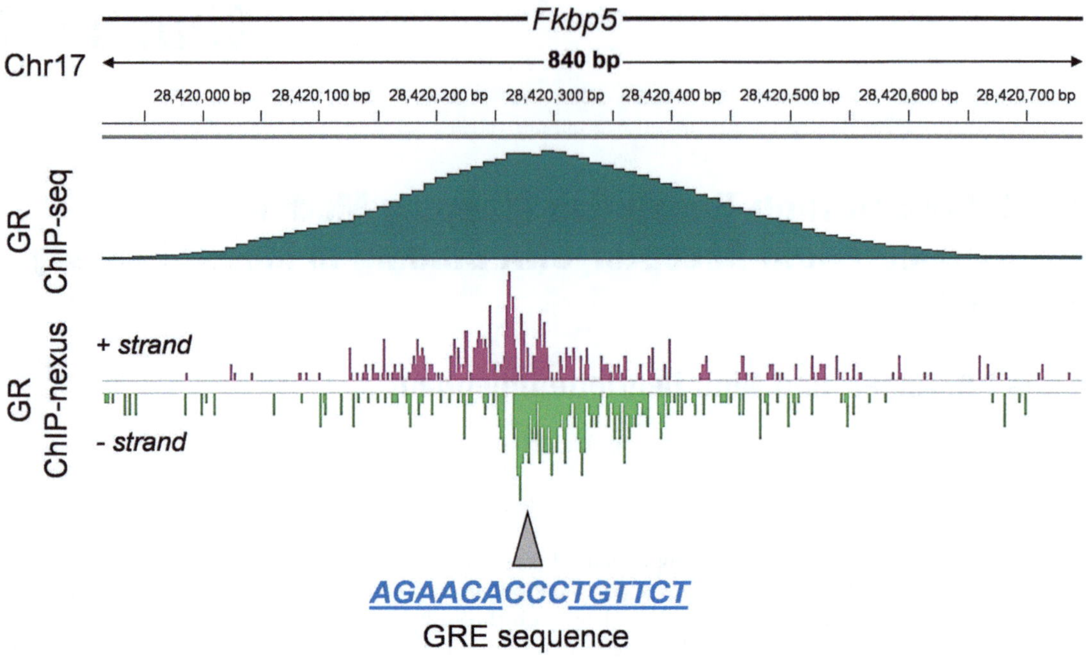

Fig. 1 Comparison of ChIP-seq and ChIP-nexus. Integrative Genomics Viewer (IGV) snapshot showing an example of the glucocorticoid receptor binding at a *Fkbp5* enhancer. The position of a glucocorticoid response element (GRE) sequence is also depicted. The underlined sequences represent the two half-sites of the GRE. ChIP-seq data for GR was retrieved from the GEO database (GSE1865112)

protein-DNA cross-linking point. However, the technical complexity of the original ChIP-exo methodology makes library preparation more difficult, frequently resulting in inefficient ligation of adapters on both ends of DNA. ChIP-nexus was developed as a simplified version of the ChIP-exo protocol in 2015, which combined the standard ChIP-exo protocol with the library preparation protocol from the iCLIP method [9, 10]. Although the procedure remains technically challenging and costly, the use of CircLigase in this method has significantly improved adapter ligation [11]. Compared to ChIP-seq, ChIP-nexus demonstrates enhanced specificity and near base-pair resolution for protein-DNA binding events (Fig. 1).

2 Materials

2.1 Preparation of Cross-Linked Cells

1. Dulbecco's phosphate-buffered saline (DPBS).
2. Cell fixation solution: 1% (v/v) formaldehyde solution in DPBS. It is better to prepare fresh solutions each time, as needed. Always use molecular biology grade and methanol-free formaldehyde.

3. 1 M glycine: Dissolve glycine in double distilled water (ddH_2O). Sterilize by filtering using a 0.22 μm filter.
4. Desired cells, e.g., freshly differentiated bone marrow-derived macrophages (BMDMs).
5. Cell scraper.
6. 2 mL tubes.
7. Cooling centrifuge for 2 mL tubes.

2.2 Preparation of Beads

1. 1 M Tris–HCl pH 8.0 (molecular biology grade).
2. 20% (w/v) Sodium Dodecyl Sulfate (SDS) solution (molecular biology grade).
3. 10% (w/v) N-Lauroylsarcosine sodium salt (molecular biology grade): Prepare a 10% (w/v) stock solution in sterile ddH_2O.
4. 10% (w/v) Sodium Deoxycholate Detergent (molecular biology grade): Prepare a 10% (w/v) stock solution in sterile ddH_2O.
5. Shearing buffer: 10 mM Tris–HCl pH 8.0, 140 mM NaCl, 1% (v/v) Tween X-100, 0.1% (w/v) sodium deoxycholate, 0.5% (w/v) N-lauroylsarcosine, 0.2% (w/v) SDS. Prepare the buffer with ddH_2O. Sterilize by filtering using a 0.22 μm filter. Store for up to 6 months at 4 °C. When using, always make an aliquot and freshly add protease inhibitor.
6. Protease inhibitor tablets (EDTA-free).
7. Glucocorticoid receptor antibody (Proteintech, Catalogue# 24050-1-AP) (*see* **Note 1**).
8. Dynabeads M-280 Sheep Anti-Rabbit IgG (Life Technologies) (*see* **Note 2**).
9. Low DNA binding microcentrifuge tubes, 1.5 mL.
10. Magnetic separation rack for 1.5/2 mL tubes.
11. Rotator for 1.5 mL microcentrifuge tubes.

2.3 Preparation of Chromatin Extracts and Shearing

1. Bioruptor with water cooler (Diagenode).
2. 1.5 mL microtubes for biorupter (Diagenode).
3. Low DNA binding microcentrifuge tubes, 2 mL.

2.4 Chromatin Immunoprecipitation (ChIP)

1. 1 M Ethylenediaminetetraacetic acid (EDTA) disodium salt solution (molecular biology grade).
2. 1 M HEPES pH 7.5 (molecular biology grade).
3. 2 M Lithium Chloride (molecular biology grade): Prepare a 2 M stock solution in sterile ddH_2O.
4. 10% (v/v) IGEPAL CA-630 (molecular biology grade): Prepare a 10% (v/v) stock solution in sterile ddH_2O.

5. 20% (w/v) Sucrose (molecular biology grade): Prepare a 20% (w/v) stock solution in sterile ddH_2O.
6. Wash buffer 1: 10 mM Tris–HCl pH 8.0, 1 mM EDTA pH 8.0, 0.1% (v/v) Triton X-100. Prepare the buffer with ddH_2O. Sterilize by filtering using a 0.22 μm filter. Store for up to 6 months at 4 °C. When using, always make an aliquot and freshly add protease inhibitor.
7. Wash buffer 2: 20 mM Tris–HCl pH 8.0, 150 mM NaCl, 5 mM EDTA pH 8.0, 5.2% (w/v) Sucrose, 1% (v/v) Triton X-100, 0.2% (w/v) SDS. Prepare the buffer with ddH_2O. Sterilize by filtering using a 0.22 μm filter. Store for up to 6 months at 4 °C. When using, always make an aliquot and freshly add protease inhibitor.
8. Wash buffer 3: 5 mM Tris–HCl pH 8.0, 25 mM HEPES, 250 mM NaCl, 0.5 mM EDTA pH 8.0, 0.5% (v/v) Triton X-100, 0.05% (w/v) sodium deoxycholate. Prepare the buffer with ddH_2O. Sterilize by filtering using a 0.22 μm filter. Store for up to 6 months at 4 °C. When using, always make an aliquot and freshly add protease inhibitor.
9. Wash buffer 4: 10 mM Tris–HCl pH 8.0, 250 mM Lithium Chloride (LiCl), 10 mM EDTA pH 8.0, 0.5% (w/v) sodium deoxycholate, 0.5% (v/v) IGEPAL CA-630. Prepare the buffer with ddH_2O. Sterilize by filtering using a 0.22 μm filter. Store for up to 6 months at 4 °C. When using, always make an aliquot and freshly add protease inhibitor.
10. Wash buffer 5: 10 mM Tris–HCl pH 8.0. Prepare the buffer with ddH_2O. Sterilize by filtering using a 0.22 μm filter. Store at 4 °C. When using, always make an aliquot and freshly add protease inhibitor.

2.5 DNA End Repair (See Note 3)

1. NEBNext End Repair Module (New England Biolabs).
2. Nuclease-free water (molecular biology grade).
3. Thermomixer.

2.6 A-Tailing (See Note 3)

1. NEBNext dA-Tailing Module (New England Biolabs).
2. Quick Ligation Kit (New England Biolabs).
3. 100X Tris-EDTA (TE) buffer (molecular biology grade).
4. ChIP-nexus adapters: The Nexus adapter is formed by annealing the oligos for the Nexus_universal_adapter and the barcoded adapters (Nexus_adapter 1–4). Each barcoded adapter includes a random barcode (AGCGG, CGCGG, GCCTG, or CCAGG) plus one of four fixed barcodes (AGTC, CAGT, GTCA, or TCAG, respectively) (Table 1). A random barcode helps to track PCR overamplification in the DNA library [12] (*see* **Note 4**). For the preparation of 20 μM Nexus adapter

Table 1
List of adapter and primer oligos used in the current optimized protocol

Oligo name	Sequence (5′–3′)	Modification	Barcode sequence
Nexus_Universal_Adapter	/5Phos/GATCGGAAGAGCACACGTCTGGATCC ACGACGCTCTTCC	5′-Phosphate	
Nexus_Adapter_1	/5Phos/AGTCAGCGGAGATCGGAAGAGCGTCGTGGATCCAGACGTGTGCTCTTCCGATCT	5′-Phosphate	AGTCAGCGG
Nexus_Adapter_2	/5Phos/CAGTCGCGGAGATCGGAAGAGCGTCGTGGATCCAGACGTGTGCTCTTCCGATCT	5′-Phosphate	CAGTCGCGG
Nexus_Adapter_3	/5Phos/GTCAGCCTGAGATCGGAAGAGCGTCGTGGATCCAGACGTGTGCTCTTCCGATCT	5′-Phosphate	GTCAGCCTG
Nexus_Adapter_4	/5Phos/TCAGCCAGGAGATCGGAAGAGCGTCGTGGATCCAGACGTGTGCTCTTCCGATCT	5′-Phosphate	TCAGCCAGG
Nexus_Universal_Primer	AATGATACGGCGACCACCGAGATCTACACTCTT TCCCTACACGACGCTCTTCCGATC*T	3′ phosphorothioate bond	
Nexus_Primer_Index_01	CAAGCAGAAGACGGCATACGAGATCGTGATGT GACTGGAGTTCAGACGTGTGCTCTTCCGATC*T	3′ phosphorothioate bond	ATCACG
Nexus_Primer_Index_04	CAAGCAGAAGACGGCATACGAGATTGGTCAGT GACTGGAGTTCAGACGTGTGCTCTTCCGATC*T	3′ phosphorothioate bond	TGACCA

(100 μL): combine 20 μL 100 μM Nexus_universal_adapter, 5 μL each of 100 μM Nexus_adapter 1–4, 10 μL 10x TE, 1 μL 5 M NaCl, and 49 μL nuclease-free water. Anneal the mixture on Thermomixer at 95 °C for 5 min. Turn the instrument off and allow for a slow cooling to approximately 25 °C temperature. The procedure may take 2–3 h. Alternatively, annealing can also be performed on a PCR cycler at 95 °C for 5 min, cool to 25 °C (1 °C/min), and then 25 °C for 30 min. Dilute this stock 1:20 to make a working stock of 1 μM Nexus_adapter_-mix and keep it at −20 °C. Aliquot them to prevent repeated freezing-thaw cycles.

2.7 Barcode Extension (See Note 3)

1. Phi29 DNA Polymerase (New England Biolabs).
2. 10 mM dNTP Mix.

2.8 Lambda Exonuclease Digestion

1. RIPA buffer: 50 mM HEPES pH 7.5, 500 mM LiCl, 1 mM EDTA pH 7.5, 0.7% (w/v) sodium deoxycholate, 1% (v/v) IGEPAL CA-630. Prepare the buffer with ddH_2O. Sterilize by filtering using a 0.22 μm filter. Store for up to 6 months at 4 °C. When using, always make an aliquot and freshly add protease inhibitor.
2. Lambda Exonuclease (New England Biolabs).
3. Dimethyl sulfoxide (DMSO) (molecular biology grade).
4. 10% (v/v) Triton X-100. Prepare a 10% (v/v) stock solution from Triton-X100 (molecular biology grade) in sterile ddH_2O.

2.9 DNA Purification

1. Elution buffer: 25 mM Tris–HCl pH 8.0, 5 mM EDTA pH 8.0, 0.5% (w/v) SDS. When using, freshly prepare the buffer with nuclease-free water.
2. 10 mg/mL Proteinase K solution: 10 mg/mL proteinase K prepared in nuclease-free water.
3. 5 M NaCl (molecular biology grade).
4. Phenol/Chloroform/Isoamyl alcohol.
5. 20 mg/mL glycogen (molecular biology grade).
6. Ethanol absolute (molecular biology grade).
7. 70% (v/v) EtOH: Absolute ethanol in nuclease-free water.
8. 0.2 mL PCR tubes.

2.10 Library Preparation

1. CircLigase ssDNA Ligase kit (Biozym).
2. 50x Tris-acetate-EDTA (TAE) buffer: To prepare 1 L, dissolve 242 g Tris-base (Molecular Weight = 121.14 g/mol) in 700 mL sterile ddH_2O. Add 100 mL 0.5 M EDTA pH 8.0

and 57.1 mL glacial acetic acid carefully and adjust the final volume of the solution to 1 L with water. The pH of the buffer should be approximately 8.5. Store the solution at 20–25 °C.

3. 6x color-less loading dye: 30% (v/v) Glycerol (molecular biology grade, v/v) in nuclease-free water.
4. GelRed Nucleic Acid Gel Stain Dye.
5. PeqGREEN or other DNA gel dye.
6. Agarose.
7. Agarose gel electrophoresis equipment: gel chambers, power supplies, transilluminator.
8. 100 bp DNA Ladder.
9. Monarch DNA Gel Extraction Kit (New England Biolabs).
10. PCR machine.
11. Q5 High-Fidelity 2X Master Mix (New England Biolabs).
12. Library amplification primers: During the PCR amplification of the libraries, universal primers (Nexus_Universal_primer) and index primers (Nexus_Primer_Index_01 and Nexus_Primer_-Index_04 for two replicates, respectively) were used (Table 1) (*see* **Note 5**). Prepare 10 μM working stocks of each primer and store at −20 °C.
13. Qubit dsDNA HS Assay Kit (Life Technologies).
14. Qubit fluorometer (Thermo Fisher Scientific).

2.11 Quality Control of ChIP-Nexus Library and Sequencing

1. High Sensitivity DNA Kit (Agilent).
2. Bioanalyzer Instrument (Agilent).

3 Methods

We describe the ChIP-nexus protocol specifically optimized for the GR antibody and BMDM cells. We provide detailed steps and tips for a reproducible ChIP-nexus library preparation (Fig. 2). In the first step, intact BMDM cells are treated with formaldehyde for in vivo cross-linking of the protein-DNA complex. Following cell lysis, chromatin is sheared to an appropriate size (100–500 bp). The soluble chromatin is immunoprecipitated with GR antibody for selective enrichment of GR and its DNA targets. Several enzyme-based steps, including digestion with lambda exonuclease, are performed, while the immunoprecipitated protein-DNA complex remains bound to the beads. Single-strand DNA is eluted and purified from the beads. Following circularization, the DNA is ready for PCR amplification to generate a sequencing library.

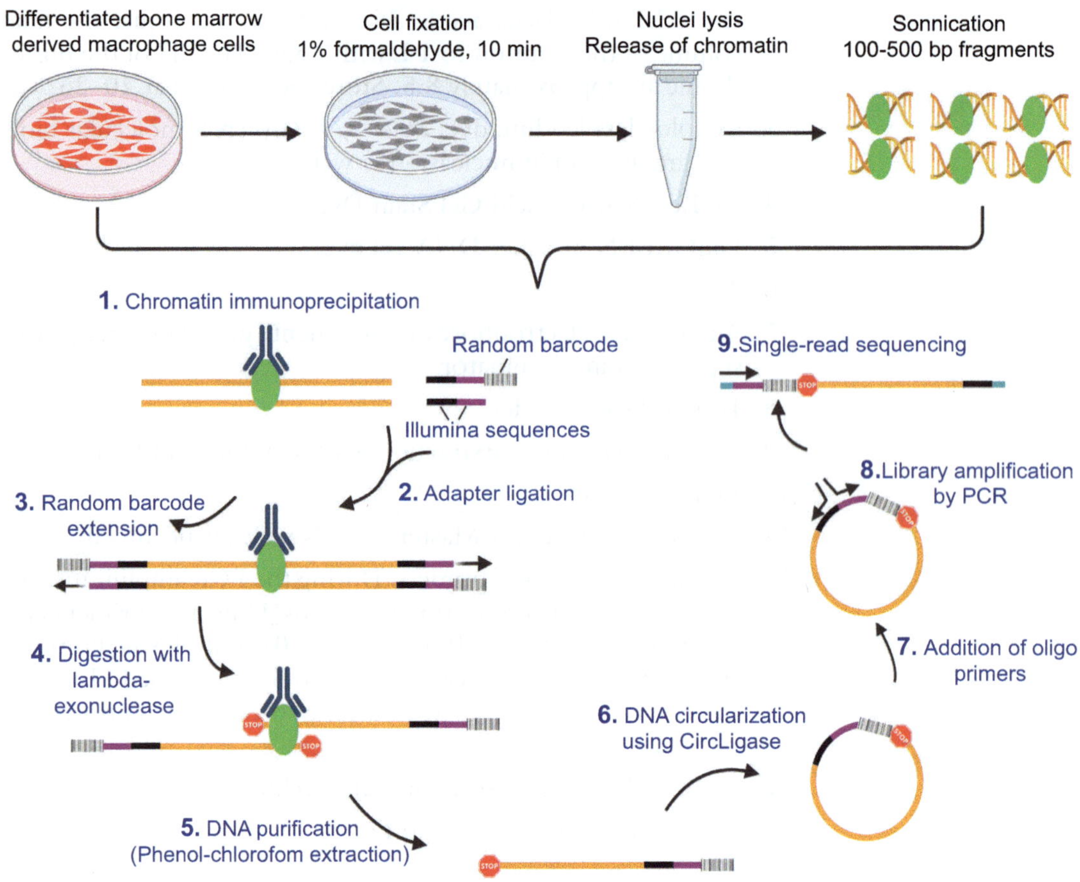

Fig. 2 Schematic illustration of the ChIP-nexus library preparation. To generate the high-resolution glucocorticoid receptor-DNA footprints, the BMDM cells are cross-linked using 1% formaldehyde. Following nuclear lysis, the chromatin is sonicated to obtain the desired fragments (100–500 bp). (**1**) A ChIP-optimized glucocorticoid receptor antibody is used to immunoprecipitate the receptor-DNA complex. (**2**) While the DNA-protein adducts are still on the beads, the DNA ends are repaired, dA-tailed, and ligated to an adaptor containing a pair of sequences for library amplification. (**3**) A nine-nucleotide barcode with five random bases and four fixed bases is used to remove reads from library DNA overamplification. After adapter ligation, 5′ overhangs are filled to generate blunt ends for efficient lambda exonuclease digestion. (**4**) Lambda exonuclease trims the DNA until it encounters a cross-linking point of the protein-DNA complex ("stop" symbol). (**5**) Single-stranded DNA is purified after reverse cross-linking. (**6**) Circularization of single-stranded DNA that places barcode sequence next to the stop point. (**7**, **8**) PCR amplification with primers that anneal to Illumina sequences on both ends of the DNA. (**9**) Single-end sequencing of a DNA library in which the barcode is sequenced first, followed by the protein-DNA cross-link stop point. (Figure created with BioRender.com)

3.1 Preparation of Cross-Linked Cells

Formaldehyde cross-linking is frequently used to detect and quantify protein-DNA interactions. Cross-linking stabilizes the interaction between target protein and DNA sequences. This modification is also reversible, enabling the isolation and amplification of ChIP-enriched DNA.

1. This protocol has been optimized with freshly cultured or frozen pellets of bone marrow–derived macrophages

(BMDMs). Twenty million cells were seeded in a 15 cm plate overnight (16–24 h) in a serum-free macrophage medium (*see* **Note 6**).

2. The differentiated macrophage cells were treated the next day with 1 μM Dexamethasone, a GR activator in combination with 100 ng/μL lipopolysaccharides (LPS) (*see* **Note** 7).
3. Aspirate the medium and wash the cells with 10 mL of DPBS. Aspirate again.
4. Add 10 mL Cell fixation solution per plate and incubate at 20–25 °C for 10 min with gentle shaking using a rocker.
5. Add 1.5 mL of 1 M glycine and incubate at 20–25 °C for 5 min with gentle shaking on a rocker.
6. Aspirate the liquid from the plate and wash the cells two times with ice-cold 10 mL of DPBS. Aspirate and leave 1 mL DPBS on the plate.
7. With a cell scraper, scrape the cells gently from the plate into the leftover DPBS.
8. Transfer the cell suspension to a 2 mL tube and place it on ice.
9. Spin the tube at 1000× *g* for 5 min at 4 °C. Remove supernatant.
10. *Optionally*: Cell pellet can be stored at −80 °C for up to 6 months.

3.2 Preparation of Beads

1. Aliquot 50 μL (50 μg) of Dynabeads slurry in a 1.5 mL low DNA binding tube per sample.
2. Wash the beads three times with 1 mL of Shearing buffer (with protease inhibitor). Every time, wash the beads quickly and vigorously with 1 mL buffer using a wrist motion. During each wash, the beads should be thoroughly resuspended. Place samples on a magnetic rack that holds the beads to remove liquid. Allow the liquid to settle for approximately 1 min. Aspirate the supernatant.
3. Add 500 μL shearing buffer (with protease inhibitor) per reaction to the Dynabeads and add 10 μg of specified GR antibody (*see* **Note 8**).
4. Incubate with rotation at 4 °C for 10–12 h.

3.3 Preparation of Chromatin Extracts and Shearing

1. Resuspend the cross-linked cells (per tube) in 1 mL Shearing buffer (with protease inhibitor). Put it on ice for 15 min. If stored and frozen cells (from −80 °C) are taken, place them on ice for at least 30 min before resuspending them in the Shearing buffer.
2. Sonicate samples for 30 s on, 30 s off on high setting in the Bioruptor. The number of sonication cycles and amplitude of

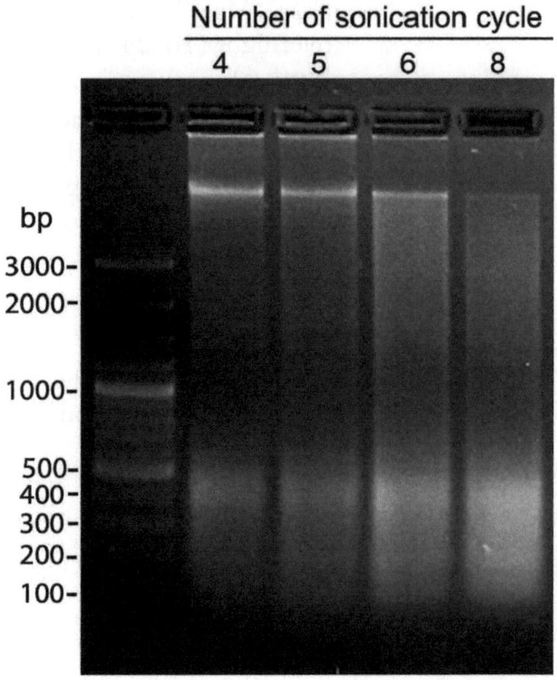

Fig. 3 Optimization of chromatin shearing. Samples are sheared for different sonication cycles (high setting, 30 s on/30 s off) with the Bioruptor. Sheared chromatin is de-cross-linked and purified using the Phenol: Chloroform extraction method. 1 μg of DNA is analyzed on a 1.5% agarose gel. The optimal sonication is 8 cycles in BMDMs with the most DNA sheared. In this condition, a 100 bp DNA molecular weight marker indicated that the sheared DNA size ranged from 100 to 500 bp on the gel

sonication must be optimized (*see* **Note 9**). We have determined that 6–8 cycles are optimal for obtaining the desired size distribution (100–500 bp fragments) from BMDM cells (Fig. 3).

3. Spin for 30 min at 13,600× *g*, and 4 °C. Transfer the supernatant containing the soluble chromatin to a new 1.5 mL low DNA binding tube.

3.4 Chromatin Immunoprecipitation (ChIP)

1. After the beads have been incubated with the antibody, place them in a new 2 mL low DNA binding tube.
2. Wash the ChIP Dynabead three times with Shearing buffer (with protease inhibitor) as mentioned in Subheading 3.2, **step 2**.
3. Add 800 μL of soluble chromatin to the Dynabead-Antibody mix and incubate at 4 °C with gentle rotation (20 rpm) for at least 16 h.
4. Wash the beads using 1 mL each of Wash buffers 1, 2, 3, 4, and 5 (*see* **Note 10**).

3.5 DNA End Repair

Following chromatin shearing, the fragmented DNA has 5′ and 3′ overhangs that require repair. The end repair process fills in 5′ overhangs with T4 DNA polymerase and eliminates 3′ overhangs with Klenov fragment. In addition, 5′ blunt ends are phosphorylated by T4 polynucleotide kinase, ensuring the 3′ ends have a hydroxyl group.

1. Prepare a master mix with 5 μL of 10x NEBNext End Repair reaction buffer, 1 μL of NEBNext End Repair enzyme mix, and 39 μL of nuclease-free water. After the last wash step, suspend the beads in the master mix and incubate at 25 °C for 30 min on the Thermomixer, without shaking (*see* **Note 11**).
2. Wash the beads using 1 mL each of Wash buffers 1, 2, 3, 4, and 5 (*see* **Note 10**).

3.6 A-Tailing and Nexus Adapter Ligation

This process incorporates a single adenosine residue at the 3′ ends of blunt-ended DNA. This reduces the likelihood that these fragments will ligate to one another and enhances the rate of adapter ligation, as adapters have a single overhanging "T" base.

1. Prepare a master mix with 5 μL of 10x NEBNext dA-tailing reaction buffer, 1 μL Klenow Fragment (3′ > 5′ exo-), and 39 μL of nuclease-free water. After the last wash step, resuspend the beads in the master mix and incubate at 37 °C for 30 min on the Thermomixer, without shaking (*see* **Note 11**).
2. After the reaction is completed, wash the beads using 1 mL each of Wash buffers 1, 2, 3, 4, and 5 (*see* **Note 10**).
3. Prepare a master mix with 25 μL of 2x Quick Ligase buffer, 3 μL of Quick Ligase, and 15.5 μL of nuclease-free water. Add 1.5 μL of 1 μM Nexus adapter mix at the end, just before adding to the beads. After the last wash step, resuspend the beads in the master mix and incubate at 25 °C for 40 min on the Thermomixer, without shaking (*see* **Note 11**).
4. Afterward, wash the beads using 1 mL each of Wash buffers 1, 2, 3, 4, and 5 (*see* **Note 10**).

3.7 Barcode Extension

1. Prepare a master mix with 5 μL of 10x Phi29 reaction buffer, 1 μL 10 mM dNTPs, 0.5 μL 20 mg/mL BSA, 1 μL of Phi29 DNA polymerase, and 37.5 μL of nuclease-free water.
2. After the last wash step, resuspend the beads in the master mix and incubate at 30 °C for 30 min on the Thermomixer, without shaking (*see* **Note 11**).
3. When the reaction is completed, wash the beads using 1 mL each of Wash buffers 1, 2, 3, 4, and 5 (*see* **Note 10**).

3.8 Lambda Exonuclease Digestion

Adaptor-ligated DNA is digested in the 5′–3′ direction by lambda exonuclease until a few nucleotides away from the cross-linked protein-DNA complex.

1. Prepare a master mix with 10 μL of 10x Lambda exonuclease buffer, 1 μL 10% (v/v) Triton X-100, 5 μL DMSO, 4 μL of Lambda exonuclease, and 75 μL of nuclease-free water.
2. After the wash step, resuspend the beads in the master mix and incubate at 37 °C and 1000 rpm for 60 min on the Thermomixer.
3. When the reaction is completed, wash the beads three times using ice-cold RIPA buffer. Wash each sample quickly and vigorously with 1 mL buffer using a wrist motion. During each wash, the beads should be thoroughly resuspended. Place samples on a magnetic rack that holds the beads to remove liquid. Allow the liquid to settle for approximately 1 min, before removing the liquid completely by aspiration.

3.9 DNA Purification

1. Resuspend the beads in 300 μL of Elution buffer, add 6 μL of 10 mg/mL Proteinase K solution, and incubate the mixture for 16 h at 65 °C, 1000 rpm in a Thermomixer for reverse cross-linking.
2. After reverse cross-linking, add 300 μL of cold phenol/chloroform/isoamyl alcohol. Mix thoroughly at least ten times with hand inversion and separate phases by centrifuging for 5 min at 13,600× *g*, and 25 °C.
3. Collect the upper aqueous phase into a new 1.5 mL low DNA binding tube.
4. Add 3 μL 20 mg/mL glycogen, 12 μL 5 M NaCl, and 750 μL ice-cold absolute ethanol for precipitate formation. Mix thoroughly and incubate overnight (16–24 h) at −20 °C.
5. Centrifuge for 30 min at 13,600× *g*, and 4 °C and discard the supernatant. Glycogen forms a white precipitate with DNA.
6. Wash the white pellet with 500 μL of ice-cold 70% ethanol, centrifuge for 5 min at 13,600× *g*, 4 °C, and discard the supernatant.
7. Allow the sample to air dry for 5–10 min and resuspend the pellet in 12 μL nuclease-free water.
8. Transfer the sample into a new 0.2 mL PCR tube for library preparation.

3.10 Library Preparation

Following the purification of single-stranded DNA, DNA is circularized in the presence of the circLigase enzyme, which positions the barcode adjacent to the exo-nuclease stop base. The circularized ssDNA fragments serve as the direct template for the PCR library amplification.

1. Denature the sample for 5 min at 95 °C in a PCR machine and then cooled on ice.
2. Prepare a master mix per sample with 1.5 μL of 10x CircLigase Buffer, 0.75 μL 50 mM $MnCl_2$, 0.75 μL 1 mM ATP, and 0.5 μL CircLigase ssDNA Ligase.
3. Add the master mix to the sample and incubate for 60 min at 60 °C in a PCR machine for single-stranded DNA intramolecular ligation (circularization). Keep the sample on ice after circularization.
4. Prepare a master mix per sample with 25 μL of 2x Q5 Master Mix, 1 μL 10 μM Nexus_Universal_Primer, 1 μL 10 μM Nexus_primer_Index_01 or Nexus_primer_Index_04, and 8 μL nuclease-free water.
5. Add the master mix to each sample and run the following PCR program: (a) Initial Denaturation for 2 min at 98 °C. (b) Denature 10 s at 98 °C. (c) Anneal Nexus primers for 30 s at 65 °C. (d) Extend DNA for 30 s at 72 °C. (e) Repeat **steps b–d** 18 times. (f) Final Extension for 5 min at 72 °C (*see* **Note 12**).
6. Hold the sample at 4 °C.

The selection of the appropriate library size is a crucial factor for optimal results. A precise size selection can improve the efficiency of sequencing. We have used agarose gel for the size selection of the ChIP-nexus library. It is a straightforward size selection strategy that, when done properly, is highly effective. Additionally, gel-based size selection helps in the efficient elimination of larger fragments from DNA library samples.

7. Prepare a 1.8% (w/v) agarose gel containing GelRed nucleic acid stain in a 1x TAE buffer. Use a gel comb of the appropriate size to load the PCR library DNA from each sample into two wells.
8. Add 10 μL of color-less 6x loading dye to 50 μL of the ChIP-nexus library.
9. Load 100 bp DNA ladder in an agarose gel well.
10. Load the library DNA into two gel lanes, one lane apart from the ladder or other sample library DNA.
11. Run the gel at 100 V for about 60 min. Examine the gel with a transilluminator (keep the UV exposure as short as possible).
12. Using a clean, sharp razor blade, carefully excise the desired library band (150–300 bp), avoiding the adapter multimer (125 bp) and primer dimer (<100 bp) (Fig. 4a).
13. Transfer the gel piece into a new 2 mL low DNA binding tube and weigh the gel slice.

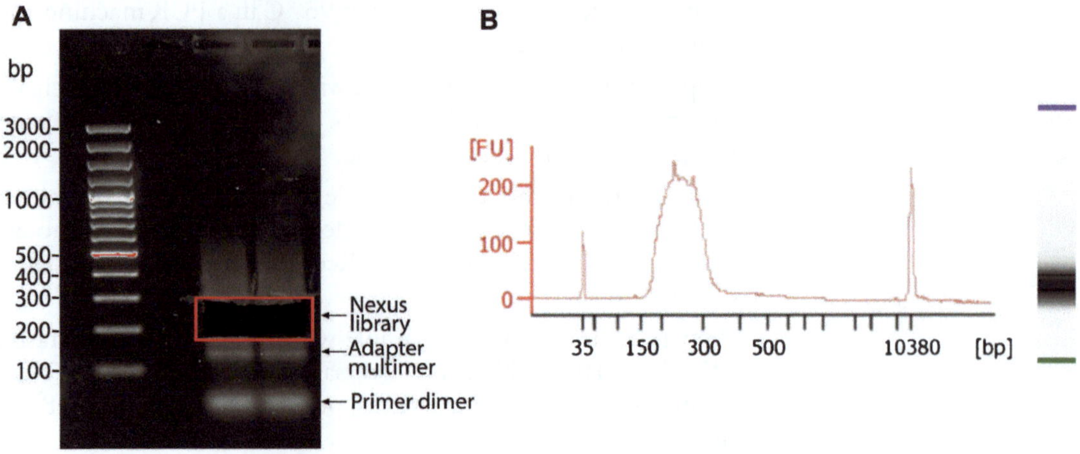

Fig. 4 Profile of GR ChIP-nexus library. (**a**) PCR amplified library was run on 1.8% agarose gel. The DNA library of the desired size was excised from the gel while avoiding adapter and primer dimers. (**b**) The Bioanalyzer profile showed a clean library size range (150–300 bp)

14. Purify the DNA from agarose gel using the Monarch DNA Gel Extraction Kit according to the manufacturer's guidelines.
15. Add four volumes of Monarch Gel Dissolving Buffer to the gel slice-containing tube (i.e., 100 mg of gel in 400 μL of buffer).
16. Incubate the sample at 32 °C at 1000 rpm in a thermomixer until the gel slices are completely dissolved (generally 20–30 min).
17. Place the column into the collection tube and load the sample.
18. Spin it at 16,000× *g* for 1 min and discard the flow-through.
19. Place the column back into the same collection tube.
20. Add 200 μL DNA Wash Buffer and spin at 16,000× *g* for 1 min. Repeat this step.
21. Transfer the column to a new collection tube and spin for 2 min at 16,000× *g* to remove any residual buffer.
22. Transfer the column to a new 1.5 mL tube. Add 10 μL of nuclease-free water at the center of the column matrix. Wait for 2 min and spin the column to elute the DNA.
23. Repeat the elution **step 22** with an additional 10 μL nuclease-free water to improve the yield.
24. Store the DNA at −20 °C.
25. Measure the library DNA concentration using the Qubit fluorometer. In our case, the optimal library yield from 20 million BMDM cells is 60–80 ng.

3.11 Quality Control of ChIP-Nexus Library and Sequencing

Before performing next-generation sequencing, run the library sample on an Agilent Bioanalyzer system to assess the purity and fragment size distribution of library DNA (Fig. 4b). In this case, the optimal library fragment size range is between 150 and 300 base pairs. Pool at least two replicates of distinct barcoded samples in equimolar concentration. We prepare 10 nM (in 10 μL) of pooled samples and perform single-end 75 bp sequencing; 40–50 million single-end reads (per replicate) are typically adequate for ChIP-nexus experiments to investigate the genome-wide profiling of glucocorticoid receptor-DNA binding in BMDM cells.

4 Notes

1. It is important to note that generating reliable ChIP-nexus data is dependent on using antibodies that have first been validated for high sensitivity and target specificity in the ChIP-seq experiment. In this instance, the specified catalog number GR antibody was validated in previous studies for generating high-quality ChIP-seq data [1, 3]. Moreover, the suitability of GR antibody for chromatin immunoprecipitation can vary among different lots of the same antibody. If using an antibody from a different lot or supplier, titration studies must be performed to determine the optimal antibody concentration and for detecting locus-specific enrichment in ChIP-qPCR (at least fivefold enrichment) or detection of enrichment peaks in ChIP-seq assay without significant background noise.
2. The specified Dynabead is selected to match with the antibody used in this optimized technique. However, other commercially available Dynabeads that match the antibody can also be utilized. Dynabeads are favored over sepharose/agarose beads for the fast separation of protein-DNA complexes due to (i) their low nonspecific binding and (ii) beads are visible within the tube, allowing for easy handling during washes.
3. We had success with the library preparation strategy for Illumina sequencing using the given kits and the protocol adapted from an existing protocol developed by Zeitlinger's lab [9].
4. The random barcode distinguishes between PCR-duplicated reads originating from the single DNA fragment (with the same random barcode) and reads from different DNA molecules (with different random barcodes).
5. For both adapters and primers, HPLC-purified oligos are recommended.

6. The optimum number of differentiated BMDM cells for a ChIP-nexus experiment is 20 million cells seeded in a 15 cm dish. While the precise number of cells needed for a ChIP-nexus experiment will vary depending on the cell type being tested, it is recommended that cells be at 80% confluency in a 15 cm culture dish at the time of drug treatment.
7. The provided protocol was established as part of our research study to evaluate the anti-inflammatory mechanism of GR. Therefore, the cells were treated with a combination of Dexamethasone and LPS.
8. If the shearing buffer is too cold, SDS may precipitate. While keeping it on ice, keep a close eye on it and make sure it has been dissolved completely before using it.
9. In general, macrophages are quite resistant to sonication. DNA must be sheared efficiently to a consistent fragment size of 100–500 bp. In the event of an inefficient shearing, it is also worth trying to perform nuclei extraction before sonication. Perform a sonication time course experiment for each cell type to determine the optimal and the least time condition for chromatin shearing. A prolonged sonication may alter antigenic properties of DNA-binding proteins and result in antigenicity loss during immunoprecipitation. Once optimal sonication time conditions have been determined, use the same parameters for every experiment to ensure reproducibility.
10. Wash each sample quickly and vigorously with wrist motion. During each wash, the beads should be thoroughly resuspended. Place samples on a magnetic rack that holds the beads to remove liquid. Allow the liquid to settle for approximately 1 min, before removing the liquid completely by aspiration. It is also recommended that any residual liquid be removed by brief centrifugation and pipetting before washing with buffer 1 and after washing with buffers 4 and 5. After the last wash step in each round of washes, briefly centrifuge the tube to collect the beads at the bottom, so that they can be resuspended in the master mix of the next enzyme reactions.
11. After a few minutes, the beads tend to settle down. The tube should be periodically (approximately every 10 min) and quickly vortexed or flicked with a finger to resuspend the beads.
12. To prevent over-amplification, minimal PCR cycling is highly desirable. The number of cycles should be determined by the amount of starting material (i.e., number of cells) used per ChIP.

References

1. Escoter-Torres L, Greulich F, Quagliarini F et al (2020) Anti-inflammatory functions of the glucocorticoid receptor require DNA binding. Nucleic Acids Res 48:8393–8407
2. Greulich F, Mechtidou A, Horn T et al (2021) Protocol for using heterologous spike-ins to normalize for technical variation in chromatin immunoprecipitation. STAR Protoc 2:100609
3. Greulich F, Wierer M, Mechtidou A et al (2021) The glucocorticoid receptor recruits the COMPASS complex to regulate inflammatory transcription at macrophage enhancers. Cell Rep 34:108742
4. Quagliarini F, Mir AA, Balazs K et al (2019) Cistromic reprogramming of the diurnal glucocorticoid hormone response by high-fat diet. Mol Cell 76:531–545.e5
5. Uhlenhaut NH, Barish GD, Yu RT et al (2013) Insights into negative regulation by the glucocorticoid receptor from genome-wide profiling of inflammatory cistromes. Mol Cell 49:158–171
6. Mir AA, Dyar KA, Greulich F et al (2019) In vivo ChIP-seq of nuclear receptors: a rough guide to transform frozen tissues into high-confidence genome-wide binding profiles. Methods Mol Biol 1966:39–70
7. Rhee HS, Pugh BF (2012) ChIP-exo method for identifying the genomic location of DNA-binding proteins with near-single-nucleotide accuracy. Curr Protoc Mol Biol Chapter 21:Unit 21.24
8. Starick SR, Ibn-Salem J, Jurk M et al (2015) ChIP-exo signal associated with DNA-binding motifs provides insight into the genomic binding of the glucocorticoid receptor and cooperating transcription factors. Genome Res 25: 825–835
9. He Q, Johnston J, Zeitlinger J (2015) ChIP-nexus enables improved detection of in vivo transcription factor binding footprints. Nat Biotechnol 33:395–401
10. König J, Zarnack K, Rot G et al (2010) iCLIP reveals the function of hnRNP particles in splicing at individual nucleotide resolution. Nat Struct Mol Biol 17:909–915
11. Rossi MJ, Lai WKM, Pugh BF (2018) Simplified ChIP-exo assays. Nat Commun 9:2842
12. Kivioja T, Vähärautio A, Karlsson K et al (2011) Counting absolute numbers of molecules using unique molecular identifiers. Nat Methods 9:72–74

Chapter 7

Sequential ChIP-Seq

Joseph D. Schinderle and Irina M. Bochkis

Abstract

ChIP-Seq has been used extensively to profile genome-wide transcription factor binding and post-translational histone modifications. A sequential ChIP assay determines the in vivo co-localization of two proteins to the same genomic locus. In this chapter, we combine the two protocols in Sequential ChIP-Seq, a method for identifying genome-wide sites of in vivo protein co-occupancy.

Key words Sequential chromatin immunoprecipitation, DNA binding, Protein co-localization, Next-generation sequencing

1 Introduction

The chromatin immunoprecipitation (ChIP) technique is a powerful tool to assess DNA-protein interactions in vivo and determine protein occupancy at a specific region [1]. ChIP, in combination with next-generation sequencing (ChIP-Seq), extends the capability to identify genome-wide DNA binding sites and is used to profile transcription factor occupancy and localization of chromatin marks, post-translational modifications on histone tails [2–5]. DNA regulatory elements, such as promoters and enhancers, contain binding sites for several transcription factors and in bivalent domains, nucleosomes are dually marked by different covalent modifications [6, 7]. Hence, identifying the co-localization of two proteins on DNA is important to studies of transcriptional and chromatin regulation. Sequential ChIP assays determine the in vivo co-occupancy of two proteins to the same genomic locus (Fig. 1) [8]. In this chapter, we provide a Sequential ChIP-Seq protocol, combining Sequential ChIP with next-generation sequencing, which identifies genome-wide sites of in vivo protein co-occupancy.

Franziska Greulich (ed.), *Chromatin Immunoprecipitation: Methods and Protocols*, Methods in Molecular Biology, vol. 2846,
https://doi.org/10.1007/978-1-0716-4071-5_7,

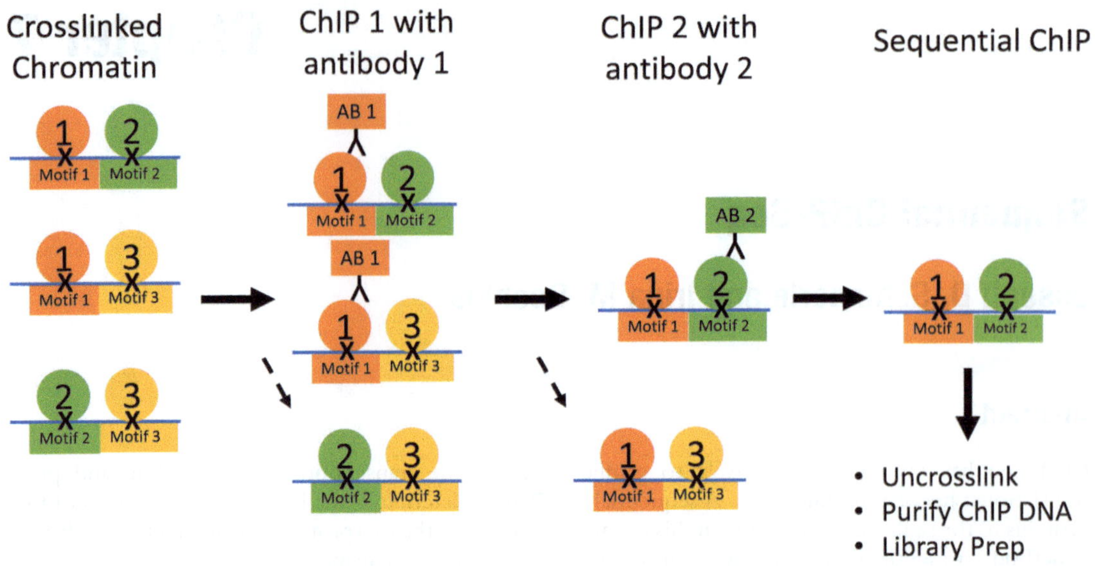

Fig. 1 Principle of sequential chromatin immunoprecipitation (ChIP) assay. The principle relies on combinatorial selection of two proteins using the first antibody in the first ChIP (ChIP 1) and a second antibody in the sequential ChIP (ChIP 2). ChIP DNA is used in library preparation for next-generation sequencing. Alternatively, ChIP DNA can be used in qPCR to assay for known regions bound by both proteins

2 Materials

2.1 First Chromatin Immunoprecipitation (ChIP 1)

1. 5 M NaCl.
2. 1 M Tris–HCl pH 8.0.
3. 10% (v/v) Triton-X100.
4. 10% (w/v) SDS.
5. 0.5 M EDTA pH 8.0.
6. 10% (v/v) NP-40.
7. 8 M LiCl.
8. 10% (w/v) Sodium Deoxycholate.
9. 1 M DTT.
10. ChIP Dilution Buffer (CDB): 0.01% (w/v) SDS, 1.1% (v/v) Triton-X 100, 167 mM NaCl, 16.7 mM Tris–HCl. To make 50 mL, add 50 μL 10% (w/v) SDS, 5.5 mL 10% (v/v) Triton-X 100, 835 μL 1 M Tris–HCl pH 8.0, and 1.67 mL 5 M NaCl.
11. TSE1: 0.1% (w/v) SDS, 1% (v/v) Triton-X 100, 2 mM EDTA, 20 mM Tris–HCl, 150 mM NaCl. To make 50 mL, add 500 μL 10% (w/v) SDS, 5 mL 10% (v/v) Triton-X 100, 200 μL 0.5 M EDTA pH 8.0, 1 mL 1 M Tris–HCl pH 8.0, 1.5 mL 5 M NaCl.
12. TSE2: 0.1% (w/v) SDS, 1% (v/v) Triton-X 100, 2 mM EDTA, 20 mM Tris–HCl, 500 mM NaCl. To make 50 mL, combine

500 μL 10% (w/v) SDS, 5 mL 10% (v/v) Triton-X 100, 200 μL 0.5 M EDTA pH 8.0, 1 mL 1 M Tris–HCl pH 8.0, 5 mL 5 M NaCl.

13. ChIP Buffer 3: 0.25 M LiCl, 1% (v/v) NP-40, 1% (w/v) Sodium Deoxycholate, 1 mM EDTA, 10 mM Tris–HCl. To make 50 mL, add 2.5 mL 5 M LiCl, 5 mL 10% (v/v) NP-40, 500 μL 10% (w/v) Sodium Deoxycholate, 100 μL 0.5 M EDTA pH 8.0, 500 μL 1 M Tris–HCl pH 8.0.
14. TE Buffer: 0.2 mL 0.5 M EDTA pH 8.0, 1 mL 1 M Tris–HCl pH 8.0, bring up to 100 mL with water.
15. Elution Buffer 1: 1% (w/v) SDS, 1 mM DTT. To make 5 mL, combine 4.495 mL of water, 0.5 mL 10% (w/v) SDS, 5 μL of 1 M DTT.
16. Fixed and isolated chromatin.
17. 1–4 μg of antibody of interest.
18. Protein G agarose (Thermo Scientific).
19. Cooling microfuge for 2 mL tubes.
20. Rotator.
21. Heating block for 2 mL tubes.

2.2 Second Chromatin Immunoprecipitation (ChIP 2)

1. Elution Buffer 2: 1% (w/v) SDS, 0.1 M $NaHCO_3$. To make 50 mL, add 420 mg $NaHCO_3$ to 45 mL water. Take 4.5 mL of this and add 0.5 mL 10% (w/v) SDS.
2. 20 mg/mL proteinase K.
3. QIAquick PCR Purification Kit (*see* **Note 1**).

2.3 Library Preparation

1. Library Elution Buffer: 10 mM Tris–HCl pH 8.0.
2. END-IT kit (Epicentre Biotechnologies).
3. Klenow (3′–5′ exo-) (NEB, *see* **Note 1**).
4. Quick Ligation Kit (NEB, *see* **Note 1**).
5. Qiagen MinElute PCR Purification Kit (*see* **Note 1**).
6. Kapa HiFi Kit (Roche, *see* **Note 1**).
7. Ampure XP beads: Remove the Ampure XP beads from 4 °C and bring them to room temperature (21–25 °C) before use.
8. 70–80% (v/v) ethanol.
9. Magnetic stand for 2 mL tubes.
10. 0.2 mL PCR tubes.
11. PCR machine.

2.4 Library Size Selection and Cleanup

1. Pippin Prep, kits and cassettes (2% (w/v) agarose).

3 Methods

This protocol starts with fixed and prepared chromatin of interest. We refer to published protocols for advice on nuclear extraction and fixation (*see* **Note 2**).

3.1 First Chromatin Immunoprecipitation (ChIP 1)

1. Thaw frozen chromatin quickly at 37 °C (*see* **Note 3**). Then place samples on ice.
2. Spin in microfuge at 20,000× *g* for 5 min at 4 °C; place supernatant in a fresh 2 mL tube and discard cellular debris.
3. Measure the chromatin concentration. Use 8–12 μg chromatin DNA and 2 μg antibody per ChIP (*see* **Note 4**).
4. Set 10 ng chromatin aside as an input sample, which is used later (Subheading 3.3, **step 1**).
5. For pre-clearing, add 125 μL of vortexed protein-G agarose to each chromatin sample (*see* **Note 5**). Dilute with room temperature (21–25 °C) ChIP Dilution Buffer (CDB) up to 1 mL and rotate for 1 h at 4 °C.
6. Spin all samples at max speed for 15 s in a microfuge at room temperature (21–25 °C) and transfer the supernatant to a fresh 2 mL tube.
7. Add 1–4 μg affinity purified or monoclonal antibody and incubate for 15 h at 4 °C while rotating.
8. To prepare the beads, wash the protein-G agarose three times with 1 mL of room temperature (21–25 °C) CDB, then rotate for 15 h at 4 °C.
9. The next day, add 200 μL of washed protein-G agarose to each chromatin sample and rotate at 4 °C for 1 h to precipitate the chromatin. Use cut-off tips as in **step 5**. Vortex the beads between withdrawals.
10. Pellet agarose beads by spinning the sample down at 20,000× *g* and room temperature (21–25 °C) for 15 s.
11. Use a pipette to discard the supernatant and add 1 mL of prepared wash buffer TSE1 (21–25 °C). Then rotate for 5 min at room temperature (21–25 °C).
12. Repeat **steps 10** and **11** using each of the following buffers in order: TSE2, ChIP Buffer 3, TE Buffer.
13. Elute into 100 μL of Elution Buffer 1.
14. Repeat the elution once more and combine eluates.

3.2 Second Chromatin Immunoprecipitation (ChIP 2)

1. After the second elution, add 1.3 mL of TE Buffer to obtain 1.5 mL total, using 2 mL tubes.
2. Add 2.5 μL of the second antibody and rotate at room temperature (21–25 °C) for 15 h.
3. Take 125 μL of well-mixed protein-G agarose and dilute with 875 μL of ChIP dilution buffer (CDB). Rotate at room temperature (21–25 °C) for 15 h.
4. The next day, set a heating block to 65 °C.
5. Add 200 μL of pre-cleared protein-G agarose (from **step 3**) to each chromatin sample to precipitate the antibody-chromatin complexes and rotate at 4 °C for 1 h. Use cut-off tips and vortex the beads between withdrawals.
6. Pellet agarose beads by spinning the sample down at 20,000× *g* and room temperature (21–25 °C) for 15 s.
7. Use a pipette to discard the supernatant and add 1 mL of prepared wash buffer TSE1 (21–25 °C). Then rotate for 5 min at room temperature (21–25 °C).
8. Repeat **step 7** using TSE1, TSE2, ChIP Buffer 3, TE Buffer, in this order.
9. Elute the chromatin by adding 100 μL freshly made Elution Buffer 2 to the final pellet and rotate 10 min at room temperature (21–25 °C). Spin the samples at 20,000× *g* at room temperature (21–25 °C) for 15 s. Transfer the supernatant to a fresh tube.
10. Repeat elution **step 9** once and combine eluates.
11. Add 8 μL 5 M NaCl per 200 μL eluate (final concentration 192 mM NaCl) and incubate on the preheated thermoblock at 65 °C for at least 4 h to reverse cross-link the chromatin.
12. Add 8 μL 1 M Tris–HCl pH 7.5, 4 μL 0.5 M EDTA, and 1 μL 10 mg/mL proteinase K. Incubate for 1 h at 45 °C.
13. Purify de-cross-linked chromatin using QIAquick PCR purification kit according to the manufacturer's manual. Elute in 50 μL EB provided by the kit.

3.3 Library Preparation

The protocol described here is a reference; alternatives are possible.

1. Use 34 μL of ChIP DNA from Subheading 3.2, **step 13**. Dilute 10 ng input DNA from Subheading 3.1, **step 4** in 34 μL nuclease-free water.
2. For DNA end repair using the END-IT Kit, prepare the following reaction mix per sample (ChIP or input):

DNA	34 μL
10x End-Repair Buffer	5 μL

(continued)

2.5 mM dNTPs	5 μL
10 mM ATP	5 μL
END-IT enzyme mix	1 μL
	50 μL

3. Incubate at room temperature (21–25 °C) for 45 min.
4. Purify the DNA using the QIAquick PCR purification kit according to manufacturer's manual. Elute in 35 μL EB provided by the kit.
5. For "A" Base addition to the 3′ End of the DNA Fragments (A-tailing), prepare the following reaction mix:

Klenow buffer (NEB buffer #2)	5 μL
1 mM dATP	10 μL
Klenow exo (3′–5′ exo minus)	1 μL
	16 μL

6. Add 16 μL of the reaction mix to 34 μL of purified DNA from **step 4**.
7. Incubate for 30 min at 37 °C on a thermoblock.
8. Purify using the MinElute Purification Kit according to the manufacturer's manual. Elute in 15 μL elution buffer provided by the kit.
9. Determine adapter concentration empirically.
10. For adapter ligation to the DNA Fragments, prepare the following reaction mix

	x1	x5
2x Quick DNA ligase buffer	15 μL	75 μL
Diluted Adapter oligo mix	1 μL	5 μL
Quick DNA ligase	1 μL	5 μL
	17 μL	

11. Add 17 μL of the reaction mix to 13 μL of purified DNA from **step 8**.
12. Incubate for 15 min at 20 °C.
13. Vortex/mix room temperature (21–25 °C) Agencourt Ampure XP beads until they are well dispersed, and keep mixing every couple of samples.

14. Add 30 μL of mixed Ampure XP beads to each sample (30 μL from **step 12**, 1x purification) and mix by pipetting.
15. Incubate at room temperature (21–25 °C) for 10–15 min.
16. Place tubes on a magnetic stand and let sit for 2 min or until the liquid appears clear.
17. Remove and discard 55 μL of the supernatant from each tube by pipetting.
18. Add 200 μL 70–80% ethanol, resuspend the beads with a pipette, and let them incubate on the magnetic stand for 1 min before discarding the ethanol.
19. Repeat ethanol wash (**step 18**) once more.
20. Let the samples sit at room temperature (21–25 °C) for 5–10 min to allow the ethanol to evaporate.
21. Remove tubes from the magnetic stand and resuspend the dried beads in 20 μL of Library Elution Buffer.
22. Mix by pipetting 10x up and down, and incubate at room temperature (21–25 °C) for 5 min.
23. Place tubes on a magnetic stand and let sit for 2 min or until the liquid appears to be clear.
24. Transfer 18 μL of the supernatant to a new PCR tube for PCR amplification.
25. Set up a PCR reaction using the Kapa HiFi kit as follows. The indicated amounts are for one sample.

5x Buffer	5 μL
dNTP	0.75 μL
Primer	0.75 μL
Polymerase	0.5 μL
DNA	18 μL
	25 μL

Run the following PCR program in a PCR machine:

Denaturation	
Temp.	*Duration*
95 °C	3 min
Cycling: 16 cycles (*see* **Note 6**)	
Temp.	*Duration*
98 °C	20 s
60 °C	15 s

(continued)

72 °C	60 s
Final extension	
Temp.	*Duration*
72 °C	60 s

3.4 Library Size Selection and Cleanup

After library preparation, you may follow one of the options below for size selection and cleanup.

Optional Cleanup 1: Pippin Prep

1. Bring DNA sample up to 30 μL with TE.
2. Bring loading solution to room temperature (21–25 °C).
3. For each sample, combine 30 μL of DNA sample with 10 μL of loading solution/marker mix (Labeled Marker L).
4. Mix samples thoroughly by vortexing. Briefly centrifuge to collect the sample.
5. In the Pippin prep software, go to the Protocol Editor Tab to program the Pippin protocol.
   - Click the "Cassette" folder, and select "2 % DF Marker L."
   - Select the collection mode for each lane (Usually "Tight" or "Range"), and enter the size selection parameters.
   - Click the "Use Internal Standards" button.
   - Make sure the "Ref Lane" values match the lane numbers.
   - Press "Save as" and name and save the protocol.
6. Calibrate the optics with the calibration fixture by placing the calibration fixture onto the optical nest. Close the lid and press "CALIBRATE" and launch the calibration window.
7. Enter 0.80 in the "Target I ph, mA" field. Press the "CALIBRATE" button in the window, and when complete, press "EXIT."
8. Remove the cassette from the foil packaging and inspect the cassettes for agarose gel breakage, buffer levels, and air bubbles (*see* **Note 7**).
9. Prepare the cassette for loading (*see* **Note 8**).
10. Dislodge bubbles from behind the elution wells. Tilt the cassette sample well side down, to release any trapped bubbles behind the elution models.
11. Place the cassette into the optical nest. Keep the cassette slightly tilted down so that the bubbles in the elution reservoirs do not return to the area behind the elution models. Be sure the cassette is fully seated into the bottom of the nest to ensure proper optical alignment.

12. Remove the white tabbed adhesive strips from the cassette. Place one hand on the cassette, and hold it firmly in the nest. Grab the white tabs of the tape and pull the strips firmly and slowly toward the front of the PippinPrep until they are removed.
13. Remove buffer from all elution models and replace with 40 μL of fresh electrophoresis buffer. Make sure that the pipette tips extend all the way to the bottom of the elution modules without sealing the elution port opening. Test tip fit using the empty rinse cassette supplied with the instrument.
14. Seal the elution wells with the adhesive tape strips. Tape for sealing the elution wells is supplied with cassette packaging. Place tape over the elution wells and rub firmly to fix the tape in position.
15. Check the buffer level in the sample wells. Immediately before loading, sample wells should be filled to the top with buffer. If any wells are underfilled, top them up with an additional buffer.
16. Perform the continuity test. Close the continuity test. Close the lid and press the "Test" button located in the lower right area of the Main screen. The test routine runs automatically and measures the current in each separation and elution channel. Moreover, it should return a "PASS" for each separation and elution channel. The cassette temperature must be above 17 °C (62 °F).
17. Recheck the buffer level in the sample's wells. Make sure that sample wells are completely full to the top with electrophoresis buffer. Fill with an additional buffer if necessary.
18. Remove 40 μL of buffer from the first sample well and load 40 μL of sample into that well (*see* **Note 9**).
19. Repeat **step 21** for the remaining four wells.
20. Close the lid, go to the Main Tab, and make sure the proper protocol is loading in the "Protocol Name" field.
21. Press "START." The run will automatically stop when every collection is complete.
22. After the run is completed, collect samples from the elution wells using a standard 100–200 μL pipette. Samples will be in a Tris-TAPS buffer at a volume of 40 μL. Samples should be suitable for amplification. Do not let samples remain in the cassette overnight.
23. Remove the cassette and dispose it properly. Do not keep used cassettes in the Pippin with the cover closed. Humidity from the cassette may cause damage to the electrodes.

Optional Cleanup 2: Pippin HT

1. For best results, samples should be de-proteinized (e.g., using proteinase K).
2. Bring the DNA sample up to 20 μL with TE.
3. Bring the internal standard/loading solution (20B) to room temperature (21–25 °C).
4. For each sample, combine the 20 μL DNA sample with 5 μL of the 20B solution.
5. Mix samples thoroughly by vortexing. Briefly centrifuge to collect the sample on the tube bottom.
6. In the PippinHT Software, go to the Protocol Editor Tab and program the desired protocol as follows.
   - Click the "Cassette" folder and select the appropriate cassette definition:

     "2 % Agarose, 100–600 bp, 20B"
   - Make sure the Cassette 1 tab is selected.
   - Enter the size selection parameters in lanes ½.
   - To copy the size selection to the remainder of lanes, press "Copy Lane 1 to Cassette."
   - Press "Use Internal Standards" to assign internal standards to each lane pair.
   - Select the cassette 2 tab. To copy the cassette 1 size selection parameters to cassette 2, press "Copy Cassette 1 to Cassette 2."
   - Press "Save As" and name and save to protocol.
7. Place the calibration fixtures onto both optical nests. Close the lid and press "Calibrate" to launch the calibration window.
8. When complete, the "Calibration Status" field will indicate "Calibration OK." Press "EXIT" to continue.
9. Remove the cassettes from the foil packaging and inspect the gel cassette for breakage of the agarose gel columns (*see* **Note 7**).
10. Dislodge bubbles from behind the elution modules. Do this by gently tapping the cassette while holding it sideways with the elution modules facing up.
11. Remove the adhesive tapes. Do this by placing the cassettes onto a benchtop and firmly pulling with an even motion.
12. Remove 80 μL of buffer from each of the upper buffer chambers.

13. Remove all buffer from the elution modules and replace them with 30 μL of spare electrophoresis buffer provided by the manufacturer.
14. Re-seal the elution modules with the adhesive tape provided with the kit.
15. Perform the continuity test. Place the cassettes onto the nests, close the lid, and press "Test" in the Main Tab.
16. Re-check the buffer level in the sample's wells. Make sure that sample wells are completely full to the top with electrophoresis buffer. Fill with additional buffer, if necessary.
17. Remove 30 μL of buffer from the first sample well and load 25 μL of sample into that well (*see* **Note 9**).
18. Repeat **step 17** for the remaining four wells, or use a 12-channel Multipipettor to load the wells at the same time.
19. Enter sample information (*Optional*). Sample information may be manually keyed, read with an external barcode reader, or imported as a text file from a USB drive. A log file name prefix may be entered in the System Options Tab.
20. Close the lid, go to the Main Tab, and make sure the proper protocol is loaded in the "Protocol Name" field.
21. Press "START." The run will automatically stop when every collection is complete.
22. After the run is completed, remove 30 μL samples from the elution wells using a standard 100–200 μL pipette. Samples will be in a Tris-TAPS buffer. Do not let samples remain in the cassette overnight.
23. Remove the cassette and dispose of it properly. Do not keep used cassettes in the PippinHT with the cover closed. Humidity from the cassette may cause damage to electrodes.

Optional Cleanup3: AmpureXP Beads

Cleanup according to manufacturer's protocol.

4 Notes

1. These are the reagents that we use in our lab and are optimized for our parameters. Other reagents may work for your experiment, but optimization will potentially be required.
2. Extraction of nuclei and chromatin and cross-linking protocols require intensive adaptation depending on the tissue, cell type, and protein of interest. We refer to domain-specific protocols for this part and start our protocol assuming cross-linked chromatin was previously prepared.

3. Thawing chromatin reliably is difficult. We have found that hand thawing works best for this step since the body temperature is about 37 °C and allows you to easily see the exact moment when the sample has thawed.
4. The amount of chromatin and antibody was determined empirically and needs to be optimized in advance.
5. When taking up agarose, it may help to cut off the tip of your pipette tip. This allows the agarose beads to be taken up evenly and will not clog up your pipette.
6. Sixteen cycles are optimized for the concentration that we are looking for in our sample. Your sample may require more or fewer cycles depending on what is optimal for your specific sample.
7. Inspect the levels of buffer in all buffer reservoirs. Reservoirs should be nearly full. If the buffer level in any reservoir appears less than 50% full, fill it with spare buffer provided by the manufacturer. Inspect the gel columns. Look for obvious breakage of the agarose column in each channel. If there is obvious breakage, do not use the lane. The remaining lanes can be used. Inspect for bubbles due to the separation of agarose from the bottom of the cassette in the region used for the optical detection of DNA. If a bubble is observed, do not use the lane. A bubble on the top of the gel column will not affect detection.
8. It is recommended to familiarize yourself with the orientation of the cassette at this point to ensure proper use.
9. Take care not to pierce the agarose with the pipette tip. There is gel on all sides and bottom of the sample well. In addition, there is an agarose chimney surrounding the top of the sample well that protrudes up through the cassette cover. When removing the buffer, some users find it useful to immerse the pipette tip just below the surface of the buffer and follow the liquid level down with the tip as the buffer is removed. When adding the sample, place the tip of the pipette just below the surface of the buffer, and follow the liquid level up with the tip as the well fills. Do not be concerned if the sample well slightly overfills. The density of the sample will allow it to sink before it can flow out of the well.

Acknowledgments

I.M.B. is supported by the National Institute of Diabetes and Digestive and Kidney Diseases R01 award DK121059.

References

1. Collas P (2010) The current state of chromatin immunoprecipitation. Mol Biotechnol 45(1): 87–100
2. Bochkis IM, Schug J, Ye DZ et al (2012) Genome-wide location analysis reveals distinct transcriptional circuitry by paralogous regulators Foxa1 and Foxa2. PLoS Genet 8(6):e1002770
3. Kain J, Wei X, Reddy NA et al (2021) Pioneer factor Foxa2 enables ligand-dependent activation of type II nuclear receptors FXR and LXRα. Mol Metab 53:101291
4. Wei X, Murphy MA, Reddy NA et al (2022) Redistribution of lamina-associated domains reshapes binding of pioneer factor FOXA2 in development of nonalcoholic fatty liver disease. Genome Res 32(11–12):1981–1992
5. Whitton H, Singh LN, Patrick MA et al (2018) Changes at the nuclear lamina alter binding of pioneer factor Foxa2 in aged liver. Aging Cell 17(3):e12742
6. Bernstein BE, Mikkelsen TS, Xie X et al (2006) A bivalent chromatin structure marks key developmental genes in embryonic stem cells. Cell 125(2):315–326
7. Price AJ, Manjegowda MC, Kain J et al (2020) Hdac3, Setdb1, and Kap1 mark H3K9me3/H3K14ac bivalent regions in young and aged liver. Aging Cell 19(2):e13092
8. Furlan-Magaril M, Rincón-Arano H, Recillas-Targa F (2009) Sequential chromatin immunoprecipitation protocol: ChIP-reChIP. In: Leblanc B, Moss T (eds) DNA-protein interactions: principles and protocols, 3rd edn. Humana Press, Totowa, pp 253–266

Chapter 8

Reverse Chromatin Immunoprecipitation (R-ChIP)

Xuejing Wen and Yucheng Wang

Abstract

DNA-protein interactions play fundamental roles in diverse biological functions. The gene-centered method is used to identify the upstream regulators of defined genes. In this study, we developed a novel method for capturing the proteins that bind to certain chromatin fragments or DNA sequences, which is called reverse chromatin immunoprecipitation (R-ChIP). This technology uses a set of specific DNA probes labeled with biotin to isolate chromatin or DNA fragments, and the DNA-associated proteins are then analyzed using mass spectrometry. This method can capture DNA-associated proteins with sufficient quantity and purity for identification.

Key words Chromatin immunoprecipitation, DNA-associated proteins, DNA probe, Hybridization, Biotin-streptavidin system, Mass spectrometry, Gene-centered approach, Upstream regulators

1 Introduction

DNA-protein interactions play fundamental roles in diverse biological processes, especially in DNA replication and transcription regulation [1, 2]. Transcription factors (TFs) can activate or repress the expression of genes by binding to DNA *cis*-regulatory elements. Therefore, interactions between TFs and DNA *cis*-regulatory elements build up the backbone of gene regulatory networks [3], and consequently become the focus of molecular regulation mechanism research.

The methods to identify DNA-protein interactions can be classified into two complementary approaches, i.e., gene-centered (DNA-to-protein) methods and TF-centered (protein-to-DNA) approaches [4]. Gene-centered methods use one or more DNA cis-elements to identify the TFs that bind to these DNA cis-elements, including yeast one-hybrid assays (Y1H). On the other hand, TF-centered approaches such as chromatin immunoprecipitation (ChIP), systematic evolution of ligands by exponential enrichment (SELEX), and DNA adenine methyltransferase identification (DamID) [5], use one or a set of TFs to identify the

Franziska Greulich (ed.), *Chromatin Immunoprecipitation: Methods and Protocols*, Methods in Molecular Biology, vol. 2846, https://doi.org/10.1007/978-1-0716-4071-5_8,

genes which are regulated by these TF(s). Until now, very limited in vivo gene-centered methods have been investigated, such as proteomics of isolated chromatin segments (PICh) [6], chromatin isolation by RNA purification (ChIRP) [7], protein-RNA interaction mapping assay (PRIMA) [8], and Y1H. ChIRP and PRIMA are both used to identify proteins that bind to RNA. PICh uses DNA probes to hybridize and isolate specific chromatin segment-protein complexes but is mainly used for Hela cells [6]. Murarka and Srivastava (2018) developed a technology that applies to microorganisms to identify the proteins that bind to specific DNA sequences [9]. Therefore, an efficient in vivo gene-centered method should be developed for plants.

In this study, we developed an approach to capture proteins from specific DNA *cis*-elements. The procedures of R-ChIP are comprised of protein/chromatin cross-linking using formaldehyde, nucleus isolation, shearing of the chromatin to 1 kb using a sonication, denaturation of the chromatin by heat, hybridization of the interested chromatin using specific probes with biotin, harvesting the chromatin of interest using streptavidin magnetic beads, washing and elution the target chromatin, and analysis by mass spectrometry.

2 Materials

2.1 Plant Material Preparation

1. Cross-linking buffer: 10 mM Tris–HCl pH 8.0, 3.0% (w/v) formaldehyde, 20 mg/L $NiSO_4$, 1 mM PMSF, 0.01% Silwet.
2. 2 M glycine.
3. Sterile deionized water.
4. Liquid nitrogen.
5. 50 mL centrifuge bottles.
6. Vacuum pump.

2.2 Nuclear Isolation

1. Nuclei isolation buffer: 10 mM Tris–HCl pH 8.0, 0.4 M sucrose, 10 mM $MgCl_2$, 5 mM β-mercaptoethanol, 1 mM PMSF, 1 μg/mL proteinase inhibitor cocktail.
2. Nuclei suspension buffer: 10 mM Tris–HCl pH 8.0, 0.25 M sucrose, 10 mM $MgCl_2$, 1% (v/v) Triton X-100, 5 mM β-mercaptoethanol, 1 mM PMSF, 1 μg/mL proteinase inhibitor cocktail.
3. Nuclei separation buffer: 10 mM Tris–HCl pH 8.0, 1.7 M sucrose, 0.15% (v/v) Triton X-100, 2 mM $MgCl_2$, 5 mM β-mercaptoethanol, 1 mM PMSF, 1 μg/mL proteinase inhibitor cocktail.

4. Nuclei lysis buffer: 50 mM Tris–HCl pH 8.0, 1% (w/v) SDS, 1 μg/mL proteinase inhibitor cocktail.
5. Prechilled mortars and pestles.
6. Miracloth.
7. High-speed cooling centrifuge for 50 mL and 1.5 mL tubes.
8. 1.5 mL centrifugation tubes.

2.3 Ultraviolet Cross-Linking

1. 96-well cell culture plates.
2. UV lamp capable of 7000 μJ/cm^2 at 254 nm.

2.4 Sonication

1. Ultrasonic processor.
2. Proteinase K solution.
3. Agarose and agarose gel materials: running buffer, DNA loading dye, DNA ladder, gel chambers, UV transilluminator.

2.5 Precleaning of the Chromatin

1. 2x hybridization buffer: 30 mM Tris–HCl pH 7.4, 20% (v/v) formamide, 40 mM KCl, 20 mM $(NH_4)_2SO_4$, 6 mM $MgCl_2$, 1 μg/mL proteinase inhibitor cocktail.
2. Dynabesds MyOne streptavidin C1 (Invitrogen).
3. Magnetic rack for 1.5 mL tubes.
4. Roller.
5. Thermostatic incubator.

2.6 Chromatin Hybridization

1. Biotinylated probes (*see* **Note 1**).

2.7 Isolation of Hybridized DNA Using Magnetic Beads

1. Low salt wash buffer: 20 mM Tris–HCl pH 7.4, 0.1% (w/v) SDS, 150 mM NaCl, 1 μg/mL proteinase inhibitor cocktail.
2. High salt wash buffer: 20 mM Tris–HCl pH 7.4, 0.1% (w/v) SDS, 500 mM NaCl, 1 μg/mL proteinase inhibitor cocktail.
3. LiCl wash buffer: 20 mM Tris–HCl pH 7.4, 250 mM LiCl, 1 μg/mL proteinase inhibitor cocktail.
4. Tris wash buffer: 10 mM Tris–HCl pH 7.4, 1 μg/mL proteinase inhibitor cocktail.

2.8 Elution

1. Biotin elution buffer: 12.5 mM biotin, 7.5 mM HEPES-KOH pH 7.5, 75 mM NaCl, 1.5 mM EDTA, 0.15% (w/v) SDS, 0.075% (w/v) Sarkosyl, and 0.02% (w/v) Na-Deoxycholate.

2.9 Concentration of Proteins

1. Protein filter (100 kDa).

2.10 Analysis of the Hybridization Efficiency

1. PCR purification kit.
2. Target region primer.
3. Quantitative PCR system and master mix.

3 Methods

3.1 Probe Design

Design 10–20 probes according to the promoter sequence of the target gene, and each DNA strand has 5–10 matched probes. Synthesize all the probes (20 nt) with biotin labeling at their 5' termini (*see* **Note 1**).

3.2 Plant Material Preparation

1. Harvest whole seedlings of plants (such as *Arabidopsis thaliana*) in 50 mL centrifuge bottles. Use 10 g of 3-week-old seedlings and divide into four equal portions (2.5 g/sample).
2. Add 30 mL of cross-linking buffer to each sample. Make sure to cover the tissue completely.
3. Poke needle size holes in the cap of the centrifuge bottles and screw on the cap.
4. Add 3–5 pipette tips (P1000) on the top of seedlings to keep them immersed during vacuum infiltration.
5. Place the centrifuge bottles into the vacuum device and keep the vacuum at 25 PSI for 10–12 min at room temperature (25 ± 3 °C) for cross-linking.
6. Repeat this **step 5** once. The total vacuum time should not be longer than 24 min (*see* **Note 2**).
7. Add 2 mL of 2 M Glycine (final concentration of 125 mM) to stop the cross-linking reaction and apply vacuum infiltration for an additional 5 min at room temperature (25 ± 3 °C).
8. Wash plant tissues three times with cold sterile deionized water, remove the water as much as possible by blotting the tissues between paper towels, and quick-freeze in liquid nitrogen.

3.3 Nuclear Isolation

1. Grind the samples into a fine powder using prechilled mortars and pestles under liquid nitrogen.
2. Resuspend each sample with 30 mL cold Nuclei isolation buffer in a 50 mL centrifuge bottle.
3. Vortex briefly and shake at 120 rpm for 5 min at 4 °C. Afterward, incubate the samples on ice for 20 min.
4. Filter the homogenized solution through two layers of Miracloth and centrifuge the filtrate at 3000× *g* for 15 min at 4 °C.
5. Discard the supernatant and resuspend the pellet in 1 mL of cold Nuclei suspension buffer and centrifuge the slurry at 12,000× *g* for 5 min at 4 °C.

6. Repeat **step 5** once (*see* **Note 3**).
7. Add 300 μL of Nuclei separation buffer in a new 1.5 mL centrifuge tube.
8. Resuspend the pellet from **step 6** in 300 μL of Nuclei suspension buffer, and overlay this layer on the top of the Nuclei separation buffer from **step** 7. Then centrifuge the slurry at 14,000× *g* for 30 min at 4 °C.
9. Discard the supernatant and resuspend the pellet in 350 μL of Nuclei lysis buffer.

3.4 Ultraviolet Cross-Linking

1. Transfer the chromatin solution to a 96-well cell culture plate.
2. Treat the chromatin solution with ultraviolet radiation of 254 nm with 7000 μJ/cm^2 for 10 min on the ice for UV cross-linking (*see* **Note 4**).
3. Transfer the chromatin solution into a new 1.5 mL centrifuge tube.
4. Importantly, set 20 μL from each sample aside to compare the extracted chromatin with the sonicated chromatin samples by gel electrophoresis.

3.5 Sonication

In this step, the chromatin is sheared into smaller fragments.

1. Shear the chromatin into DNA fragments of around 750 bp (500–1000 bp) by sonicating in an ice bath for 30 min (10 s sonication/20 s pause) at 90 W power setting using a contact ultrasonic processor or for 20 min (5 s sonication/10 s pause) at 15% power setting using a noncontact ultrasonic processor. The position of the sonicator probe should be a few millimeters from the bottom of the tube to avoid foaming.
2. Spin the chromatin solution at 14,000× *g* for 10 min at 4 °C to pellet debris, and collect supernatants (sonicated chromatin). Set aside 20 μL sonicated chromatin as the input control (sample 1).
3. Importantly, set aside 20 μL from each sample to compare with the aliquot from Subheading 3.4, **step 4**.
4. Add a final concentration of 1 mg/mL proteinase K and 0.2 M NaCl to each aliquot from Subheading 3.4, **step 4** and Subheading 3.5, **step 3** and incubate at 55 °C for at least 4 h to overnight (16–24 h) for reverse cross-linking and protein digestion.
5. Mix DNA samples with DNA loading buffer and load in the wells of a 1% (w/v) agarose gel containing ethidium bromide solution. Perform electrophoresis at 100–120 V for 20–30 min and visualize DNA fragments with UV transillumination (260–340 nm range) to test the sonication efficiency (*see* **Note 5**).

3.6 Preclearing of the Chromatin (See Note 6)

1. Add 35 μL streptavidin magnetic beads (1 μm superparamagnetic particles covalently coupled to a highly pure form of streptavidin) to a 1.5 mL centrifuge tube and harvest the magnetic beads using the magnetic rack.
2. Discard the supernatant.
3. Add 350 μL of hybridization buffer. Resuspend the magnetic beads and position them back in the magnetic rack. Let the beads settle and discard the supernatant.
4. Preclear the sonicated chromatin by adding 350 μL of the sonicated chromatin to the washed beads and incubate at 37 °C for 30 min using a roller at 10 rpm.
5. Collect beads using the magnetic rack and transfer the supernatant to a fresh 1.5 mL tube (= precleaned chromatin).
6. Repeat the preclearing process **steps 1–5** once.

3.7 Chromatin Hybridization

1. Add an equal volume of 2x hybridization buffer and 0.6 μM of the mixed probes to the precleaned chromatin, then mix and divide the solution into two equal portions (300–400 μL/centrifuge tube).
2. Incubate the solution at 90 °C for 2 min in a water bath or metal bath to denature the DNA (*see* **Note 7**).
3. Afterward, immediately place the tube in a mixture of ice and water.
4. Incubate the solution using a roller at 37 °C with 10 rpm rotation for 4–6 h for hybridization (hybridization product).

3.8 Isolation of Hybridized DNA Using Magnetic Beads

1. Transfer streptavidin magnetic beads (1 μL/pmol of probe) to a clean centrifuge tube.
2. Harvest the beads using the magnetic rack and remove the supernatant.
3. Wash the beads once with 1 mL of hybridization buffer.
4. Add the hybridization product to the magnetic beads and incubate with 10 rpm rotation at 37 °C for 2–3 h.
5. Harvest the beads using the magnetic rack and remove the supernatant (*see* **Note 8**).
6. Wash the magnetic beads with 1 mL of each of the following buffers for 5 min: 1x Low salt wash buffer, 1x High salt wash buffer, 1x LiCl wash buffer, 2x Tris wash buffer. Rotate samples at 10 rpm at room temperature (25 ± 3 °C) on a roller and place the tubes on the magnetic stand for 1 min. Afterward, remove the supernatant.

3.9 Elution

1. Elute the DNA-protein complexes from the magnetic beads by adding 300 μL of Biotin elution buffer to the beads and incubate at room temperature (25 ± 3 °C) for 30 min while rotating at 10 rpm on a roller.
2. Harvest the beads using the magnetic rack and collect the supernatant into a fresh tube.
3. Repeat the elution step by adding another 200 μL of Biotin elution buffer to the beads and incubate at 65 °C for 20 min.
4. Centrifuge at 10,000× *g* for 1 min, place the tubes in the magnetic rack, and collect the supernatant into the tube containing the eluate from above (**step 2**). The total eluate is 500 μL.
5. Put 20 μL of the supernatant as sample 2 aside for later analysis (*see* **Note 9**).

3.10 Concentration of Proteins

1. Add the elution product to a protein filter (100 kDa cutoff).
2. Spin at 3000× *g* for 10 min at 4 °C.
3. Repeat **steps 1** and **2** three to seven times to concentrate the products to 30 μL for mass spectrometry analysis (*see* **Note 10**).

3.11 Analysis of the Hybridization Efficiency

1. Add a final concentration of 1 mg/mL proteinase K, 0.2 M NaCl, and 0.5% (w/v) SDS to samples 1 (input from Subheading 3.5, **step 2**) and 2 (Subheading 3.9, **step 5**); incubate at 55 °C for at least 3 h to reverse cross-link and digest the proteins.
2. Purify the DNA fragments using a PCR purification kit, and use the purified DNA fragments as a PCR template to determine the hybridization efficiency.

4 Notes

1. Analyze the cis-acting element distribution of the target gene promoter using PlantCARE database (https://bioinformatics.psb.ugent.be/webtools/plantcare/html/), then design 10–20 probes according to cis-acting element enrichment region to cover more cis-acting element. Each DNA strand should have the same number of probes. For example, we designed 12 probes according to the promoter sequence of AtCAT3 (AT1G20620), each DNA strand had six matched probes, and their distribution is shown in Fig. 1.
2. After cross-linking, seedlings should become water-soaked and translucent, because formaldehyde penetrates plant cell walls. The formaldehyde concentration and the cross-linking time are very important for cross-linking efficiency. A short time of

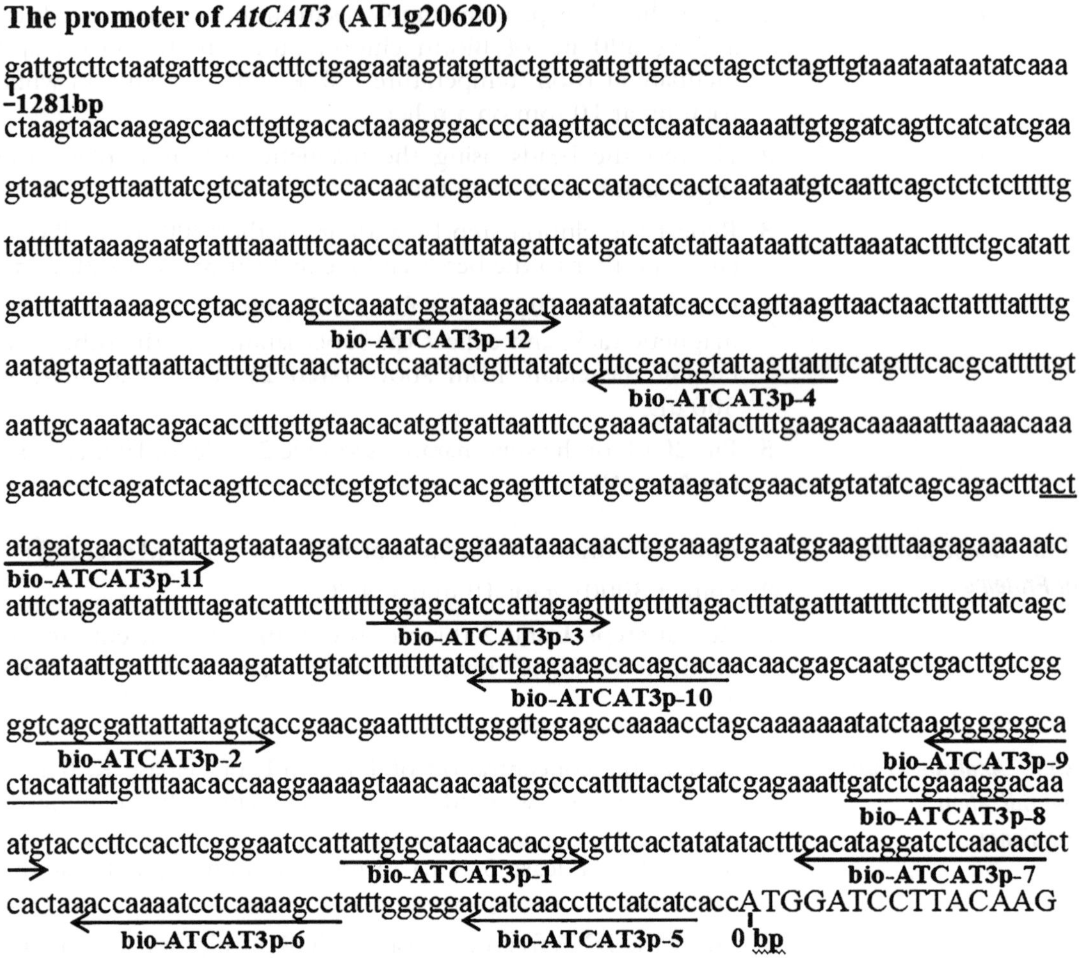

Fig. 1 The distribution of probes on promoter of AtCAT3 (AT1G20620)

vacuum infiltration can lead to inefficient cross-linking, whereas a long time (>24 min) can cause excessive cross-linking, which affects chromatin shearing, reversal of cross-linking, and DNA precipitation.

3. Nuclei suspension buffer contains enough detergents to burst open the chloroplasts but not the nuclei. Therefore, we can repeat the wash step once if the solution appears dark green due to the large sample size. If the sucrose density gradient centrifugation is working well, this pellet should have a white fleck (nuclei).
4. This step is employed for enhancing DNA-protein cross-linking to avoid de-cross-linking during sonication and hybridization. Keep the plate on ice to protect the DNA-protein complex from heat generated during UV cross-linking and prevent melting water from entering the plate to contaminate the sample.

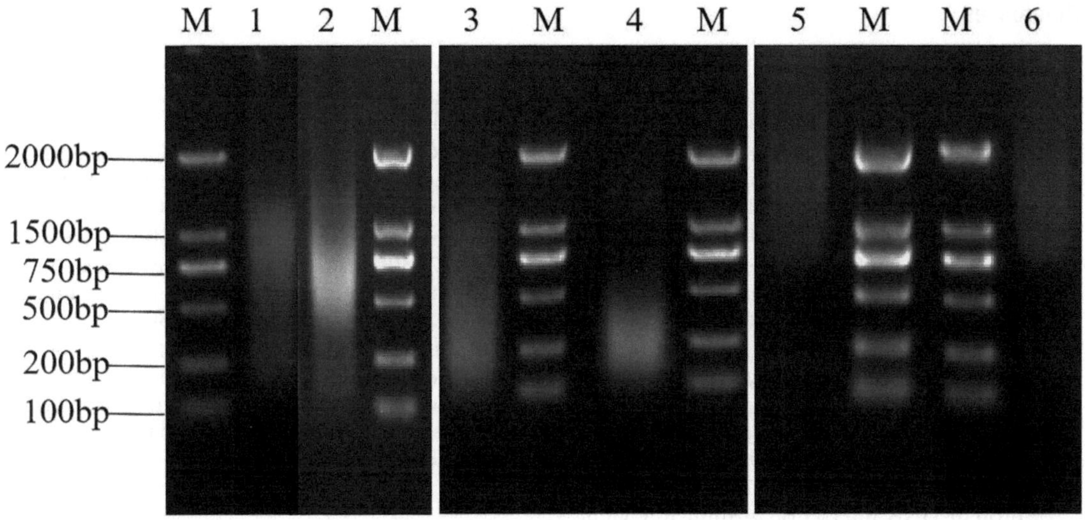

Fig. 2 Agarose gel picture with the optimal fragment size (500–1000 bp) and under-/over-sonicated chromatin. M: DNA ladder; 1, 2: optimal fragment size chromatin; 3, 4: over-sonicated chromatin; 5, 6: under-sonicated chromatin

5. The output power and the sonication time are very important for sonication efficiency. Under- or over-sonicated chromatin could reduce the hybridization efficiency and needs to be avoided. Therefore, always check the size of the chromatin fragments (Fig. 2).
6. To reduce the nonspecific binding of magnetic beads, we introduce the precleaning step to remove the DNA-protein complex that is easily bound to magnetic beads before hybridization.
7. The heating temperature and time are very important for DNA denaturing, overheating could stimulate reverse cross-linking, and the heating time should not exceed 2 min. Underheating could decrease the DNA denaturing efficiency. The liquid volume should be less than 400 μL/tube to ensure be heated uniformly within 2 min.
8. The hybridization product should not be shaken vigorously. Every harvest step using the magnetic rack should be held on for at least 3 min, due to a small number of magnetic beads are not visible to the naked eye.
9. Subheading 3.5, **step 2** and Subheading 3.9, **step 4** are employed for the analysis of the hybridization efficiency.
10. Keep an eye on the fluid flow rate when using a protein filter, and if the fluid flow rate suddenly increases, the protein filter may leak. The liquid volume in the protein filter should be greater than 30 μL until the last spin, and mix the liquid remaining in the filter with a pipette.

References

1. Chang TC, Lin JJ, Lin KC et al (2002) Investigation of DNA–protein recognition by satellite hole spectra of labeling dye. J Lumin 98(1):149–152
2. Malonga H, Neault JF, Arakawa H et al (2006) DNA interaction with human serum albumin studied by affinity capillary electrophoresis and FTIR spectroscopy. DNA Cell Biol 25(1):63–68
3. Walhout AJM (2006) Unraveling transcription regulatory networks by protein-DNA and protein-protein interaction mapping. Genome Res 16(12):1445–1454
4. Reece-Hoyes JS, Barutcu AR, McCord RP et al (2011) Yeast one-hybrid assays for gene-centered human gene regulatory network mapping. Nat Methods 8(12):1050–1052
5. Kim TH, Ren B (2006) Genome-wide analysis of protein-DNA interactions. Annu Rev Genomics Hum Genet 7(1):81–102
6. Déjardin J, Kingston RE (2009) Purification of proteins associated with specific genomic loci. Cell 136(1):175–186
7. Chu C, Zhang Qiangfeng C, da Rocha Simão T et al (2015) Systematic discovery of Xist RNA binding proteins. Cell 161(2):404–416
8. Tamburino AM, Kaymak E, Shrestha S et al (2017) PRIMA: a gene-centered, RNA-to-protein method for mapping RNA-protein interactions. Translation 5(1):e1295130
9. Murarka P, Srivastava P (2018) An improved method for the isolation and identification of unknown proteins that bind to known DNA sequences by affinity capture and mass spectrometry. PLoS One 13(8):e0202602

Chapter 9

Estrogen Receptor Chromatin Profiling by CUT&RUN

Bruno Gegenhuber and Jessica Tollkuhn

Abstract

Gonadal steroid hormones, namely, testosterone, progesterone, and estrogens, influence the physiological state of an organism through the regulation of gene transcription. Steroid hormones activate nuclear hormone receptor (HR), transcription factors (TFs), which bind DNA in a tissue- and cell type–specific manner to influence cellular function. Identifying the genomic binding sites of HRs is essential to understanding mechanisms of hormone signaling across tissues and disease contexts. Traditionally, chromatin immunoprecipitation followed by sequencing (ChIP-seq) has been used to map the genomic binding of HRs in cancer cell lines and large tissues. However, ChIP-seq lacks the sensitivity to detect TF binding in small numbers of cells, such as genetically defined neuronal subtypes in the brain. Cleavage Under Targets & Release Under Nuclease (CUT&RUN) resolves most of the technical limitations of ChIP-seq, enabling the detection of protein-DNA interactions with as few as 100–1000 cells. In this chapter, we provide a stepwise CUT&RUN protocol for detecting and analyzing the genome-wide binding of estrogen receptor α (ERα) in mouse brain tissue. The steps described here can be used to identify the genomic binding sites of most TFs in the brain.

Key words CUT&RUN, Estrogen receptors, Transcription factor, Chromatin, Epigenetics, Sex differences, Social behavior

1 Introduction

Gene regulation involves the dynamic recruitment of a large class of chromatin-associated proteins, including TFs, cofactors, and chromatin remodeling complexes. TFs recognize and bind select regions in the genome through the specificity encoded in their DNA binding domains, and are thus essential for maintaining genetic identity and responding to extracellular cues. Steroid HRs are a family of TFs that bind to DNA in a ligand-dependent manner to regulate transcription [1]. Nuclear HRs can be divided into those that respond to gonadal hormones—estrogen receptors α (ERα) and β (ERβ), androgen receptor (AR), and progesterone receptor (PR) - and those that bind glucocorticoids - glucocorticoid receptor (GR) and mineralocorticoid receptor (MR). For decades,

Franziska Greulich (ed.), *Chromatin Immunoprecipitation: Methods and Protocols*, Methods in Molecular Biology, vol. 2846, https://doi.org/10.1007/978-1-0716-4071-5_9,

mechanistic insights into HRs have been driven by studies in cancer cell lines [2–7], in which HRs contribute to malignancy. However, in addition to their role in disease, HRs influence the physiological state of reproductive and nonreproductive tissues through regulating transcription in a cell type–specific manner [8–11].

Chromatin immunoprecipitation followed by sequencing (ChIP-seq) has long been the gold-standard approach for identifying the genomic targets of HRs and other DNA-binding proteins. ChIP-seq involves fixation of cells or tissue to preserve protein-DNA interactions, followed by chromatin sonication and immunoprecipitation to extract DNA fragments bound to a protein of interest (Fig. 1a). Subsequent preparation of next-generation sequencing libraries from immunoprecipitated DNA enables the detection of protein binding sites throughout the genome. Although many important insights have been generated with this approach, ChIP-seq often requires a high amount of input material (~ten million cells) to overcome excessive epitope loss arising from protein masking and degradation during fixation and sonication, respectively. Moreover, because ChIP-seq fragments DNA by sonication, the size of DNA fragments limits the resolution of TF binding events, often resulting in considerably broader peaks (~1 kilobase) than protein-DNA contact sites (~10–50 bp). Collectively, these technical limitations have prohibited the use of ChIP-seq for profiling TF binding within specific cell types in vivo and identifying the precise location of TF-DNA contacts on chromatin.

The recently developed in situ chromatin profiling method Cleavage Under Targets & Release Under Nuclease (CUT&RUN) resolves most of the technical limitations of ChIP-seq, enabling the detection of TF binding with as few as 100–1000 cells [12, 13]. CUT&RUN utilizes a fusion protein consisting of immunoglobulin-binding proteins A or G linked to micrococcal nuclease (pA-MNase or pA/G-MNase), which is recruited to a protein of interest in permeabilized cells or nuclei by a target-specific antibody (Fig. 1b). The enzymatic activity of MNase requires Ca^{2+}. By delivering Ca^{2+} after pA/G-MNase has bound its target, DNA is selectively cleaved at sites bound by the protein of interest. Performing this digestion reaction on ice and for a limited period of time (30–60 min) minimizes nonspecific cleavage of accessible chromatin following the release of pA/G-MNase from DNA. Upon completion of the digestion reaction, DNA fragments are separated from undigested chromatin by centrifugation or magnetic separation of permeabilized cells or nuclei. DNA fragments in the supernatant are then prepared for paired-end sequencing using a commercial library preparation kit. Following sequencing, standard command line tools are implemented to align paired-end fragments to the genome and identify protein binding sites.

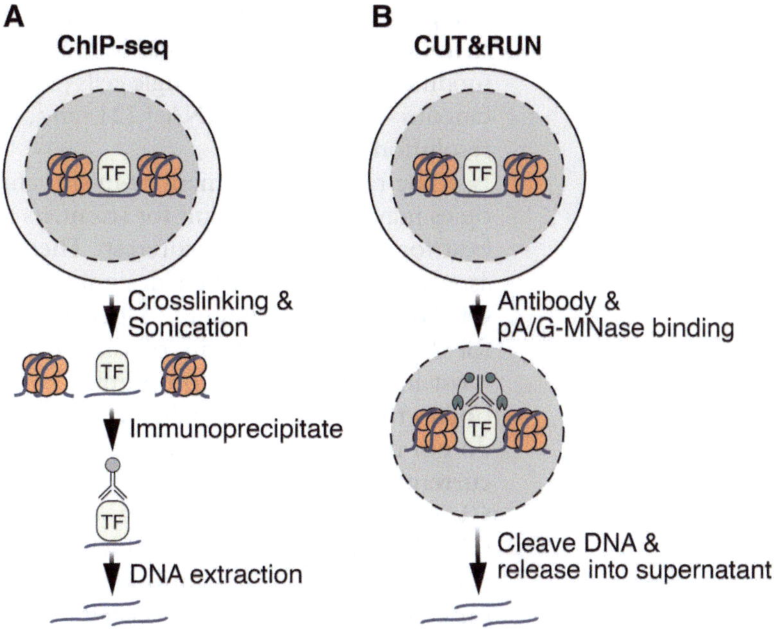

Fig. 1 Comparison of ChIP-seq and CUT&RUN methodology. (**a**) In ChIP-seq, cells are fixed to cross-link protein-DNA and/or protein-protein interactions. Subsequently, cells are lysed and sonicated to solubilize chromatin prior to immunoprecipitation with a target-specific antibody coupled to protein A– or protein G–coated magnetic beads. Following several washes, the protein-bound DNA fragments are eluted from the beads with proteinase K and SDS and extracted with a spin column or PCI precipitation. (**b**) In CUT&RUN, native cells or nuclei are directly immobilized on concanavalin A–coated magnetic beads throughout the protocol. Samples are first incubated with an antibody recognizing the protein of interest and then incubated with the pA/G-MNase fusion protein. Following several washes, selective cleavage of target-bound DNA is initiated with the addition of Ca^{2+} to the buffer, which activates the pA/G-MNase. Performing this step on ice minimizes the diffusion of pA/G-MNase upon target cleavage, which reduces nonspecific fragmentation of accessible chromatin. Cleaved DNA fragments are separated from undigested chromatin by centrifugation and extracted with PCI precipitation. In both ChIP-seq and CUT&RUN, DNA is prepared into Illumina-compatible sequencing libraries using standard ligation-based DNA library preparation kits

Since its inception, several modifications to the original CUT&RUN protocol have been described, including a low NaCl/high Ca^{2+} wash buffer that minimizes nonspecific pA-MNase cleavage [14], a spike-in normalization strategy that utilizes carryover of *Escherichia coli* DNA from bacterial pA-MNase expression [14], and a single-cell implementation [15]. The fusion of proteins A or G to Tn5 transposase (pA-Tn5 or pA/G-Tn5), which inserts sequencing adapters into DNA in a "cut-and-paste" manner rather than cutting, has led to the development of Cleavage Under Targets & Tagmentation (CUT&Tag) [16]. The direct, in situ tagging of DNA bound to a target protein has enabled a legion of pA/G-Tn5-based methods that measure either individual or multiple histone post-translational modifications (PTMs) in single cells [17–21], with or without simultaneous detection of RNA [22], and, most recently, with spatial resolution [23, 24].

The rapid development of in situ chromatin profiling technologies may make it difficult for scientists to assess which approach is best for their protein of interest. There are a number of technical aspects to consider when designing and optimizing a successful CUT&RUN experiment. In general, CUT&RUN works better for TFs and is the focus of this chapter. For CUT&RUN, the most important factor to optimize is the primary antibody. Antibodies that work well in ChIP-seq or immunofluorescence staining do not necessarily give specific and robust CUT&RUN signal, and currently, only certain commercial vendors validate antibodies for CUT&RUN. Some primary antibodies only work well in CUT&RUN with light formaldehyde fixation and/or in combination with a secondary antibody that amplifies the signal. At the onset of targeting a new protein with CUT&RUN, we highly recommend performing a large-scale optimization experiment (Fig. 2), in which several antibodies and parameters are tested in

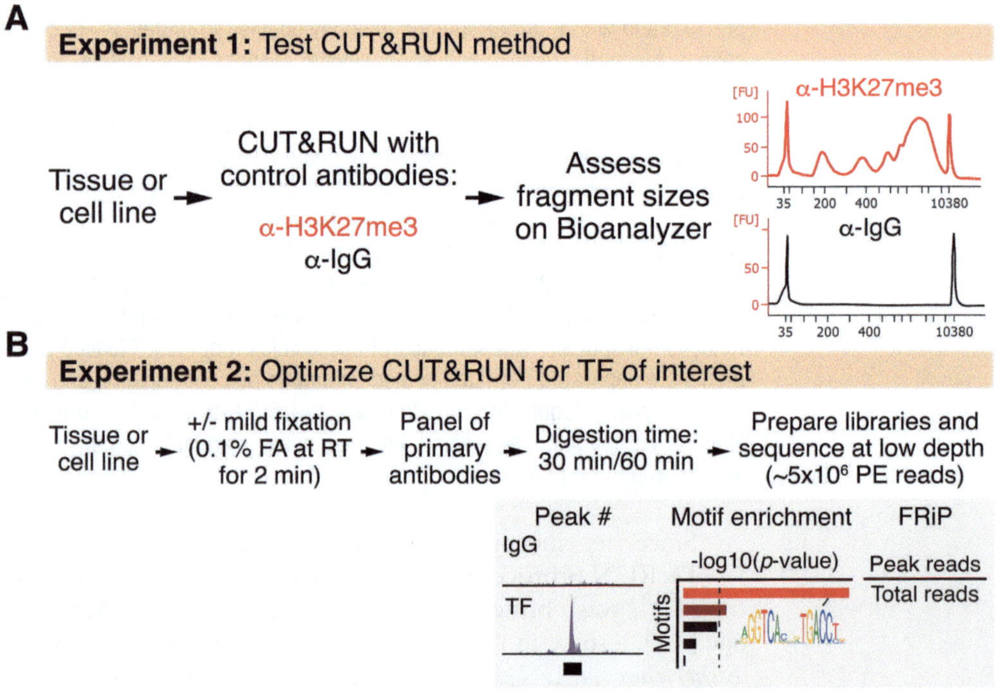

Fig. 2 Guidelines for optimizing CUT&RUN conditions. (**a**) Experiment 1 validates the CUT&RUN protocol in the tissue or cell line of interest using positive and negative control antibodies. (**b**) Experiment 2 optimizes CUT&RUN for the protein of interest by testing several critical parameters in parallel (*see* **Note 1**) and sequencing at low-depth coverage

parallel and sequenced at low-depth coverage (*see* **Note 1**). After sequencing, the optimal CUT&RUN condition can be identified by comparing the peak number, target motif enrichment, and fraction of reads in peaks (FRiP) across samples.

All next-generation sequencing experiments involve the processing and analysis of sequencing reads to derive biological insights. Although the basic steps of CUT&RUN analysis are largely similar to ChIP-seq, there are several key modifications to account for pA/G-MNase-derived DNA fragments as opposed to sonicated DNA. We provide a Linux data processing script (https://github.com/brunogegenhuber/cut-n-run) compatible with a SLURM job scheduler that accomplishes the basic steps of adapter trimming, read alignment, duplicate read removal, fragment size filtering, and peak-calling (Fig. 3). Downstream analysis

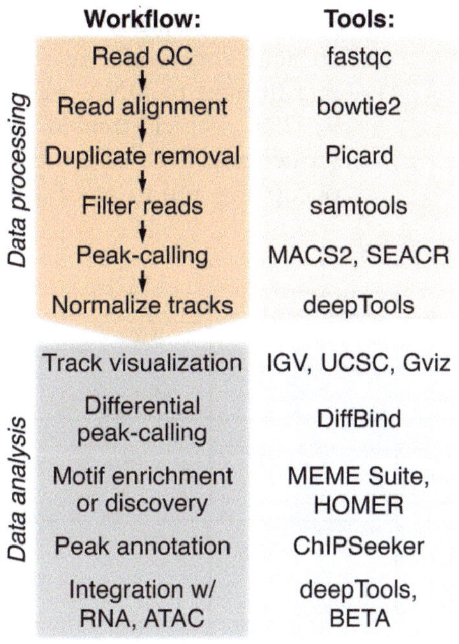

Fig. 3 Workflow of CUT&RUN data processing and analysis. The processing of CUT&RUN fastq files involves (1) quality assessment; (2) alignment to species genome; (3) removal of PCR and optical duplicate fragments; (4) filtering of multi-mapping reads, low-quality reads, and fragments larger than 120 bp (TFs only); (5) peak-calling; and (6) track generation. These steps are executed on a high-performance computing cluster with the indicated bioinformatic tools and can be performed with the provided script file (https:/github.com/brunogegenhuber/cut-n-run). After read alignment and processing, CUT&RUN data may be analyzed to address a range of biological questions. Popular downstream analysis methods include (1) track visualization, (2) differential peak-calling, (3) motif enrichment and/or discovery, (4) genomic annotation of peaks and functional enrichment of nearest genes, and (5) integration with other sequencing modalities, such as RNA-seq and ATAC-seq

of CUT&RUN datasets are user-specific and beyond the scope of this chapter, although we recommend reading the extensive literature covering best practices in bioinformatic analysis [25–27] or consulting with a bioinformatician.

The CUT&RUN protocol described herein was successfully used to determine the genomic binding sites of ERα within MCF7 breast cancer cells and a sexually dimorphic neural circuit of the developing and adult mouse brain [28]. Comparison with prior ERα ChIP-seq datasets revealed that over 60% of the sites occupied by ERα in the brain were not targets in other tissues, highlighting the specificity of HR gene regulation. CUT&RUN led to the discovery of ERα-regulated genes that coordinate sexual differentiation of neuronal wiring and survival during an early-life critical period of brain development (Fig. 4). This protocol allows for reproducible profiling of ERα genomic binding, although, with minor modifications, it may be generally applied to measure the genomic binding of most TFs in the brain. We note that we utilize an acute hormone treatment paradigm, to synchronize receptor recruitment to DNA, as is standard for HR ChIP-seq studies [7–9, 29]. Vehicle-treated samples should also be prepared and sequenced, both to act as a control for ligand-dependent receptor occupancy and to reveal sites of ligand-independent binding [30].

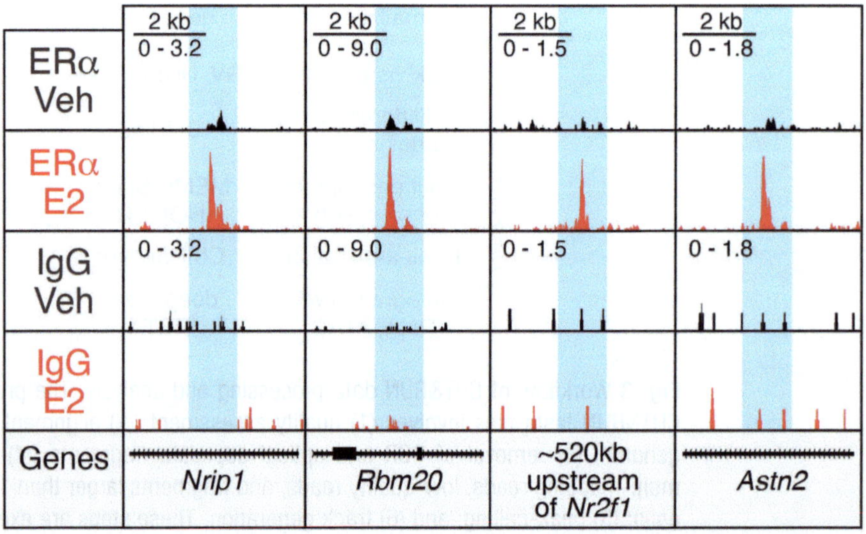

Fig. 4 Example ERα-bound sites in the posterior bed nucleus of the stria terminalis (BNSTp), medial amygdala (MeA), and medial pre-optic area (MPOA) of mice treated for 4 h with estradiol (E2) or vehicle control (Veh). ERα CUT&RUN data were generated with the EMD Millipore Sigma 06–935 antibody and pooled across two biological replicates and across females and males. The IgG control was generated with the ABIN101961 antibody from the same animals. A 5 kb window surrounding each ERα-bound site was plotted using Gviz [31]. The y-axis scale, corresponding to the counts per million (CPM)-normalized read coverage at each base pair, is the same across the four tracks and plotted in the top-left corner for each condition

2 Materials

2.1 Isolation of Nuclei from Brain Tissue

1. Animals: Adult (P48-P56) gonadectomized C57/Bl6J female and male mice from Jackson Laboratory (000664); 3 weeks following gonadectomy, animals were injected subcutaneously with 5 μg of β-Estradiol 3-benzoate (E2, Sigma E8515) or vehicle control (corn oil, Sigma C8267) and harvested at 4 h following injection.
2. Homogenization buffer: 250 mM sucrose, 25 mM KCl, 20 mM Tricine-KOH pH 7.8, 5 mM $MgCl_2$. Adjust pH to 7.8 with KOH and a sterile filter. Homogenization buffer is stable for 6 months at 4 °C.
3. Diluent buffer: 150 mM KCl, 30 mM $MgCl_2$, 120 mM Tricine-KOH pH 7.8. Adjust the final pH to 7.8 with KOH and sterile filter. Diluent buffer is stable for 6 months at 4 °C.
4. 50% OptiPrep solution: 30 mL OptiPrep Density Gradient Medium, 6 mL Diluent buffer. This buffer should be prepared on the day of the experiment.
5. 30% OptiPrep solution: 12 mL 50% OptiPrep solution, 8 mL Homogenization buffer. This buffer should be prepared on the day of the experiment.
6. 40% OptiPrep solution: 16 mL 50% OptiPrep solution, 4 mL Homogenization buffer. This buffer should be prepared on the day of the experiment.
7. Trypan blue.
8. 5% (v/v) IGEPAL CA-630.
9. 0.15 M spermine.
10. 0.5 M spermidine.
11. 1 M DTT.
12. cOmplete, EDTA-free protease inhibitor cocktail (PIC) tablets.
13. 38.5 mL Ultra-Clear tubes (Beckman-Coulter C14292).
14. 40 μm cell strainers.
15. 50 mL conical tubes
16. 1.5 and 2 mL tubes.
17. 3 mL syringes.
18. 18G needles.
19. Adult mouse brain metal dissection matrix.
20. Dissection microscope.
21. 1 mL tissue grinders.
22. Beckman Coulter SW-28 rotor and ultracentrifuge.
23. Automated Cell Counter or hemocytometer.

2.2 Bind Nuclei to Concanavalin A Beads and Incubate with Primary Antibody

1. CUT&RUN Wash buffer: 20 mM HEPES-KOH pH 7.5, 150 mM NaCl, 0.5 mM spermidine, 0.1% (w/v) BSA, 1 cOmplete protease inhibitor cocktail tablet. Sterile filter. This buffer should be prepared on the day of the experiment.
2. CUT&RUN Antibody buffer: CUT&RUN Wash buffer, 2 mM EDTA. This buffer should be prepared on the day of the experiment.
3. CUT&RUN Binding buffer: 20 mM HEPES-KOH pH 7.5, 10 mM KCl, 1 mM $CaCl_2$, 1 mM $MnCl_2$. Sterile filter. CUT&RUN Binding buffer is stable for 6 months at 4 °C.
4. Concanavalin A magnetic beads (Bangs Laboratories BP531).
5. ERα antibody (validate for tissue/cells of interest using optimization strategy in Fig. 2).
6. H3K27me3 antibody (Cell Signaling Technology 9733).
7. Guinea pig anti-rabbit IgG antibody (Antibodies-online ABIN101961).
8. 1.5 mL low-binding tubes.
9. Magnetic stand for 1.5 mL tubes.
10. Tube rocker or nutator.
11. Tabletop microcentrifuges at 20–22 °C and 4 °C.

2.3 Wash Samples and Incubate with pA/G-MNase Fusion Protein

1. pA/G-MNase fusion protein (EpiCypher 15-1016).

2.4 Wash Samples and Activate Chromatin Digestion with $CaCl_2$

1. CUT&RUN 2X Stop buffer: 340 mM NaCl, 20 mM EDTA, 4 mM EGTA, 50 μg/mL RNase A, 50 μg/mL glycogen, 100 pg/mL *E. coli* spike-in DNA (*see* **Note 2**). Sterile filter. This buffer should be prepared on the day of the experiment.
2. 100 mM $CaCl_2$.
3. Metal heat block.

2.5 Extract DNA with Phenol-Chloroform-Isoamyl Alcohol

1. 0.1X TE buffer: 1 mM Tris–HCl pH 8.0, 0.1 mM EDTA. Sterile filter. This buffer should be prepared on the day of the experiment.
2. Phenol:chloroform:isoamyl alcohol.
3. 10% (w/v) SDS.
4. 20 mg/mL proteinase K.
5. Chloroform.
6. 20 mg/mL glycogen.
7. Absolute ethanol.
8. 1.5 mL MaXtract High-Density tubes (Qiagen 129046).

2.6 Prepare Sequencing Library

1. ThruPLEX DNA-seq Library Prep Kit (Takara R400676).
2. Barcoded primers (Table 1), ordered at 50 or 100 nM scale with standard desalting purification.

2.7 Size-Select the Final Libraries

1. AMPure XP beads.
2. 80% (v/v) ethanol freshly prepared.

2.8 Quality Control

1. Bioanalyzer or TapeStation and reagents.
2. Qubit 3.0 Fluorometer and reagents.

3 Methods

3.1 Isolation of Nuclei from Brain Tissue (Day 1)

Keep all samples, buffers, tissue grinders, and ultracentrifuge tube holders on ice throughout the duration of the protocol.

1. Anesthetize the mouse with ketamine/dexmedetomidine and remove the brain.
2. Place the brain into a metal brain dissection matrix on ice and collect 0.5 mm brain sections with razor blades.
3. Under a dissection microscope, collect brain regions of interest and place them in 1 mL of Homogenization buffer containing 0.15 mM spermine, 0.5 mM spermidine, 1 mM DTT, and 1X PIC in a 1.5 mL tube.
4. After completing all of the dissections, transfer the tissue to a 1 mL glass tissue grinder with a wide-bore P1000 pipette tip.
5. Dounce the tissue 15 times with the loose glass pestle.
6. Add 60 μL of 5% IGEPAL CA-630 to a final concentration of 0.3%. Dounce the tissue an additional 5 times with the tight glass pestle.
7. Filter the tissue homogenate through a 40 μm cell strainer into a 50 mL conical tube.
8. Mix the homogenate with 1 mL of 50% OptiPrep solution and transfer the sample to a 38.5 mL Ultra-Clear tube within a Beckman SW-28 tube holder on ice (*see* **Note 3**).
9. Carefully underlay the homogenate with 4.8 mL of 30% Opti-Prep solution. Next, underlay the 30% OptiPrep layer with 4.8 mL of 40% OptiPrep solution.
10. Centrifuge the sample in a Beckman SW-28 swinging-bucket rotor at 10,000× *g* for 18 min at 4 °C.
11. After centrifugation, puncture the tube at the 30%/40% OptiPrep interface with a 3 mL syringe connected to an 18G needle. Aspirate ~1.5 mL at the interface and transfer the solution to a 2 mL tube.
12. Dilute 10 μL of the sample with 10 μL of trypan blue and count the nuclei with an automated cell counter or hemocytometer.

Table 1
Primer sequences

Oligo	Sequence	Barcode
TSPE1.0	AATGATACGGCGACCACCGAGATCTACACTCTTTCCC TACACGACGCTCTTCCGATCT	
TSPE2.1	CAAGCAGAAGACGGCATACGAGATCGTGATGTGACTGGAG TTCAGACGTGTGCTCTTCCGATC	ATCACG
TSPE2.2	CAAGCAGAAGACGGCATACGAGATACATCGGTGACTGGAG TTCAGACGTGTGCTCTTCCGATC	CGATGT
TSPE2.3	CAAGCAGAAGACGGCATACGAGATGCCTAAGTGACTGGAG TTCAGACGTGTGCTCTTCCGATC	TTAGGC
TSPE2.4	CAAGCAGAAGACGGCATACGAGATTGGTCAGTGACTGGAG TTCAGACGTGTGCTCTTCCGATC	TGACCA
TSPE2.5	CAAGCAGAAGACGGCATACGAGATCACTGTGTGACTGGAG TTCAGACGTGTGCTCTTCCGATC	ACAGTG
TSPE2.6	CAAGCAGAAGACGGCATACGAGATATTGGCGTGACTGGAG TTCAGACGTGTGCTCTTCCGATC	GCCAAT
TSPE2.7	CAAGCAGAAGACGGCATACGAGATGATCTGGTGACTGGAG TTCAGACGTGTGCTCTTCCGATC	CAGATC
TSPE2.8	CAAGCAGAAGACGGCATACGAGATTCAAGTGTGACTGGAG TTCAGACGTGTGCTCTTCCGATC	ACTTGA
TSPE2.9	CAAGCAGAAGACGGCATACGAGATCTGATCGTGACTGGAG TTCAGACGTGTGCTCTTCCGATC	GATCAG
TSPE2.10	CAAGCAGAAGACGGCATACGAGATAAGCTAGTGACTGGAG TTCAGACGTGTGCTCTTCCGATC	TAGCTT
TSPE2.11	CAAGCAGAAGACGGCATACGAGATGTAGCCGTGACTGGAG TTCAGACGTGTGCTCTTCCGATC	GGCTAC
TSPE2.12	CAAGCAGAAGACGGCATACGAGATTACAAGGTGACTGGAG TTCAGACGTGTGCTCTTCCGATC	CTTGTA
TSPE2.13	CAAGCAGAAGACGGCATACGAGATTTGACTGTGACTGGAG TTCAGACGTGTGCTCTTCCGATC	AGTCAA
TSPE2.14	CAAGCAGAAGACGGCATACGAGATGGAACTGTGACTGGAG TTCAGACGTGTGCTCTTCCGATC	AGTTCC
TSPE2.15	CAAGCAGAAGACGGCATACGAGATTGACATGTGACTGGAG TTCAGACGTGTGCTCTTCCGATC	ATGTCA
TSPE2.16	CAAGCAGAAGACGGCATACGAGATGGACGGGTGACTGGAG TTCAGACGTGTGCTCTTCCGATC	CCGTCC
TSPE2.17	CAAGCAGAAGACGGCATACGAGATCTCTACGTGACTGGAG TTCAGACGTGTGCTCTTCCGATC	GTAGAG
TSPE2.18	CAAGCAGAAGACGGCATACGAGATGCGGACGTGACTGGAG TTCAGACGTGTGCTCTTCCGATC	GTCCGC

(continued)

Table 1
(continued)

Oligo	Sequence	Barcode
TSPE2.19	CAAGCAGAAGACGGCATACGAGATTTTCACGTGACTGGAG TTCAGACGTGTGCTCTTCCGATC	GTGAAA
TSPE2.20	CAAGCAGAAGACGGCATACGAGATGGCCACGTGACTGGAG TTCAGACGTGTGCTCTTCCGATC	GTGGCC
TSPE2.21	CAAGCAGAAGACGGCATACGAGATCGAAACGTGACTGGAG TTCAGACGTGTGCTCTTCCGATC	GTTTCG
TSPE2.22	CAAGCAGAAGACGGCATACGAGATCGTACGGTGACTGGAG TTCAGACGTGTGCTCTTCCGATC	CGTACG
TSPE2.23	CAAGCAGAAGACGGCATACGAGATCCACTCGTGACTGGAG TTCAGACGTGTGCTCTTCCGATC	GAGTGG
TSPE2.24	CAAGCAGAAGACGGCATACGAGATGCTACCGTGACTGGAG TTCAGACGTGTGCTCTTCCGATC	GGTAGC

3.2 Bind Nuclei to Concanavalin A Beads and Incubate with Primary Antibody (Day 1)

1. Calculate the quantity of nuclei needed for each CUT&RUN sample and controls (*see* **Note 4**).
2. Transfer the desired amount of nuclei per sample into 1.5 mL low-binding tubes. If the volume is less than 1 mL, bring the volume to 1 mL with CUT&RUN Wash buffer. Add 5% IGEPAL CA-630 to a final concentration of 0.4% to increase affinity to concanavalin A beads.
3. Calculate the volume of concanavalin A beads required for all CUT&RUN samples (20 uL beads per sample multiplied by the number of samples). Transfer the desired volume to a 1.5 mL low-binding tube.
4. Add 1 mL of CUT&RUN Binding buffer to the beads and mix by gentle pipetting. Place the tube on a magnetic rack for ~1 min to separate the beads and then remove the liquid with a P1000 pipette.
5. Repeat **step 4** once for a total of two washes.
6. Resuspend the beads to their original volume in CUT&RUN Binding buffer.
7. Add 20 uL of beads to each sample and incubate the tubes on a rotator at 4 °C for 30 min to bind nuclei to the beads.
8. Place the tubes on a magnetic rack and let the beads separate for ~1 min.
9. Wash the samples with 1 mL of CUT&RUN Wash buffer by gentle pipetting or tube inversion.
10. Repeat **steps 8** and **9** once for a total of two washes.

11. Resuspend the beads in 198 μL of CUT&RUN Antibody buffer and add 2 μL of primary antibody to a final concentration of 1:100 (*see* **Note 5**).
12. Incubate the samples overnight (16–24 h) at 4 °C on a tube rocker or nutator (*see* **Note 6**). Do not use a tube rotator.

3.3 Wash Samples and Incubate with pA/G-MNase Fusion Protein (Day 2)

1. Briefly spin the tubes (~1–2 s) in a tabletop microcentrifuge to remove any beads from the tube cap.
2. Place the samples on a magnetic rack and let the beads separate for ~1 min.
3. Wash the samples with 1 mL of CUT&RUN Wash buffer by gentle pipetting or tube inversion.
4. Repeat **steps 2** and **3** once for a total of two washes.
5. Resuspend the beads in 47.5 μL of CUT&RUN Wash buffer and add 2.5 uL of EpiCypher pA/G-MNase fusion protein (*see* **Note** 7).
6. Incubate the samples for 1 h at 4 °C on a tube rocker or nutator.

3.4 Wash Samples and Activate Chromatin Digestion with $CaCl_2$ (Day 2)

1. Briefly spin the tubes (~1–2 s) in a tabletop microcentrifuge to remove any beads from the tube cap.
2. Place the samples on a magnetic rack and let the beads separate for ~1 min.
3. Wash the samples with 1 mL of CUT&RUN Wash buffer by gentle pipetting or tube inversion.
4. Repeat **steps 2** and **3** once for a total of two washes.
5. Resuspend the beads in 150 μL of CUT&RUN Wash buffer and place the tubes in a metal block on ice. Incubate for 5 min to chill the tubes.
6. To each sample, add 3 μL of 100 mM $CaCl_2$, quickly mix by low-speed vortexing, and immediately place back into the chilled metal block (*see* **Note 8**).
7. Incubate the digestion reaction for 30–90 min, depending on the protein of interest (*see* **Note 9**).
8. Stop the digestion reaction by adding 150 μL of CUT&RUN 2X Stop buffer and thoroughly mixing by gentle pipetting.
9. Incubate the samples at 37 °C for 10 min without shaking to release the digested chromatin fragments from insoluble chromatin.
10. Centrifuge the samples at 16,000× *g* for 5 min at 4 °C.
11. Place the samples on a magnetic rack and let the beads separate for ~1 min.
12. Transfer the supernatant containing the digested fragments to fresh 1.5 mL low-binding tubes.

3.5 Extract DNA with Phenol-Chloroform-Isoamyl Alcohol (Day 2)

1. To each sample, add 3 μL of 10% SDS and 2 μL of proteinase K (20 mg/mL). Mix by tube inversion and incubate for 1 h at 50 °C. During this incubation, bring the UltraPure phenol: chloroform:isoamyl alcohol (PCI) to 20–22 °C.
2. In a biosafety cabinet, add 200 μL of PCI to each sample and mix by vortexing for ~2 s.
3. Transfer the solution to a 1.5 mL MaXtract High Density tube. Centrifuge at 16,000× *g* for 5 min at 20–22 °C.
4. In a biosafety cabinet, add 200 μL of chloroform to each sample, invert ~10 times to mix, and centrifuge at 16,000× *g* for 5 min at 20–22 °C.
5. Transfer the top phase to a fresh 1.5 mL low-binding tube containing 1 μL of 20 mg/mL glycogen and 500 μL of ice-cold absolute ethanol. Mix by vortexing.
6. Incubate the samples at −20 °C for ≥1 h (*see* **Note 10**).
7. Centrifuge the samples at 16,000× *g* for 10 min at 4 °C. A small white glycogen pellet should form.
8. Carefully pour the liquid into a glass beaker without disturbing the pellet.
9. Add 1 mL of ice-cold absolute ethanol and invert ~10 times to mix. Centrifuge at 16,000× *g* for 10 min at 4 °C.
10. Carefully pour off the liquid into a glass beaker and drain the excess liquid on a Kimwipe.
11. Leave the tubes open to air-dry for approximately 5–10 min, or until the ethanol has completely evaporated.
12. Once the pellet is dry, dissolve it in 10 μL of 0.1X TE Buffer.
13. Store the DNA samples at −20 °C until library preparation. The DNA can be stored at this temperature for several months before proceeding.

3.6 Prepare Sequencing Library (Day 3)

1. Prepare DNA sequencing libraries using the ThruPLEX DNA-seq library preparation kit, with the following modifications:
   (a) In the Library Amplification Reaction Mixture, exclude the Indexing reagent and instead add 2.5 μL of 10 μM TSPE1.0 primer and 2.5 μL of 10 μM barcoded TSPE 2.X primer. Use different barcoded TSPE 2.X primers for samples that will be multiplexed for sequencing.
   (b) To increase the representation of short DNA fragments during library amplification, shorten the Stage 4 extension time from 40 s to 30 s and the Stage 5 extension time from 50 s to 15 s.

3.7 Size-Select the Final Libraries with 0.5X-1.7X AMPure XP Beads

1. Add 25 μL of AMPure XP beads to 50 μL of library and pipette to mix.
2. Incubate at 20–22 °C for 10 min, then place samples on a magnetic stand for 5 min.
3. Transfer the supernatant to a fresh 1.5 mL low-binding tube and discard the beads.
4. Add 60 μL of AMPure XP beads and pipette to mix.
5. Incubate at 20–22 °C for 10 min, then place samples on a magnetic stand for 5 min.
6. Wash twice with 200 μL of freshly prepared 80% ethanol for 30 s. Do not remove the tubes from the stand during the wash steps.
7. Remove the supernatant and air-dry the beads until they are fully dry and have a matte finish.
8. Resuspend the beads in 17 μL of 0.1X TE buffer by gentle pipetting.
9. Incubate at 20–22 °C for 5 min, then place on a magnetic stand for 3 min.
10. Transfer 16 μL of 0.1X TE buffer to a fresh 1.5 mL low-binding tube and avoid transferring beads.

3.8 Quality Control

1. Measure the library fragment size distribution on a Bioanalyzer or TapeStation (*see* **Note 11**).
2. Measure the library concentration on a Qubit Fluorometer (*see* **Note 12**).

3.9 Sequence and Process Data

1. Pool CUT&RUN libraries at equimolar concentration and sequence the library pool on an Illumina instrument. We recommend sequencing each library to a depth of 20–30 million paired-end reads using a 75 Cycle kit (*see* **Note 13**).
2. Process the fastq files on a high-performance computing cluster using the provided script file (https://github.com/brunogegenhuber/cut-n-run).

4 Notes

1. When performing CUT&RUN for the first time, we recommend validating the protocol using antibodies recognizing an abundant histone PTM (α-H3K27me3) and a nontargeting control (guinea pig α-rabbit IgG). The success of this experiment can be evaluated without library preparation or sequencing by examining the DNA fragment size distribution on a Bioanalyzer or TapeStation. The α-H3K27me3 control should

contain a ladder of nucleosome-sized fragments, while the α-IgG control should contain little DNA prior to library amplification. After completing this validation experiment, we recommend optimizing CUT&RUN for the protein of interest, assuming it has not been previously profiled with this assay. Critical parameters for optimization include the following: 1) native vs. light fixation, 2) primary antibody, and 3) digestion time. The samples from this large-scale experiment may be pooled and sequenced at low-depth coverage to minimize cost and can be evaluated for several quality metrics, including peak number, motif enrichment *p*-value, and FRiP. After identifying the optimal CUT&RUN conditions for the protein of interest, we recommend applying these conditions to the desired experimental paradigm and collecting multiple biological replicates.

2. *E. coli* spike-in DNA may be purchased from a commercial vendor (EpiCypher 18-401) or generated in-house by shearing *E. coli* genomic DNA to mononucleosome-sized fragments (~200 bp) with MNase treatment.
3. If the SW-28 rotor is not available, the SW-41 rotor may be used instead. For the SW-41 rotor, load the gradient in 13.2 mL Ultra-Clear tubes (Beckman 344059) instead of 38.5 mL Ultra-Clear tubes.
4. The number of nuclei required to achieve high-quality CUT&RUN data depends on the expression level of the TF and the proportion of cells expressing the TF within the tissue dissection. In general, the nuclei quantity can range from 10,000 to 1,000,000 for each CUT&RUN sample. For ERα, we used 400,000 nuclei isolated from mouse BNSTp, MeA, and MPOA, although only approximately 20–30% of nuclei within these tissues express ERα. During target optimization, we recommend using a high number of nuclei (500,000-1,000,000) and reducing the number only if sufficient nuclei cannot be isolated from the tissue or cell type of interest.
5. To validate the efficiency and specificity of CUT&RUN, it is essential to always profile positive (α-H3K27me3) and negative control (guinea pig α-rabbit IgG) epitopes in parallel to the protein of interest. Although the α-H3K27me3 control does not need to be sequenced, the α-IgG control must be sequenced to distinguish true protein binding sites from non-specific pA/G-MNase cutting events.
6. To avoid sample loss, the concanavalin A beads should not stick to the tube walls and dry out during the overnight incubation. It is normal for the beads to clump together at the bottom of the tube during the incubation steps.
7. If a high number of CUT&RUN reactions are necessary, to save on cost, we recommend purifying a large batch of pA/G-

MNase in-house following a published protocol [14]. The purified protein is stable in 50% glycerol at −20 °C for several years. The plasmid is available on Addgene (Plasmid #123461).

8. This step should be performed within ~10 s for each sample to prevent tube warming, which increases nonspecific cleavage of accessible chromatin.
9. The length of the digestion time should be optimized for the target of interest. We have previously performed 90 min digestions for ERα CUT&RUN and 30 min digestions for abundant histone PTMs.
10. At this step, DNA samples can be incubated overnight (16–24 h) at −20 °C for a convenient stopping point.
11. Optimal CUT&RUN libraries for TFs should contain sub-nucleosome-sized fragments (150–200 bp) as well as mono- and poly-nucleosome-sized fragments (300–1000 bp) with minimal adapter dimers at ~120 bp. Example of a successful and failed CUT&RUN library traces are shown in Fig. 5.

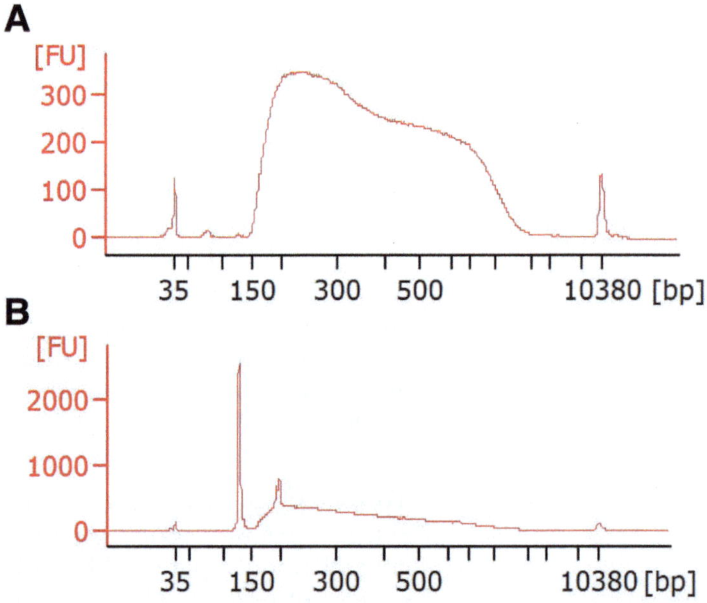

Fig. 5 (**a**) An example Bioanalyzer trace of a successful ERα CUT&RUN library prepared from the BNSTp, MeA, and MPOA of female mice treated for 4 h with estradiol benzoate. The library predominantly contains sub-nucleosome-sized fragments (150–200 bp), which correspond to direct TF-DNA contact sites, as well as larger mono- and poly-nucleosome-sized fragments (300–1000 bp). (**b**) An example Bioanalyzer trace of a failed ERα CUT&RUN library. In the absence of sufficient starting material, sequencing adapters tend to self-anneal and form dimers that migrate at ~120 bp. A library with an abundance of adapter dimers will not sequence well, because adapter dimers bind efficiently to the flow cell and occupy a high proportion of reads

12. To accurately calculate the molarity of CUT&RUN libraries prior to next-generation sequencing, we recommend inputting the DNA concentration, as determined by Qubit, and the mean fragment size, as determined by Bioanalyzer or TapeStation, into the following equation (C = Qubit concentration in ng/μL, S = mean fragment size): molarity (nM) = $(C / 660 \times S) \times 10^6$.

13. The number of reads necessary to generate high-quality CUT&RUN data depends on the signal-to-noise ratio of each individual reaction, which varies by protein of interest, antibody, and input material (tissue vs. cell lines). For profiling highly abundant histone PTMs (H3K27me3, H3K4me2) or TFs (CTCF) in cell lines, published FRiP values often range between 25% and 40% [14, 16], whereas profiling ERα from heterogeneous brain tissue results in FRiP values between 5% and 10%. Accordingly, sequencing to a higher depth than previously recommended for CUT&RUN (20–30 million reads vs. 5–8 million reads) is essential for optimal peak detection using this protocol.

References

1. Aranda A, Pascual A (2001) Nuclear hormone receptors and gene expression. Physiol Rev 81: 1269–1304
2. Shang Y, Hu X, DiRenzo J et al (2000) Cofactor dynamics and sufficiency in estrogen receptor-regulated transcription. Cell 103: 843–852
3. Carroll JS, Liu XS, Brodsky AS et al (2005) Chromosome-wide mapping of estrogen receptor binding reveals long-range regulation requiring the forkhead protein FoxA1. Cell 122:33–43
4. Hurtado A, Holmes KA, Ross-Innes CS et al (2011) FOXA1 is a key determinant of estrogen receptor function and endocrine response. Nat Genet 43:27–33
5. John S, Sabo PJ, Thurman RE et al (2011) Chromatin accessibility pre-determines glucocorticoid receptor binding patterns. Nat Genet 43:264–268
6. Mohammed H, Russell IA, Stark R et al (2015) Progesterone receptor modulates ERα action in breast cancer. Nature 523:313–317
7. Swinstead EE, Miranda TB, Paakinaho V et al (2016) Steroid receptors reprogram FoxA1 occupancy through dynamic chromatin transitions. Cell 165:593–605
8. Hewitt SC, Li L, Grimm SA et al (2012) Research resource: whole-genome estrogen receptor α binding in mouse uterine tissue revealed by ChIP-seq. Mol Endocrinol 26: 887–898
9. Gertz J, Savic D, Varley KE et al (2013) Distinct properties of cell-type-specific and shared transcription factor binding sites. Mol Cell 52: 25–36
10. Droog M, Mensink M, Zwart W (2016) The estrogen receptor α-cistrome beyond breast cancer. Mol Endocrinol 30:1046–1058
11. Yao G, Hu S, Yu L et al (2017) Genome-wide mapping of in vivo ERα-binding sites in male mouse efferent ductules. Endocrinology 158: 3724–3737
12. Skene PJ, Henikoff S (2017) An efficient targeted nuclease strategy for high-resolution mapping of DNA binding sites. elife 6
13. Skene PJ, Henikoff JG, Henikoff S (2018) Targeted in situ genome-wide profiling with high efficiency for low cell numbers. Nat Protoc 13:1006–1019
14. Meers MP, Bryson TD, Henikoff JG et al (2019) Improved CUT&RUN chromatin profiling tools. elife 8
15. Hainer SJ, Bošković A, McCannell KN et al (2019) Profiling of pluripotency factors in single cells and early embryos. Cell 177:1319–1329.e11
16. Kaya-Okur HS, Wu SJ, Codomo CA et al (2019) CUT&Tag for efficient epigenomic

profiling of small samples and single cells. Nat Commun 10:1930
17. Bartosovic M, Kabbe M, Castelo-Branco G (2021) Single-cell CUT&Tag profiles histone modifications and transcription factors in complex tissues. Nat Biotechnol 39:825–835
18. Wu SJ, Furlan SN, Mihalas AB et al (2021) Single-cell CUT&Tag analysis of chromatin modifications in differentiation and tumor progression. Nat Biotechnol 39:819–824
19. Bartosovic M, Castelo-Branco G (2022) Multimodal chromatin profiling using nanobody-based single-cell CUT&Tag bioRxiv 2022.03.08.483459
20. Meers MP, Llagas G, Janssens DH et al (2022) Multifactorial profiling of epigenetic landscapes at single-cell resolution using MulTI-Tag. Nat Biotechnol
21. Stuart T, Hao S, Zhang B et al (2022) Nanobody-tethered transposition enables multifactorial chromatin profiling at single-cell resolution. Nat Biotechnol
22. Zhu C, Zhang Y, Li YE et al (2021) Joint profiling of histone modifications and transcriptome in single cells from mouse brain. Nat Methods 18:283–292
23. Deng Y, Bartosovic M, Kukanja P et al (2022) Spatial-CUT&Tag: spatially resolved chromatin modification profiling at the cellular level. Science 375:681–686
24. Lu T, Ang CE, Zhuang X (2022) Spatially resolved epigenomic profiling of single cells in complex tissues. Cell 185:4448–4464.e17
25. Zhu Q, Liu N, Orkin SH et al (2019) CUT&RUNTools: a flexible pipeline for CUT&RUN processing and footprint analysis. Genome Biol 20:192
26. Yan F, Powell DR, Curtis DJ et al (2020) From reads to insight: a hitchhiker's guide to ATAC-seq data analysis. Genome Biol 21:22
27. Nakato R, Sakata T (2021) Methods for ChIP-seq analysis: a practical workflow and advanced applications. Methods 187:44–53
28. Gegenhuber B, Wu MV, Bronstein R et al (2022) Gene regulation by gonadal hormone receptors underlies brain sex differences. Nature 606:153
29. La Greca A, Bellora N, Le Dily F et al (2022) Chromatin topology defines estradiol-primed progesterone receptor and PAX2 binding in endometrial cancer cells. elife 11
30. Le Dily F, Vidal E, Cuartero Y et al (2019) Hormone-control regions mediate steroid receptor-dependent genome organization. Genome Res 29:29–39
31. Hahne F, Ivanek R (2016) Visualizing genomic data using Gviz and bioconductor. In: Mathé E, Davis S (eds) Statistical genomics: methods and protocols. Springer, New York, pp 335–351

Chapter 10

Cleavage Under Targets and Release Using Nuclease (CUT&RUN) in Macrophages

Antonia Babl and Franziska Greulich

Abstract

Cleavage Under Targets and Release Using Nuclease (CUT&RUN) is a method to detect specific interactions between DNA and DNA-associated proteins. It is valuable for the characterization of the binding of transcription factors or co-regulators genome wide. Furthermore, it can be used for epigenetic profiling, chromatin accessibility assessment, and identification of regulatory elements. Compared to the more commonly used chromatin immunoprecipitation (ChIP), CUT&RUN has several advantages including an in situ approach as well as no need for sonication. However, the biggest advantage is the reduced cell amounts that are required for CUT&RUN, which makes it more attractive for experiments with limited cell numbers. In this chapter, we describe a reliable CUT&RUN protocol for macrophages that can be performed within 2 days and includes a library preparation so that the sample can be directly sequenced.

Key words CUT&RUN, Chromatin immunoprecipitation (ChIP), DNA-protein interactions, Transcription factors, Epigenetic profiling, Regulatory elements

1 Introduction

The CUT&RUN technique, formally known as Cleavage Under Targets and Release Using Nuclease, represents a significant advancement in molecular biology for chromatin profiling. It operates in a manner akin to chromatin immunoprecipitation followed by sequencing (ChIP-seq) but offers several distinct advantages over traditional ChIP-seq methodologies.

Like ChIP-seq, CUT&RUN employs antibodies to identify genome-wide sequences associated with antigens. These antibodies commonly target proteins or protein modifications to elucidate their localization on DNA across various conditions, thereby enabling investigations into their roles in gene regulation, whether for transcription factors, histone marks, or co-regulators.

For nearly two decades, ChIP stood as the gold standard for detecting protein-DNA interactions, assessing chromatin structure,

Franziska Greulich (ed.), *Chromatin Immunoprecipitation: Methods and Protocols*, Methods in Molecular Biology, vol. 2846, https://doi.org/10.1007/978-1-0716-4071-5_10,

and exploring epigenetic modifications. However, in 2017, Henikoff and Skene introduced CUT&RUN, which addresses similar questions as ChIP but with enhanced efficacy and without the limitations associated with traditional ChIP methods [1].

CUT&RUN operates on a similar principle as ChIP but implements a cell permeabilization strategy in conjunction with a protein A/G-Micrococcal Nuclease (pA/G-MNase) fusion protein. This approach eliminates the need for intensive cross-linking and sonication, which are typical in ChIP protocols. Cross-linking and sonication may inadvertently mask or destroy epitopes, thereby affecting the performance of antibodies. Consequently, antibodies effective in ChIP may not perform optimally in CUT&RUN and vice versa.

In CUT&RUN, whole cells are bound to Concanavalin A (ConA) beads, permeabilized, and subjected to specific DNA cleavage by pA/G-MNase following the binding of the primary antibody to its target protein. This process substantially reduces background signal, as DNA fragments are generated only in proximity to the protein of interest. These fragments diffuse out of the permeabilized cells, leaving larger, uncut DNA fragments behind, further diminishing background noise. This improvement in signal-to-noise ratio facilitates more accurate identification of true binding events while simultaneously reducing the required sample material and sequencing costs.

While a ChIP-seq experiment in primary murine macrophages typically demands approximately 20 million cells [2], a CUT&RUN reaction can be executed with as few as 100,000 to 1 million cells [3, 4].

Moreover, CUT&RUN boosts a streamlined protocol, requiring fewer steps and less labor compared to ChIP-seq, rendering it particularly efficient for high-throughput experiments.

Although CUT&RUN protocols can function entirely without cross-linking [3], a brief 2 min cross-linking step with 0.1% formaldehyde is employed in this protocol to stabilize the interaction between the protein of interest and DNA. Following this cross-linking, cells are bound to ConA beads and permeabilized using digitonin. Antibodies targeting the protein of interest diffuse into the nucleus and bind to their respective targets. Subsequently, cells are incubated with the fusion protein A/G-MNase, comprising the chimeric protein A/G, which binds to the constant region of various IgG-type antibodies [5], and a calcium-dependent Micrococcal Nuclease (MNase) that cleaves nucleic acids [6]. Initially added without calcium, the pA/G-MNase binds to the antibody-antigen complex. Unbound pA/G-MNase is then washed away, and the remaining bound pA/G-MNase is activated by calcium addition, facilitating specific DNA cleavage around the protein of interest. The released DNA fragments diffuse out of the cell into the supernatant, where they are subsequently purified for library preparation and Illumina sequencing.

2 Materials

2.1 Preparation of Macrophages/Cells

1. Dulbecco's Phosphate-Buffered Saline (D-PBS).
2. Accutase (thaw freshly on ice).
3. 1 M HEPES-KOH pH 7.9.
4. 5 M NaCl.
5. 20% (w/v) SDS.
6. 1 M Spermidine.
7. 0.1% (v/v) formaldehyde: Dilute 16% (w/v) methanol-free formaldehyde in D-PBS. The solution should be used within 2 days. Store in the dark and at 8 °C.
8. 1 M glycine: Dissolve 75.07 g glycine in 1× D-PBS. Sterilize by filtering through a 0.22 μM filter. The solution is stable at room temperature (18–22 °C) for up to 1 year.
9. Wash Buffer (50 mL): 20 mM HEPES-KOH pH 7.9, 150 mM NaCl, 1% (v/v) Triton X-100, 0.05% (w/v) SDS, 0.5 mM Spermidine, 1 Roche Complete Protease Inhibitor-EDTA-free. Filter sterile. Stable for 2 days at 8 °C (*see* **Note 1**).
10. Cell scraper.
11. 15 mL conical tubes.
12. 1.5 mL DNA low-binding tubes.
13. Cell counter.
14. Centrifuge for 15 mL conical tubes.
15. Cooling centrifuge for 1.5/2 mL tubes.

2.2 Activation of Concanavalin A Beads

1. 1 M KCl.
2. 100 mM $CaCl_2$.
3. 100 mM $MnCl_2$.
4. Bead Activation Buffer: 20 mM HEPES-KOH pH 7.9, 10 mM KCl, 1 mM $CaCl_2$, 1 mM $MnCl_2$. Filter sterilize and store at 8 °C for up to 6 months.
5. Concanavalin A magnetic beads.
6. Magnetic rack for 1.5 mL tubes.
7. Low-retention pipette tips.

2.3 Binding of Antibody

1. 0.5 M EDTA pH 8.0.
2. 5% (w/v) Digitonin: Dissolve 50 mg digitonin in 1 mL DMSO (*see* **Note 2**). Store aliquots at −20 °C. Digitonin is toxic. Wear protective clothing and avoid dust formation. Dispose of digitonin-containing solution according to your local regulations.

3. Digitonin buffer: Add Digitonin to the Wash buffer from Subheading 2.1, **item 9** to reach a final concentration of 0.1% (w/v) (*see* **Note 3**). This buffer needs to be prepared freshly on the day. Store on ice.
4. Antibody binding buffer: Add 2 mM EDTA pH 8.0 to Digitonin Buffer from **item 3**. Prepare freshly on the day and store on ice.
5. Antibodies of interest including an isotype control antibody (IgG, *see* **Note 4**).
6. Nutator.

2.4 pA/G-MNase Binding

1. pA/G-MNase mix (per reaction): 50 μL Digitonin buffer and 1.5 μL 20× pA/G-MNase.
2. Thermoblock.

2.5 DNA Cleavage and Diffusion

1. 0.5 M EGTA pH 8.0.
2. 10 mg/mL RNAse A.
3. 15 mg/mL GlyocoBlue™
4. 2× STOP buffer: 340 mM NaCl, 20 mM EDTA pH 8.0, 4 mM EGTA pH 8.0. Sterile filter and store at 8 °C for up to 6 months. Add 0.1% (w/v) Digitonin, 50 μg/mL RNAse A, 50 μg/mL GlycoBlue™, and 1 ng/500,000 cells of spike-in DNA freshly on the day of use (*see* **Note 5**).
5. 10 mg/mL proteinase K.

2.6 Input Sample Preparation

1. DNA extraction buffer: 100 mM Tris–HCl pH 8.5, 200 mM NaCl, 5 mM EDTA pH 8.0, 0.5% (w/v) SDS. Sterile filter and store at room temperature (18–22 °C) for up to 6 months.
2. Bioruptor or other sonicator with appropriate tubes.

2.7 DNA Purification

1. Zymo ChIP DNA Cleanup and Concentrator kit or similar from another vendor.

2.8 Library Preparation

1. Nuclease-free water.
2. KAPA HyperPrep Kit or library preparation kit for Illumina sequencing from another vendor.
3. 80% (v/v) EtOH in nuclease-free water.
4. DNA elution buffer.
5. Adapter (*see* **Note 6**).
6. 0.2 mL PCR tubes.
7. Thermocycler.
8. AMPure XP beads or other SPRI beads.
9. Magnetic rack for 0.2 mL PCR tubes.

2.9 Quality Control

1. Bioanalyzer or similar device with respective reagents (e.g., High Sensitivity DNA kit).
2. Qubit and respective reagents.
3. qPCR equipment, master mix, and primers for positive and negative genomic regions (*see* **Note 7**).

2.10 Illumina Sequencing

1. Illumina sequencer with respective chemistry.

3 Methods

3.1 Preparation of Macrophages/Cells

The cells should be handled carefully. Harsh mixing procedures and longer incubation times at 4 °C must be avoided to maintain cell integrity. Work at room temperature (18–22 °C).

1. Grow 500,000–1 M cells for each reaction and an additional 500,000–1 M cells for the input sample (*see* **Note 8**).
2. Aspirate the medium and wash cells with warm (37 °C) D-PBS. To detach the cells, add 1 mL of Accutase per well and incubate for 10 min at 37 °C. Follow the detachment of cells under the microscope (*see* **Note 9**).
3. Harvest cells by carefully scratching them off the bottom using a cell scraper. Collect all cells of one condition in a 15 mL conical tube.
4. Spin cells for 3 min at 500× *g* at room temperature (18–22 °C) and aspirate supernatant.
5. Wash with 1 mL warm (37 °C) D-PBS, transfer to 1.5 mL DNA low-binding reaction tube, and count the cells with an automated cell counter or hemacytometer.
6. Aliquot the desired number of cells in DNA low-binding tubes and add 10% of spike-in cells (*see* **Note 10**).
7. Spin cells for 3 min at 500× *g* at room temperature (18–22 °C) and aspirate the supernatant.
8. Add 1 mL of 0.1% (v/v) formaldehyde/D-PBS and suspend the pellet carefully. Incubate at room temperature (18–22 °C) for 2 min (*see* **Note 11**).
9. Stop cross-linking by adding 125 μL of 1 M glycine/PBS (125 mM final concentration). Incubate at room temperature (18–22 °C) for 5 min.
10. Then spin the cell suspension for 3 min at 600× *g* at room temperature (18–22 °C) and remove the supernatant.
11. For each reaction or input sample, add 100 μL of Wash Buffer and resuspend the cell pellet by gently pipetting up and down.
12. Repeat **steps 10** and **11** once for a total of two washes.

13. Transfer 100 μL of cells to a new 1.5 mL DNA low-binding reaction tube and store at 4 °C. This is the input sample (*see* **Note 12**).

3.2 Activation of Concanavalin A (ConA) Beads (see Note 13)

This step may be performed during Subheading 3.1.

1. Carefully resuspend ConA beads by vortexing them for less than 1 s (*see* **Note 14**). Transfer 15 μL of the bead suspension per CUT&RUN reaction to a new 1.5 mL DNA low-binding reaction tube.
2. Add 200 μL ice-cold Bead Activation Buffer per 15 μL beads. Gently mix beads by vortexing them for less than 1 s. Place the tube on a magnetic rack until the solution becomes clear (30 s to 2 min) and then remove the supernatant with a pipette.
3. Then remove the tube from the magnetic rack and repeat **step 2** with 1 mL ice-cold Bead Activation Buffer.
4. Add 15 μL of ice-cold Bead Activation Buffer per reaction and resuspend by vortexing them for less than 1 s and place beads in on ice.

3.3 Binding of Antibody

1. Bring ConA beads into the solution by brief vortexing.
2. Add 15 μL of ConA bead suspension per reaction to the washed cell suspension from Subheading 3.1, **step 12**. Mix the samples well by pipetting with low-retention pipette tips.
3. Incubate for 10 min at room temperature (18–22 °C) on a nutator.
4. Briefly centrifuge the sample at 100× *g* for 2 s to remove the suspension from the cap of the tube.
5. Place the tube on the magnetic rack until the solution turns clear and then remove the supernatant (*see* **Note 15**).
6. Remove the tube from the stand. Add 100 μL of Antibody Binding Buffer per reaction and place on ice (*see* **Note 3**).
7. Aliquot 100 μL of the ConA bead-cell suspension per CUT&RUN reaction into a separate 1.5 mL DNA low-binding reaction tube and place on ice (*see* **Note 16**).
8. Add the appropriate amount of antibody to each reaction and mix gently by pipetting up and down (*see* **Note 4**).
9. Incubate tubes at 4 °C for 2 h to overnight (16–24 h) on a nutator (*see* **Note 17**).

3.4 pAG-MNase Binding

1. Precool a thermoblock to 4 °C in a cold room.
2. Briefly centrifuge samples at 100× *g* for 2 s to remove all liquid and beads from the caps of the reaction tubes, and place the tubes on the magnetic rack until the solution turns clear. Then remove the supernatant.

3. Add 200 μL of precooled Digitonin Buffer while the tube remains in the magnetic rack on ice, and resuspend beads by gently pipetting up and down using low-retention pipette tips.
4. Wait until the solution turns clear and then remove the supernatant.
5. Repeat **steps 3** and **4** once more for a total of two washes.
6. Remove tubes from the magnetic rack and add 50 μL of pA/G-MNase mix to each tube. Gently mix the sample by pipetting it up and down.
7. Incubate tubes at 4 °C for 30 min using a precooled thermoblock in a cold room to ensure a temperature of 4 °C.

3.5 DNA Cleavage and Diffusion

1. Briefly centrifuge samples at 100× *g* for 2 s to remove all liquid and beads from the caps of the reaction tubes. Place the tubes on the magnetic rack on ice until the solution turns clear and remove the supernatant.
2. Wash beads twice with 200 μL of precooled Digitonin buffer as in Subheading 3.4, **steps 3** and **4**.
3. Remove tubes from the magnetic rack. Add 50 μL of ice-cold Digitonin Buffer to each tube and mix by pipetting up and down with low-retention pipette tips. Place tubes in a precooled thermoblock in a cold room for 5 min to ensure a temperature of 4 °C.
4. Activate pA/G-MNase by adding 1 μL ice-cold 100 mM $CaCl_2$ to each tube and mix by pipetting up and down.
5. Incubate samples at 4 °C for 2 h in a thermoblock in the cold room (*see* **Note 18**).
6. Add 33 μL of 2× Stop Buffer to each sample and mix by pipetting up and down (*see* **Note 5**).
7. Incubate the tubes at 37 °C for 10 min without shaking to release DNA fragments into the solution.
8. Afterward centrifuge at 4 °C for 2 min at 16,000× *g*. Place the tubes on a magnetic rack until the solution is clear and transfer the supernatants to a new 1.5 mL DNA low-binding reaction tube.
9. Only for cross-linked cells: To reverse the protein-DNA cross-links add 0.8 μL of 10% SDS and 2 μL of 10 mg/mL proteinase K to each sample. Vortex each sample and incubate at 56 °C overnight (16–24 h).
10. After incubation, quickly spin samples at 10,000× *g* for 10 s to collect evaporation from the cap of the tubes. Equilibrate samples to room temperature (18–22 °C) and proceed to DNA purification in Subheading 3.7.

3.6 Input Sample Preparation (see Note 11)

1. Equilibrate the input sample to room temperature (18–22 °C) and add 5 μL of 20% SDS.
2. Transfer the sample to 1.5 mL to the Bioruptor tube and sonicate for 20 cycles (30 s on /30 s off) at high settings (*see* **Note 19**).
3. Afterward transfer the sheared input to a 1.5 mL DNA low-binding reaction tube.
4. Add 200 μL of DNA extraction Buffer to the 100 μL input sample. Mix by pipetting up and down and place the tube at 56 °C for overnight incubation (16–24 h) with the samples from Subheading 3.5, **step 10**.

3.7 DNA Purification (see Note 20)

DNA purification is performed at room temperature (18–22 °C) using the Zymo ChIP DNA Clean&Concentrator kit. All buffers are included in the kit.

1. Add 1200 μL DNA Binding Buffer to each input and enriched chromatin sample and mix by pipetting up and down.
2. Transfer 700 μL of each sample to a DNA spin column in a collection tube. Centrifuge at 18,500× *g* for 30 s. Discard the liquid from the reaction tube.
3. Repeat **step 2** until the entire sample has been loaded to the spin column.
4. Add 200 μL of DNA Wash Buffer to the spin column. Centrifuge at 18,500× *g* for 30 s. It is not necessary to empty the spin column at this point.
5. Repeat **step 4** once more for a total of two washes. Discard the flow-through.
6. Place the spin column into a fresh collection tube and spin at 18,500× *g* for 1 min.
7. Discard the collection tube. Transfer the spin column to a fresh 1.5 mL DNA low-binding tube.
8. Add 22 μL of prewarmed (37 °C) DNA Elution Buffer to each spin column. Incubate for 1 min.
9. Spin at 18,500× *g* for 1 min to elute DNA.
10. Repeat the elution **steps 8** and **9** once more. Collect the flow-through in the same elution tube as in **step 9**.

3.8 Library Preparation

We are using the KAPA HyperPrep Kit provided by Roche for library amplification. Most buffers and enzymes are provided within the kit. Other vendors and kits may be used. Adapted the protocol according to the manufacturer's instructions.

1. Bring AMPure XP beads to room temperature (18–22 °C).
2. Transfer 40 μL purified DNA to a PCR tube and dilute to a final volume of 50 μL with nuclease-free water.

3. For end repair, add 7 μL of End Repair & A-Tailing buffer and 3 μL of End Repair & A-Tailing enzyme mix. Mix thoroughly by vortexing and spin down shortly in a benchtop centrifuge. Incubate the reaction 30 min at 20 °C and heat inactivate 30 min at 60 °C in a thermocycler with the lid set to 60 °C.
4. To ligate adapters, add 9 μL nuclease-free water, 30 μL Ligation Buffer, 10 μL DNA Ligase, and 1 μL 0.5 μM Adapters (*see* **Note 6**). Mix thoroughly by vortexing and spin down shortly in a benchtop centrifuge. Incubate the ligation reaction for 15 min at 20 °C in a thermocycler with the lid set to 60 °C.
5. For size selection, we perform a double-sided SPRI cleanup. First, we remove large fragments with a 0.5× SPRI cleanup: Importantly, the desired DNA fragments stay in solution at this step. Add exactly 55 μL well-mixed AMPure XP beads at room temperature (18–22 °C) (*see* **Note 21**). Mix beads by pipetting up and down ten times with a pipette. Incubate for 10 min at room temperature (18–22 °C), place PCR tubes in the magnetic rack, and wait until the liquid turns clear (approx. 2 min, *see* **Note 22**).
6. Transfer 150 μL of the *supernatant* to new 0.2 mL PCR tubes.
7. Now smaller fragments are removed by a 0.9× SPRI cleanup. In this step, the desired fragments (150–660 bp) will bind to the AMPure XP beads, and smaller fragments such as leftover adapters are washed away. Add 60 μL well-mixed AMPure XP beads to the supernatant from **step 6**. Mix beads by pipetting up and down several times. Incubate 10 min at room temperature (18–22 °C). Place PCR tubes in a magnetic rack and wait until the liquid turns clear (30 s to 2 min). Remove the liquid with a pipette.
8. Keep the PCR tubes in the magnetic rack and add 200 μL freshly prepared 80% EtOH without disturbing the beads. Then remove the ethanol carefully with a pipette without touching the beads. Small leftovers are acceptable.
9. Repeat **step 8** once more for a total of two washes.
10. Remove all the ethanol with a pipette, spin the PCR tubes, and place them back in the magnetic rack. Remove the remaining liquid with a P20 pipette, and let the beads dry (*see* **Note 23**). Do not overdry the beads.
11. Add 21 μL DNA elution buffer. Incubate 5 min at room temperature (18–22 °C) and place tubes in the magnetic rack. Wait until the liquid turns clear (1–2 min) and transfer 20 μL of the *clear* supernatant into a new 0.2 mL PCR tubes.
12. For amplification of the size-selected library, add 25 μL 2× KAPA HiFi HotStart ReadyMix and 5 μL 10× Library Amplification Primer Mix to 20 uL of purified samples from **step 11**.

13. Incubate the reaction in a thermocycler with the following protocol: 1× 45 s 98 °C; 10× (15 s 98 °C, 30 s 60 °C, 30 s 72 °C); 1× 60 s 72 °C. The sample can be kept at 4 °C for up to 2 h. The lid should be heated to 110 °C.
14. For post-amplification cleanup, add 50 μL well-mixed AMPure XP beads (1× SPRI cleanup). Mix beads by pipetting up and down several times (*see* **Note 22**). Incubate 10 min at room temperature (18–22 °C). Place PCR tubes in a magnetic rack and wait until the liquid turns clear (1–2 min). Remove the supernatant with a pipette.
15. Keep the PCR tubes in the magnetic rack and add 200 μL freshly prepared 80% EtOH without disturbing the beads. Then remove the Ethanol using a pipette without touching the beads. Small remains are acceptable.
16. Repeat **step 15** once more for a total of two washes.
17. Spin the PCR tubes in a benchtop centrifuge to collect the remaining liquid at the bottom of the tube, place it back in the magnetic rack, and remove the remaining liquid with a P20 pipette.
18. Let the beads dry (*see* **Note 23**). Do not overdry the beads.
19. Add 15 μL DNA elution buffer. Incubate 5 min at room temperature (18–22 °C). Then place the PCR tubes back in the magnetic rack. Wait until the liquid turns clear (2 min) and transfer 14 μL of the *clear* supernatant into new PCR tubes.
20. Take a 2.5 μL aliquot for Quality Control (Subheading 3.9) and take a 2 μL aliquot for qPCR analysis.

3.9 Quality Control

1. Run 1 μL of CUT&RUN library from Subheading 3.8, **step 20** on a Bioanalyzer using a High Sensitivity DNA kit to check for adapter contamination and fragment size distribution of the library (*see* **Note 24**). For some example traces, *see* Fig. 1.
2. Run a qPCR with positive and negative genomic loci to check for specific enrichment of DNA (*see* **Note 25**) diluting the aliquote from Subheading 3.8, **step 20**. An example result is shown in Fig. 2a.
3. Measure 1 μL of the CUT&RUN library from Subheading 3.8, **step 20** with a Qubit to determine the DNA content.

3.10 Library Pooling and Sequencing

For pooling strategies and Illumina sequencing, we recommend contacting your sequencing vendor or facility. For mouse or human genomes, we recommend sequencing 10–15 M reads per transcription factor or co-regulator with low genome coverage. Broader histone modifications can be sequenced with 5–8 M reads.

Figure 2b shows an example of the transcription factor GR and the co-regulator BRG1.

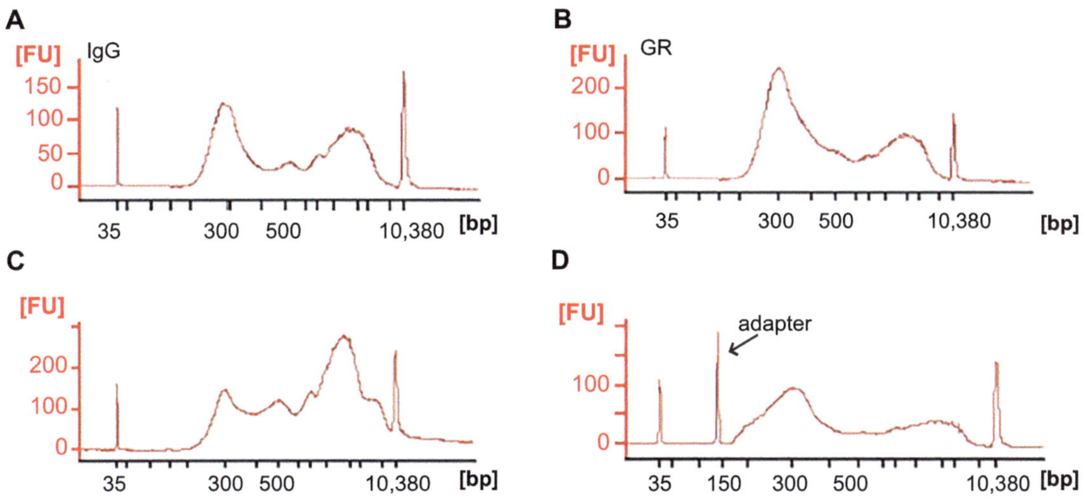

Fig. 1 Example Bioanalyzer traces. (**a**) CUT&RUN library against the isotype control (rabbit IgG) in human peripheral monocyte–derived macrophages (500,000 cells) treated with 100 ng/mL LPS and 1 μM dexamethasone for 3 h. (**b**) CUT&RUN library against the glucocorticoid receptor (GR) in human peripheral monocyte–derived macrophages (500,000 cells) treated with 100 ng/mL LPS and 1 μM dexamethasone for 3 h. (**c**) CUT&RUN library against the GR in human peripheral monocyte–derived macrophages (500,000 cells) that is underdigested by MNase. This library needs to be redone with longer incubation time for $CaCl_2$ (Subheading 3.5, **step 5**). (**d**) CUT&RUN library against the GR in human peripheral monocyte–derived macrophages (500,000 cells) that still contain adapter dimers. This library can be purified another round using a 1× SPRI cleanup (*see* Subheading 3.8, **steps 14–19**)

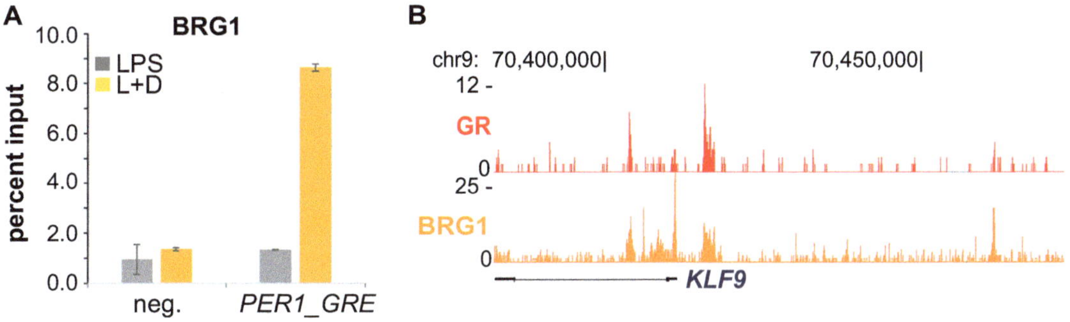

Fig. 2 CUT&RUN results. (**a**) QPCR against a genomic region not binding *Brahma-related gene 1* (BRG1) (neg.—negative) and the glucocorticoid receptor enhancer (GRE) at the *PER1* locus in human peripheral monocyte–derived macrophages after LPS (gray) or LPS and dexamethasone (L + D, orange) stimulation. The mean percent of input from two independent libraries is shown as bar, and error bars show the standard deviation. (**b**) Example genome browser track of the *KLF9* locus for CUT&RUN for the glucocorticoid receptor (GR) and BRG1 in human peripheral monocyte–derived macrophages after LPS and Dex treatment. Libraries from 500,000 cells

4 Notes

1. The addition of Triton-X100 and SDS improves the permeabilization of nuclei after cross-linking. If cells are not cross-linked, the addition of Triton-X100 or SDS is not required.
2. To avoid precipitation of digitonin, prepare a 5% digitonin stock solution in DMSO rather than in water, which requires heating for solubilization. This improves reproducibility.
3. Digitonin permeabilizes cells and allows the antibody to bind its target in situ. The digitonin concentration required for CUT&RUN varies between samples and must be optimized for every cell type and fixation condition. In general, perform a digitonin titration (0.1% down to 0.001%) and monitor cell lysis using Trypan blue staining. Find the minimum digitonin concentration needed to achieve >90% permeabilized cells. It is important to use the minimum amount of digitonin needed for efficient permeabilization to avoid cell lysis and digitonin precipitation.
4. The choice of the antibody is probably the most important one. It is advised to work with ChIP-validated antibodies. Before using it in CUT&RUN, the specificity of the antibody must be checked. We find western blotting ***and*** immunofluorescence with appropriate controls (siRNA ***knockdown*** or ***knockout*** cells, competing peptides for the epitope (especially for histone modifications), isotype controls) the most suitable assays to test antibody specificity for CUT&RUN. Additionally, an isotype matching IgG control is necessary to determine an unspecific signal in CUT&RUN, which may come from unspecific binding of DNA to beads/tubes or from off-target cleavage of the MNase. If you are new to CUT&RUN, we also advise the use of a positive control antibody such as H3K27ac or H3K4me3. Antibody concentration should be optimized. We recommend starting with 0.5 μg antibody.
5. Optional: You may add spike-in DNA into the STOP buffer to calibrate your CUT&RUN for loss during DNA purification and library preparation, especially if the use of heterologous spike-in cells is not possible. Make sure your spike-in DNA can later be distinguished from your sample DNA by using a species with sufficient evolutionary distance. You can test for cross-mappability by mapping the genome of your spike-in species against the genome of your target species. The cross-mappability should be below 10%. Commonly, 1 ng *Escherichia coli* spike-in DNA per 500,000 cells is used.
6. The adapter identifies each sample. Accordingly, use different adapters for each sample and record the adapters used, to enable computational demultiplexing after sequencing. Briefly,

an adapter contains the primer binding sequence for the universal amplification primers in Subheading 3.8, **step 12** that allow for amplification of the library. Additionally, adapters harbor the unique barcode (index) on one (single index) or both strands (dual index), which are required for the identification of each sample. Dual indices are recommended for sequencing on Illumina NovaSeq6000 machines (or higher), to avoid index hopping. Additionally, dual indices allow for more multiplexing possibilities, which are necessary for CUT&RUN on high-capacity sequencing devices due to the reduced sequencing depth per sample. Adapters might be ordered from any oligonucleotide vendor.

7. To test the specificity of the CUT&RUN, design at least one oligo primer pair for a region bound by the protein of interest (positive locus). Additionally, design one oligo primer pair for a region not bound by the protein of interest (negative locus) to evaluate the background signal of the library. For histone marks, use publicly available data from ENCODE (https://www.encodeproject.org/ [7]), public track hubs on UCSC (https://www.genome.ucsc.edu/ [8]), or available data on the Gene Expression Omnibus (GEO) [9].
8. The number of cells depends on the protein of interest. Higher cell numbers are required for transcription factors and co-regulators occupying small fractions of the genome. The broader the binding pattern of the protein interest, the fewer cells can be used.
9. Accutase is only used to detach adherent cells. The duration of Accutase treatment can vary depending on the cell type but should not extend to 10 min. Always thaw Accutase freshly at 4 °C or room temperature (18–22 °C). Observe the cellular detachment process with a microscope. Stop the reaction as soon as the cells float. If you scratch cells from the plate, do not use force. This damages the cells. If you are working with low cell numbers, you can rinse the wells afterward with D-PBS to collect all cells.
10. To account for variations in sample preparation, we employ a spike-in strategy. Of note, a calibration strategy using *E. coli* DNA, which is carried over from the production of pA/G-MNase [10], did not work in our hands as we did not receive sufficient bacterial reads. As we aim to compare different treatment conditions, we add 10% (50,000–100,000) of spike-in cells, yielding around 5% of total sequencing reads. Of note, spike-in cells need to be chosen carefully depending on the cross-mappability of the respective genomes and the recognition by the antibody of interest. We have successfully used *Drosophila* S2 cells for histone modifications [2].

11. If these cross-linking conditions are not sufficient, increase the formaldehyde concentration to 0.5% or 1%. We do not recommend extending the incubation time as this may reduce the ability of MNase to efficiently cleave DNA into sufficiently small fragments for diffusion out of the nucleus.
12. In contrast to ChIP-seq, the input control is not sequenced and only serves to normalize quantitative PCR (qPCR) results for variations in cell material.
13. Lectins play a major role in the innate immune system, and immune cells may be stimulated via binding to ConA beads. To circumvent this potential problem in CUT&RUN, you may work with nuclei rather than cells [11].
14. Concanavalin A beads exhibit strong adherence; hence, we advise using low-retention pipette tips for handling. When in an unbound state, brief vortexing is sufficient to mix the beads effectively. However, after cell binding, avoid vortexing the suspension; instead, gently mix it by pipetting. Regularly inspect the distribution of beads within the reaction tube to ensure they remain evenly dispersed and free from clumping or adherence to the tube wall. Minimize unnecessary pipetting to prevent bead loss, and exercise caution to avoid splashing the beads while pipetting.
15. You may check the efficient binding of cells to Concanavalin A beads by taking a 10 μL sample of the supernatant after Subheading 3.3, **step 5** (unbound fraction) and after Subheading 3.3, **step 6** (bound fraction). Stain with Trypan blue and observe under the microscope. The majority of stained cells should be in the bound fraction. If you observe a high amount (>30% of the initial cells) in the unbound fraction, make sure you have stored your ConA beads properly (not frozen) and all buffers are prepared correctly.
16. For batch processing, use 0.2 mL PCR tubes compatible with multichannel pipettes. PCR tubes additionally lower the risk of sample loss due to the smaller tube surface.
17. The antibody incubation time depends on your epitope of interest and must be optimized. We use shorter incubation times for transcription factors with low residence times to avoid losing them during the incubation. For histone modifications, antibody incubations may extend to overnight (16–24 h) and pose a possible break point in the protocol.
18. The digestion time after the addition of $CaCl_2$ is an important factor. If incubation times are too short, DNA may be insufficiently cut. Longer incubation times may lead to unspecific DNA cleavage and increased background signal. The incubation time depends on the epitope of interest. Histone modifications may require shorter incubation times (e.g., 30 min),

whereas transcription factors require longer times up to 90 min. This step might be fine-tuned, if you experience only mononucleosomal or large fragments (>3 kb) in the bioanalyzer analysis of the CUT&RUN library.

19. We use the Bioruptor to fragment the input DNA. Sonication conditions must be optimized for each cell type and fixation. For optimization, solubilize fixed cells of interest at 5 M cells/mL in Wash Buffer with 1% SDS and transfer into a 1.5 mL Bioruptor tube. Do not exceed 500 μL per tube. Sonicate for 2 cycles at high settings, and take a 20 μL aliquots into a fresh 1.5 mL tube. Mix well before taking the aliquot. Sonicate another 2 cycles and take another sample. Repeat until 10–-20 cycles are reached. Perform reverse cross-linking by adding 80 μL 1× TE buffer and 5 μL of 5 M NaCl per test sample. Incubate 15 min at 99 °C and 1000 rpm using a thermoblock. Bring tubes to room temperature (18–22 °C) and add 1 uL 10 mg/mL RNaseA per tube. Incubate 30 min at 37 °C 600 rpm in a thermoblock. Add 2 μL 10 mg/mL proteinase K and incubate 1 h at 56 °C 600 rpm in a thermoblock. Purify the DNA according to Subheading 3.7. Elute in 20 μL elution buffer (provided by the DNA purification kit) and add 6 μL of colorless loading dye (50% glycerol in 1× TE buffer). Run samples on a 0.6% agarose gel to obtain a sonication test gel, as in Fig. 3. The fragmented DNA should be between 200 and 700 bp (black box). Note that the number of cycles and intervals depends on your sonication device, the cell type, and cell density and needs to be adapted. For a start, try 2, 5, 10, 15, 20, 25, and 30 cycles.

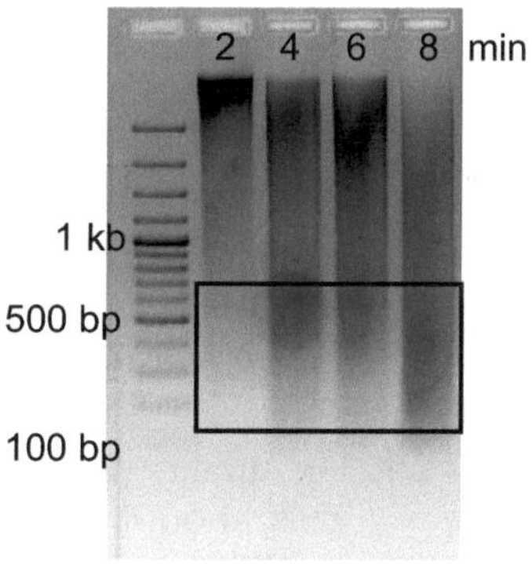

Fig. 3 Sonication conditions for the input sample. The given examples show bone marrow–derived macrophages fixed in 0.1% formaldehyde. The cell density was 2 M cells/100 μL. Sonication was performed at high settings with 30 s on/off cycling using a Bioruptor PLUS. The ideal size of 200–700 bp (black box) was reached after 8 sonication cycles. 20 μL DNA aliquots were purified as described in **Note 19**

20. We perform DNA purification using the Zymo ChIP DNA Clean&Concentrator kit. Other DNA purification kits might be used according to the manufacturer's manuals. Ensure that the lower size limit goes down to 100 bp. Alternatively, Phenol-Chloroform extraction might be performed. The advantage of Phenol-Chloroform extractions is the higher yield of DNA and the purification of fragments smaller than 100 bp.

21. The binding of DNA is dependent on the buffer concentration of AMPure beads. Therefore, always pipette precisely and make sure to work with the right amounts. Also, avoid drops hanging on the outside of the pipette. If you notice drops of cell suspension on the lid, wait until the incubation time is finished before spinning them down to avoid a high local concentration of beads. If you work with a high number of samples, you can work in batches to avoid the drying of beads and differences in incubation times. We recommend to purify a maximum of 8 to 16 libraries in one go.

22. For batch processing (more than two libraries), we recommend using a multichannel pipette to avoid variations in the incubation times. When working with a multichannel pipette always check the amount of aspirated liquid in each tip.

23. As long as the beads are still humid, the beads are dark brown and shiny but later turn matt and lighter brown. This is the best time point to add the elution buffer. Overdried beads start to crack. To avoid cracking, the elution buffer can be added first to all samples and later the samples can be mixed depending on the sample-specific drying state.

24. The desired fragment size is about 300 bp (mononucleosomes, Fig. 1a, b). Too many large fragments could indicate an incomplete digest (Fig. 1c). In this case, an adjustment in the incubation time of the digest is advised (Subheading 3.5, **step 5**). A sharp peak at about 150 bp indicated adapter contamination (Fig. 1d), which required an additional 1× AMPure bead cleanup as described in Subheading 3.8, **steps 14–19**. For this, it is advised to dilute the samples to a volume of 50 μL before the cleanup.

25. For quality control before sequencing, a qPCR is carried out. This should include a negative locus as control as well as multiple positive loci. A clear difference between the negative and some positive loci should be detectable. The pooled Input samples can be used as Standard. Although a successful qPCR does not guarantee a successful CUT&RUN, however sequencing of failed samples can be avoided.

References

1. Skene PJ, Henikoff S (2017) An efficient targeted nuclease strategy for high-resolution mapping of DNA binding sites. elife 6:e21856
2. Greulich F, Wierer M, Mechtidou A et al (2021) The glucocorticoid receptor recruits the COMPASS complex to regulate inflammatory transcription at macrophage enhancers. Cell Rep 34(6):108742
3. Skene PJ, Henikoff JG, Henikoff S (2018) Targeted in situ genome-wide profiling with high efficiency for low cell numbers. Nat Protoc 13(5):1006–1019
4. Wang L, Oh TG, Magida J et al (2021) Bromodomain containing 9 (BRD9) regulates macrophage inflammatory responses by potentiating glucocorticoid receptor activity. Proc Natl Acad Sci USA 118(35)
5. Choe W, Durgannavar TA, Chung SJ (2016) Fc-binding ligands of immunoglobulin G: an overview of high-affinity proteins and peptides. Materials (Basel) 9(12):994
6. Cuatrecasas P, Fuchs S, Anfinsen CB (1967) Catalytic properties and specificity of the extracellular nuclease of Staphylococcus aureus. J Biol Chem 242(7):1541–1547
7. Davis CA, Hitz BC, Sloan CA et al (2018) The Encyclopedia of DNA elements (ENCODE): data portal update. Nucleic Acids Res 46(D1): D794–D801
8. Kent WJ, Sugnet CW, Furey TS et al (2002) The human genome browser at UCSC. Genome Res 12(6):996–1006
9. Edgar R, Domrachev M, Lash AE (2002) Gene Expression Omnibus: NCBI gene expression and hybridization array data repository. Nucleic Acids Res 30(1):207–210
10. Meers MP, Bryson TD, Henikoff JG et al (2019) Improved CUT&RUN chromatin profiling tools. elife 8
11. Hainer SJ, Boskovic A, McCannell KN et al (2019) Profiling of pluripotency factors in single cells and early embryos. Cell 177(5): 1319–1329. e1311

Chapter 11

Cleavage Under Targets and Release Using Nuclease (CUT&RUN) of Epigenetic Regulators

Andrew D. McCray and Xiaofeng Wang

Abstract

Chromatin immunoprecipitation followed by sequencing (ChIP-Seq) allows for the identification of genomic targeting of DNA-binding proteins. Cleavage Under Targets and Release Using Nuclease (CUT&RUN) modifies this process by including a nuclease to digest DNA around a protein of interest. The result is a higher signal-to-noise ratio and decreased required starting material. This allows for high-fidelity sequence identification from as few as 500 cells, enabling chromatin profiling of precious tissue samples or primary cell types, as well as less abundant chromatin-binding proteins: all at significantly increased throughput.

Key words CUT&RUN, ChIP-Seq, Chromatin, Epigenetics, Chromatin remodeling, Transcription factor, TF, Protocol

1 Introduction

ChIP-Seq has proven to be an invaluable resource for profiling DNA-binding proteins, and its widespread use has led to many discoveries in the field of epigenetics and gene regulation. ChIP-Seq typically uses 1 M–20 M fixed cells to cross-link DNA with interacting proteins followed by shearing of the DNA by sonication. Immunoprecipitation of a protein of interest allows the bound DNA to be isolated, amplified, and sequenced. CUT&RUN was developed to circumvent the need for millions of cells and to assess native chromatin states [1]. To accomplish this, a micrococcal nuclease (MNase) is coupled to a protein A/G fusion construct that binds the primary antibody. Upon the addition of Ca^{2+}, the MNase cleaves DNA adjacent to the primary antibody, which is subsequently released. These fragments of DNA can be isolated through either column-based purification or phenol extraction followed by a bead-based size-selection step to separate desired

Franziska Greulich (ed.), *Chromatin Immunoprecipitation: Methods and Protocols*, Methods in Molecular Biology, vol. 2846, https://doi.org/10.1007/978-1-0716-4071-5_11,

DNA fragments from genomic DNA. These modifications allow CUT&RUN to be performed without fixation on as few as 500 cells [1].

The increased signal-to-noise ratio permits low input starting material, and thus makes this assay more feasible for use on precious tissue samples or primary cells. In addition to tissue, CUT&RUN can be performed on cell lines and more complex in vitro systems such as organoids [2, 3]. The low cell number requirement enables this assay to be relatively high-throughput and decreases the cost spent on cell culture reagents.

Recently, another similar method was developed, also from Steve Henikoff's lab, called CUT&TAG. Instead of using the Protein A/G–MNase in CUT&RUN, CUT&TAG uses a protein A/G–Tn5 to tagment DNA with an adaptor or index. This allows the assay to be done at the single-cell level [4]. CUT&TAG is a great tool for profiling histone modifications, but noncovalently linked DNA-binding proteins such as transcription factors or chromatin remodelers are more challenging to apply due to the stringent washing conditions of CUT&TAG [4].

MNase is inactive until the addition of Ca^{2+}, circumventing the need for stringent wash conditions in CUT&RUN. This allows the profiling of indirect DNA-binding proteins, such as members of chromatin remodeling complexes. These proteins may contribute to the catalytic activity or protein interaction network of the complex without directly binding to DNA.

Furthermore, one can determine nucleosome occupancy at sites of CUT&RUN fragments. EChO is an R package that measures the size of the fragments generated from a CUT&RUN experiment [5]. MNase is unable to cleave DNA wrapped around histones and instead cleaves in linker regions. Thus, nucleosome-free regions will tend to have smaller DNA fragments. EChO calculates the distribution of fragment sizes to determine the fraction of reads occurring in nucleosome-free versus bound sites [5]. This can be used in a time course to determine the kinetics of nucleosome eviction [2], among many other applications of the technology.

These benefits have spurred the commercialization of this assay. Epicypher, Cell Signaling Technology, and Sigma-Aldrich have all developed their own CUT&RUN kits. While this makes the assay more accessible, we describe here a detailed in-house method that is more affordable.

2 Materials

2.1 Preparing Cells

1. Binding Buffer: 20 mM HEPES-KOH pH 7.5, 10 mM KCl, 1 mM $CaCl_2$, 1 mM $MnCl_2$. Binding Buffer can be stored at 4 °C for up to 4 weeks.

2. Wash Buffer: 20 mM HEPES-KOH pH 7.5, 150 mM NaCl, 0.5 mM Spermidine, Roche Complete Protease Inhibitor EDTA-Free. Wash Buffer should be made fresh on the day of the experiment.
3. Accutase.
4. Trypan blue.
5. Concanavalin A–coated beads (Such as Bang Labs, Epicypher, and CST).
6. 1.5 mL low-binding tubes.
7. Benchtop microcentrifuge.
8. Benchtop cooling centrifuge.
9. 1.5 mL tube magnetic rack.
10. 15 mL tube rotator.

2.2 Primary Antibody Incubation

1. Digi-Wash Buffer: Wash Buffer, 0.025–0.1% (w/v) Digitonin. Digi-Wash Buffer should be stored at 4 °C for up to 1 day. Vortex before use. *See* **Note 1** for picking the appropriate Digitonin concentration.
2. Antibody Buffer**:** Digi-Wash Buffer, 2 mM EDTA.
3. Antibody against the protein of interest as well as IgG. *See* **Note 2** for choosing the right antibodies.
4. Plate rocker.

2.3 Binding of Protein A/G MNase

1. 10% (w/v) SDS.
2. 20 mg/mL proteinase K.
3. Low-Salt Rinse Buffer: 20 mM HEPES-KOH pH 7.5, 0.5 mM Spermidine, 0.025–0.1% (w/v) Digitonin. Use the same Digitonin concentration as in the Digi-Wash Buffer.
4. Incubation Buffer: 3.5 mM HEPES-KOH pH 7.5, 10 mM $CaCl_2$, 0.025–0.1% (w/v) Digitonin. Use the same Digitonin concentration as in the Digi-Wash Buffer. Incubation Buffer should be made fresh on Day 2.
5. 2× Stop Buffer: 170 mM NaCl, 20 mM EGTA, 0.025–0.1% (w/v) Digitonin, 50 μg/mL RNAse A, 25 μg/mL Glycogen, 0.1% (w/v) SDS, 400 μg/mL Proteinase K. Stop Buffer should be made fresh on Day 2.
6. Protein A/G MNase fusion protein (Such as: Epicypher or CST): We dilute our 70 μg/mL Protein A/G MNase stock 100× in Digi-Wash Buffer to obtain 700 ng/mL.
7. 1.5 mL tube cold rack.

2.4 Phenol-Chloroform Extraction

1. Phenol:Chloroform:Isoamyl alcohol (25:24:1).
2. Chloroform.
3. Absolute ethanol.
4. 20 mg/mL glycogen.
5. 0.1× Tris-EDTA.
6. Phase Maker tubes.
7. Thermomixer.

2.5 Size Selection

1. ProNex (or comparable) size selection beads (e.g., Promega NG2001).

2.6 Library Preparation

1. A DNA library prep kit such as NEB Next Ultra II DNA library prep kit (e.g., NEB #E7645).
2. Qubit fluorometer and appropriate kits.
3. Bioanalyzer and appropriate kits.
4. PCR machine.

3 Methods

This protocol has been adapted from Wolf et al., used to profile chromatin remodeler SWI/SNF subcomplexes in conjunction with AP-1 transcription factors [2]. This protocol can easily be multiplexed, which will allow researchers to profile all DNA-binding proteins of interest in one batch. *See* **Notes 3** and **4** for batching recommendations, as well as a discussion of appropriate experimental controls.

CUT&RUN allows the use of minimal starting material. As few as 500 cells can be reliably profiled. We recommend using more cells if they are available (up to 500 K cells for cell lines; 100 K for primary cells such as organoids). We observe more consistent results when we assay 100,000–500,000 cells per sample. All cells from one experimental condition and biological replicate can usually be grown in a single 100 mm dish, but this could vary depending on the size of the cells being profiled.

The health of the cells is critical to the success of the experiment. Cells should not be over-confluent before the start of the experiment.

To minimize the stress on the cells, all steps before antibody incubation are performed at room temperature (20–22 °C). Because DNA breakage must be minimized throughout the protocol, cavitation during resuspension and vigorous vortexing must be avoided.

The most common reason for failure in our hands seems to be antibody-dependent. A failed experiment can be observed as either a lack of DNA following library preparation (likely a failure in library preparation) or as little to no enrichment over the IgG control at known binding sites.

3.1 Preparing Cells

1. Gently harvest cells with Accutase at room temperature (20–22 °C). Depending on the cells, this can take up to 15 min. Once digested, use 10 μL of cells for trypan blue staining to check for cell viability and counting. Cells should have viability >90% to proceed with the experiment.
2. Transfer the desired number of cells (500 K / sample) to a new 1.5 mL tube. If multiple CUT&RUN reactions are being done on the same cell line, they can all be processed as one at this point, i.e., 500 K cells/sample * 10 rxns → transfer 5 M cells.
3. Centrifuge for 3 min at 600× *g* at room temperature (20–22 °C), and discard supernatant.
4. Resuspend cells in 1.0 mL room temperature (20–22 °C). Wash buffer (without digitonin) by gentle pipetting.
5. Centrifuge for 3 min at 600× *g* at room temperature (20–22 ° C), and discard supernatant.
6. Repeat **steps 4** and **5** two more times for a total of three washes (*see* **Note 5**).
7. While washing in **steps 3–6** begin preparing Concanavalin A (ConA) beads solution:
   (a) Gently resuspend the beads and withdraw enough slurry such that there will be 10 μL for each final sample plus ~20 μL extra. Transfer the appropriate volume of ConA beads to a new 1.5 mL tube containing 1 mL of Binding Buffer. Mix by inversion.
   (b) Remove liquid from the cap with a quick pulse on a microcentrifuge. Place on a magnet stand until the solution clears (1–2 min). Withdraw the liquid and remove the tube from the magnet stand.
   (c) Add 1 mL of Binding Buffer and mix by inversion, before repeating **step** 7 resuspend beads in a volume of Binding Buffer equal to the original volume of ConA bead slurry.
8. Resuspend cells in 1 mL Wash Buffer by gentle pipetting at room temperature (20–22 °C) and add an appropriate volume of activated ConA beads from **step** 7 (10 μL / sample). Rotate this cell and bead mixture for 10 min at room temperature (20–22 °C) to allow cells to bind the beads.
9. Mix well by gentle resuspension (if possible, usage of low-binding pipet tips to minimize cell loss) to ensure bead-bound cells are in a homogenous suspension, and then quickly divide into aliquots in 1.5 mL tubes, one per sample/antibody to be used.

10. Place each aliquot on the magnet stand and withdraw the supernatant once clear. Remove the tube from the magnet.

3.2 Primary Antibody Incubation

1. Place each tube at a low angle and add 100 μL Antibody Buffer per sample directly to the beads. Tap to dislodge the remaining beads. *See* **Note 6** to help reduce the background signal at this step.
2. Place on ice and add the antibody to the manufacturer's recommendation (generally ~1:100). Mix by gently flicking (*see* **Note 7**).
3. Incubate on ice for a minimum of 2 h, or overnight (12–16 h) in a cold room (4–8 °C). *See* **Note 8** for tips to ensure proper antibody uptake.

3.3 Binding of Protein A/G MNase

See **Note 9** for tips on batching samples for this section. If performing CUT&RUN in bulk, refer to **Note 9** before starting this section.

1. Remove liquid from tube caps and sides with a quick pulse on a microcentrifuge. Make sure not to pellet the cells at this step, pulse at very low speeds just to collect the liquid. Place on the magnet stand, withdraw the supernatant once clear, and remove it from the magnet.
2. Wash with 1 mL Digi-Wash Buffer. Mix by inverting to disrupt clumps. Small clumps may persist when handling primary cells, which will not hurt the outcome. Place back on the magnet and remove the supernatant.
3. Repeat **step 2** two times for a total of three washes. Before aspirating the final wash, perform a quick spin in a microcentrifuge to collect the liquid from the walls of the tube.
4. Remove the samples from the magnet and add 100 μL of the Protein A/G MNase fusion protein at 700 ng/mL. Adding directly to the pellet helps to dislodge the beads. Gently tap or flick the tube to resuspend the remaining beads. Alternatively, gently resuspend the solution using low-binding pipet tips to minimize cell loss.
5. Incubate at 4 °C for 2–4 h.
6. Repeat **steps 2** and **3** for a total of three washes in Digi-Wash Buffer.
7. Wash with 1 mL Low-Salt Rinse Buffer. Mix by inverting to disrupt clumps. Place back on the magnet and remove the supernatant.
8. Remove samples from the magnet and add 100 μL of ice-cold Incubation Buffer. Adding directly to the pellet helps to dislodge the beads. Gently tap or flick the tube to resuspend the

remaining beads. Alternatively, gently resuspend the solution using low-binding pipet tips to minimize cell loss.

9. Incubate at 0 °C in a cold tube rack on ice for 30 min to allow the MNase to digest DNA. This step is critical for the success of the experiment, *see* **Note 10** to limit background.
10. Add 100 μL of 2× Stop Buffer to chelate the Ca^{2+} ions and quench the reaction. Mix by gentle vortexing. This allows digested chromatin fragments to diffuse out of the cells.
11. For small fragments from histone marks or transcription factors: incubate at 37 °C for 30 min to allow the diffusion of chromatin fragments into the solution.
12. Put tubes on a magnetic rack. Once the solution becomes clear, transfer the supernatant into a new tube.
13. Add 2 μL of 10% SDS and 2.5 μL of Protease K and incubate at 50 °C for 1 h. Cell pellets can be saved in case large fragments remain in the nucleus for future analysis. Proceed to Subheading 3.4. *See* **Note 11** for isolation of large fragments, or fragments from first-time trying targets.

3.4 Phenol-Chloroform Extraction

See **Note 12** for alternative extraction methods.

1. Centrifuge PhaseLock tubes at 16,000× *g* to ensure all gel is at the bottom of the tube. Additionally, prepare a new 1.5 mL tube for each sample containing 1 μL glycogen at the bottom.
2. Subsequent steps should be performed in a chemical fume hood, and pipette tips should be properly disposed of after use. Make sure samples are at room temperature (20–22 °C). Add 200 μL phenol:chloroform: isoamyl alcohol (25:24:1) and shake vigorously for approximately 20 s.
3. Give a quick spin in a microcentrifuge to remove liquid from the walls and cap of the tube. Then transfer samples to PhaseLock tubes. Pipette slowly using a P200 to ensure your sample does not stick to the pipette tip.
4. Centrifuge samples for 5 min at 16,000× *g* at room temperature (20–22 °C) for aqueous phase separation.
5. Add 200 μL chloroform; invert a few times to mix. Centrifuge samples for 5 min at 16,000× *g* at room temperature (20–22 °C).
6. Transfer the upper aqueous phase (~200 μL) to the prepared new tubes with glycogen (in **step 1**). Try to remove as much sample as possible without transferring any phenol.
7. Add 500 μL of absolute ethanol to each sample, and invert to mix. Samples should be transferred to a −80 °C freezer for overnight precipitation (12–16 h). At this stage, the DNA is stable for a few days/weeks.

8. The following day, thaw the samples and centrifuge at 4 °C for 10 min at 16,000× *g* to pellet the DNA. Aspirate with a P1000 being careful not to disturb the pellet. The pellet often appears spherical. *See* **Note 13** on visualizing the pellet.
9. Add 1 mL of 80% ethanol and invert to mix.
10. Centrifuge at 4 °C for 2 min at 16,000× *g*. Aspirate with a P1000 as much as possible without disturbing the pellet.
11. Briefly collect residual ethanol at the bottom of the tube with a pulse on a microcentrifuge. Carefully remove the remaining ethanol using a P20 (*see* **Note 14**).
12. With the cap open, allow the pellet to air-dry for 10 min. No ethanol should be visible when flicking the tube.
13. Reconstitute the DNA in 30 μL of 0.1× Tris-EDTA buffer. Incubate at 37 °C for 10–15 min. DNA at this point is stable at −20 °C or can proceed directly to library preparation if started from Subheading 3.3, **step 11**.

3.5 Size Selection

This step is only required if large fragments from Subheading 3.3, **step 11** are used.

1. Add 33 μL PRONEX beads to each sample and vortex to mix. Allow to sit for 5 min at room temperature (20–22 °C) to bind DNA to the beads. *See* **Note 15** for further information on size selection.
2. While beads are incubating, prepare the next bead ratio by adding 60 μL of beads (Final ratio of 3×) to a new tube. *See* **Note 16** for batch processing size selection.
3. Apply the sample from **step 1** to the magnet rack and let sit for 1–2 min until clear. Transfer the supernatant to the tube prepared in **step 2** and mix by pipetting at least 10×. Allow to sit for 5 min at room temperature (20–22 °C) to bind DNA to the beads (*see* **Note 15**).
4. Apply sample to magnet rack and let sit for 1–2 min until clear. Pipette off supernatant and discard.
5. Add 200 μL PRONEX wash buffer while beads remain on the magnet. Incubate for ~30 s and then discard wash.
6. Repeat **step 5** once for a total of 2 washes. Quickly pulse on a microcentrifuge to collect residual wash buffer, apply to the magnet, and remove using a P20. It is important to not pellet the beads but remove as much wash buffer as possible.
7. Remove from the magnet and allow to air-dry for 3–5 min. Once beads begin to visibly crack, they are ready to be eluted.
8. Elute in 27 μL of 0.1× Tris-EDTA buffer. Mix well to fully resuspend beads. Incubate for 2 min before returning to the magnet.

9. Once samples are clear of beads, transfer 25 μL of the supernatant to a new tube. Samples can now be stored at −20 °C before library preparation.

3.6 Library Preparation

For DNA library preparation, we follow the NEB protocol here with one modification during the end-repair step: 30 min @ 20 °C and 60 min @ 50 °C. We use a Qubit fluorometer to determine the DNA concentration of samples before library preparation, and generally proceed with 10–50 ng DNA depending on the abundance of the protein of interest, which corresponds to 7–8 PCR cycles to avoid overamplification.

3.7 Sequencing

Finished libraries should be subjected to paired-end 50 bp sequencing to a depth of five to ten million reads per sample.

4 Notes

1. Digitonin permeabilization varies between cell lines and batches of digitonin. The concentration to use for each line + batch should be experimentally determined. Titrate digitonin from 0.1% and monitor permeabilization with trypan blue. Use the minimum concentration required to permeabilize >95% of cells [6].
2. Generally, we find that ChIP-Seq-validated antibodies are generally effective for CUT&RUN. That said, not all ChIP-Seq-validated antibodies will work for CUT&RUN. We recommend profiling multiple potential antibodies when performing CUT&RUN on a new factor.
3. Batch effects are prominent in next-generation sequencing assays [7], and while in our hands CUT&RUN has proven incredibly reproducible, any direct comparisons should still be performed in parallel. We have found that with two sets of hands, we can comfortably run 48 CUT&RUN reactions at once (e.g., profile 11 factors with IgG across biological replicates from two experimental conditions).
4. The appropriate negative control for CUT&RUN is up for debate, but the most widely accepted control is the use of IgG as a nontargeting, but still enzymatically active control [1, 2]. IgG should not target the pA/G-MNase to specific genomic loci, so the areas of enrichment in this sample should be considered for peak-calling of other experimental samples. One positive control that should be considered is H3K4Me3, a histone mark located at active promoters. This mark should be present near transcription start sites in any cell type.

5. These washes serve to remove free sugars and other molecules that can compete for binding to Concanavalin A–coated beads (ConA beads). Removing these molecules is critical for the efficient recovery of cells.
6. At this step, the cells are permeabilized with digitonin, and divalent cations are chelated with EDTA, quickly halting metabolic processes and endogenous DNAse activity. This helps to preserve the native chromatin state and ultimately reduce noise. Thus, it is recommended to work quickly to get cells into the antibody buffer.
7. It becomes challenging after this step to remove beads from the walls of the tubes. It is recommended to flick gently to avoid splashing, while still fully resuspending the beads. Alternatively, gently resuspend the solution using low-binding pipet tips to minimize cell loss.
8. We recommend incubating on a rocker overnight (12–16 h). The cells will settle even with rocking, so we usually incubate them on ice for 1–2 h, gently flicking every 10 min, and then allow them to rock overnight (12–16 h). This ensures all the cells have the opportunity to uptake antibodies. We have not observed any differences with shorter or longer antibody incubations, but we generally opt for ~16 h for consistency.
9. We have found that when using large sets of samples, batching in sets of 16 is effective. The length of pAG/MNase incubation is in excess, so the slightly different timing of the batches will not be of consequence. Instead, this minimizes the time that samples are sitting in the wash buffer.
10. MNase binds DNA but only cleaves in the presence of Ca^{2+}. After cleavage, the MNase-bound particles diffuse and can result in cleavage of open chromatin regions, resulting in a background signal. To limit this, the temperature should remain as close to 0 °C as possible. We use a cold tube rack to ensure consistently low temperatures. If high levels of background are observed, the incubation time can be shortened to as low as 5 min. The Low-Salt rinsing and incubation buffer will compact chromatin to limit diffusion out of the nucleus.
11. For large fragments from other epigenetic factors or first-time trying targets, it is always safe to use whole cell for DNA extraction: incubate tubes at 37 °C for 15 min; then add SDS and Protease K, incubate at 65 °C for 45 min, proceed to Subheading 3.4.
12. Phenol-chloroform is the preferred method for DNA extraction in CUT&RUN, although column-based methods can also be applied. Note that if using a column, you run the risk of losing smaller fragments (such as 50-80 bp fragments, which are often signatures of profiling transcription factors).

Additionally, when working with small quantities of DNA, column-based methods risk losing a substantial percentage of your yield. Thus, we recommend using the phenol-chloroform extraction shown in Subheading 3.4.

13. The addition of glycogen facilitates in visualizing the pellet, inducing a white color. The pellet should be relatively spherical yet compact.
14. We have found that we can remove almost all ethanol by tilting the tube and aspirating from above the pellet. If this is not possible without sample loss, increase the drying time in Subheading 3.4, **step 12** up to 30 min.
15. The ratio of beads: to sample in Subheading 3.5, **step 1** allows the beads to bind DNA fragments longer than 700 bp genomic DNA. The binding is determined by the viscosity of the bead + sample solution. The beads are in a highly viscous buffer. At a higher ratio of bead: sample, the high viscosity facilitates the binding of smaller DNA fragments. The CUT&RUN fragments will not bind the beads at this bead ratio in this step, so make sure not to discard the flow through. The ratio of beads to sample in Subheading 3.5, **step 3** allows the beads to bind DNA fragments longer than 100 bp. The beads now bind desired CUT&RUN fragments.
16. When batch processing multiple samples, at this point we switch to using 96 well-strip tubes for use with multichannel pipetting. Note that this requires a PCR tube magnet as well.

References

1. Skene PJ, Henikoff S (2017) An efficient targeted nuclease strategy for high-resolution mapping of DNA binding sites. eLife 6:6
2. Wolf BK, Zhao Y, McCray A et al (2023) Cooperation of chromatin remodeling SWI/SNF complex and pioneer factor AP-1 shapes 3D enhancer landscapes. Nat Struct Mol Biol 30: 10–21
3. Ringel T, Frey N, Ringnalda F et al (2020) Genome-scale CRISPR screening in human intestinal organoids identifies drivers of TGF-β resistance. Cell Stem Cell 26:431–440.e8
4. Kaya-Okur HS, Wu SJ, Codomo CA et al (2019) CUT&tag for efficient epigenomic profiling of small samples and single cells. Nat Commun 10: 1930
5. Meers MP, Janssens DH, Henikoff S (2019) Pioneer factor-nucleosome binding events during differentiation are motif encoded. Mol Cell 75:562–575.e5
6. Tramontina F, Karl J, Gottfried C et al (2000) Digitonin-permeabilization of astrocytes in culture monitored by trypan blue exclusion and loss of S100B by ELISA. Brain Res Brain Res Protoc 6:86–90
7. Meyer CA, Liu XS (2014) Identifying and mitigating bias in next-generation sequencing methods for chromatin biology. Nat Rev Genet 15: 709–721

Chapter 12

CUT&Tag for Efficient Epigenomic Profiling of Frozen Tissues

Qi Yin, Yang Li, and Yuxin Yin

Abstract

Cleavage Under Targets and Tagmentation (CUT&Tag) provides high-resolution sequencing libraries for profiling diverse chromatin components. This protocol details the steps to generate CUT&Tag libraries from fresh or frozen tissues. This CUT&Tag workflow has nine main steps: isolation of nuclei from tissues, binding of nuclei to Concanavalin A–coated beads, binding of the primary antibody, binding of the secondary antibody, binding pA-Tn5 adapter complex, tagmentation, DNA extraction, PCR, and post-PCR cleanup and size selection. This protocol enabled us to generate and sequence CUT&Tag libraries across a broad range of fresh and frozen tissue types.

Key words CUT&Tag, Protein-DNA binding, Epigenomics, High-throughput sequencing

1 Introduction

CUT&Tag, described by the Henikoff group in 2019, is an enzyme-tethering strategy in which antibodies are bound to chromatin proteins in situ in permeabilized nuclei [1]. In CUT&Tag, a hyperactive Tn5 transposase-Protein A/G (pA/G-Tn5) fusion protein loaded with sequencing adapters binds to the specific chromatin-bound antibody in situ. The following activation of pA/G-Tn5 with Mg^{2+} results in factor-targeted tagmentation, generating fragments ready for PCR enrichment and DNA sequencing. CUT&Tag provides efficient high-resolution, low--background sequencing libraries for profiling diverse chromatin components in a small number of cells [2]. This easy, highly efficient, low-cost method has been widely used to profile histone modifications, transcription factors, and chromatin remodeling factors [3–7]. Although CUT&Tag generates high-fidelity epigenomic libraries from different cell types, it remains a challenge to generate CUT&Tag libraries from complex tissues, including flash-frozen tissues.

Franziska Greulich (ed.), *Chromatin Immunoprecipitation: Methods and Protocols*, Methods in Molecular Biology, vol. 2846,
https://doi.org/10.1007/978-1-0716-4071-5_12,

In our recent studies, we developed a broadly applicable CUT&Tag protocol, that is suitable for diverse tissue types, and archival frozen samples. Using this protocol, we performed CUT&Tag analysis of endogenous RPA1, HNF4A, histone modification, and RNA polymerase II in both fresh and frozen mouse liver and generated high-quality libraries. We demonstrate that RPA1 is an important epigenetic regulator of NAFLD by controlling chromatin accessibility [8]. In this chapter, we provide a detailed protocol to generate and sequence CUT&Tag libraries across a broad range of fresh and frozen tissue types.

2 Materials

2.1 Isolation of Nuclei from Frozen Tissues for Cut&Tag

These buffers are stable at 4 °C. Sterile filtration is recommended, especially for sucrose.

1. Homogenization Buffer Stable Master Mix: 250 mM Sucrose, 25 mM KCl, 5 mM $MgCl_2$, 20 mM Tricine-KOH pH 7.8. Freshly add 0.1% (v/v) NP40, 1 mM DTT, 0.15 mM spermine, 0.5 mM spermidine, and cOmplete™ Protease Inhibitor Cocktail.
2. Solution B: 150 mM KCl, 30 mM $MgCl_2$, 120 mM Tricine-KOH pH 7.8.
3. 50% Iodixanol Solution: Dilute 60% Iodixanol Solution (OptiPrep™ Density Gradient Medium (Sigma)) to 50% with Solution B.
4. 30% Iodixanol Solution: Dilute 50% Iodixanol Solution to 30% with Homogenization Buffer Stable Master Mix.
5. 35% Iodixanol Solution (600 μL per sample): Dilute 50% Iodixanol Solution to 35% with Homogenization Buffer Stable Master Mix.
6. Nuclei resuspension buffer (NRSB): 10 mM Tris–HCl pH 7.4, 10 mM NaCl, 3 mM $MgCl_2$.
7. Cooling centrifuge for 1.5 mL tubes precooled to 4 °C.
8. 2 mL Dounce Homogenizer.
9. Automatic cell counter.
10. 2 mL LoBind tubes.
11. 100 μm nylon mesh filter.

2.2 Bind Nuclei to Concanavalin A–Coated Beads

1. Binding buffer: 20 mM HEPES-KOH pH 7.9, 10 mM KCl, 1 mM $CaCl_2$, 1 mM $MnCl_2$. Store the buffer at 4 °C for 6 months.
2. Wash buffer: 20 mM HEPES-KOH pH 7.5, 150 mM NaCl, 0.5 mM Spermidine. Freshly add 1 cOmplete™ Protease Inhibitor EDTA-Free tablet in 50 mL wash buffer.

3. Concanavalin A–coated beads (e.g., from Bangs Laboratories # BP531).
4. 1.5 mL LoBind tubes.
5. Magnetic stand for 1.5 mL tubes.
6. End-over-end rotator.

2.3 Bind Primary and Secondary Antibody

1. 0.5 M EDTA.
2. 30% (w/v) Bovine Serum Albumin (BSA).
3. 5% (w/v) Digitonin: Dissolve 50 mg digitonin in 1 mL DMSO.
4. Dig-wash buffer: Add 100 μL 5% (w/v) Digitonin to 10 mL Wash buffer. Store the buffer at 4 °C for up to 2 days.
5. Antibody buffer: Add 8 μL 0.5 M EDTA, 6.7 μL 30% (w/v) BSA to 2 mL Dig-wash buffer.
6. Primary and secondary antibodies (e.g., guinea pig α-rabbit antibody from Antibodies online #ABIN101961 or Rabbit Anti-Mouse IgG H&L from Abcam #ab6709) of interest.
7. Nutator.

2.4 Bind pA-Tn5 Adapter Complex

1. Dig-300 buffer: 20 mM HEPES-KOH pH 7.5, 300 mM NaCl, 0.5 mM Spermidine, 0.01% (w/v) Digitonin. Freshly add 1 Roche Complete Protease Inhibitor EDTA-Free tablet in 50 mL Dig-300 buffer.
2. NovoNGS pAG-Transposome (e.g., from Novoprotein # M059-YH01-01A).

2.5 Tagmentation

1. 1 M $MgCl_2$.
2. Tagmentation buffer: Add 50 μL 1 M $MgCl_2$ to 5 mL Dig-300 buffer.
3. Thermoblock.

2.6 DNA Extraction

1. 10 mM Tris–HCl pH 8.0.
2. Absolute ethanol.
3. 80% (v/v) ethanol: Mix 8 mL abs. ethanol with 2 mL nuclease-free water.
4. Tagment DNA Extract Beads (e.g., from Novoprotein #N245-01B, alternative AMPure beads).
5. 0.2 mL PCR tubes.

2.7 PCR

1. NEBNext® Ultra™ II Q5® Master Mix (NEB #M0544S).
2. Index primer with barcodes at 10 μM (e.g., from Novoprotein #N239).
3. PCR machine.

2.8 Post-PCR Cleanup and Size Selection

1. NovoNGS DNA clean Beads (e.g., Novoprotein #N240-01A, alternative AMPure beads).
2. Bioanalyzer or similar with appropriate reagents.
3. Qubit with appropriate reagents.
4. Illumina sequencer.

3 Methods

3.1 Isolation of Nuclei from Frozen Tissues for Cut&Tag

1. Place frozen tissue (about 20 mg) (*see* **Note 1**) into a prechilled 2 mL Dounce Homogenizer with 2 mL cold Homogenization Buffer Stable Master Mix.
2. Allow frozen tissue to thaw for 2–5 min on ice.
3. Tissues were homogenized in a Dounce homogenizer (with a loose and then a tight pestle) with 10–30 strokes on ice. We use 10 strokes with a loose pestle and then 15 strokes with a tight pestle when homogenizing mouse liver tissue.
4. Filtration via a 100 μm nylon mesh filter.
5. Pellet nuclei by spinning for 5 min at 4 °C at 350× *g* in a fixed-angle centrifuge.
6. Pipette the supernatant-off.
7. Resuspend nuclei in a total volume of 400 μL Homogenization Buffer Stable Master Mix.
8. Add 1 volume (400 μL) of 50% Iodixanol solution to give a final concentration of 25% Iodixanol and mix by pipetting.
9. Add 600 μL of 35% Iodixanol solution in a new round bottom 2 mL LoBind tube.
10. Layer 600 μL of 30% Iodixanol solution over the 35% mixture. Avoid mixing of layers.
11. Layer 800 μL of 25% Iodixanol and sample mixture (from **step 8**) over the 30% Iodixanol solution. Avoid mixing of layers.
12. In a swinging bucket centrifuge, spin for 40 min at 5000× *g* at 4 °C with the brake off.
13. Using 200 μL pipet tips, collect the nuclei band (at the 30–35% interface) and transfer it to a fresh low-binding tube.
14. Count nuclei using automatic cell counter and aliquot nuclei for Cut&Tag reaction. *See* Fig. 1 for an example picture of good and bad nuclei preparation.
15. If the nuclei are too concentrated to accurately count (>5 × 10^6/mL), dilute in NRSB.
16. Transfer 100,000 nuclei (per Cut&Tag reaction) into a fresh low-binding tube containing 1 mL of NRSB.

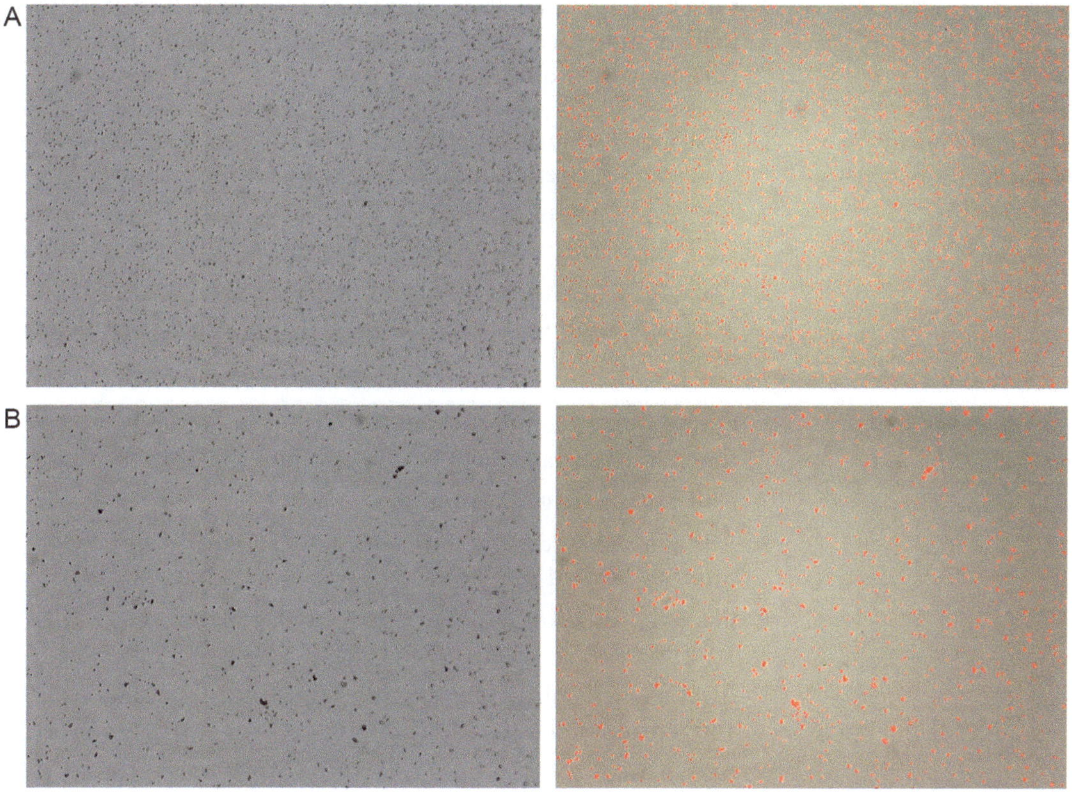

Fig. 1 Nuclei after purification. Phase contrast images before (left) and after (right) automatic counting of a good (**a**) and a bad (**b**) nuclei preparation

17. Centrifuge nuclei for 5 min at 500× *g* at 4 °C and aspirate the supernatant. The nuclei are ready for Cut&Tag.

3.2 Bind Nuclei to Concanavalin A (ConA)-Coated Beads (0.5–1 h)

1. Transfer 10 μL ConA bead slurry for each reaction of 100,000 nuclei into 100 μL Binding buffer in a 1.5 mL LoBind tube and mix by pipetting. Place the tube on a magnetic stand for 2 min to clear.
2. Aspirate the liquid, and remove it from the magnetic stand. Add 100 μL Binding buffer and mix by pipetting. Place the tube on a magnetic stand for 2 min to clear.
3. Aspirate liquid and resuspend in 10 μL Binding buffer.
4. Resuspend nuclei prepared in Subheading 3.1, **step 17** in 500 μL Wash buffer prechilled on ice, centrifuge 5 min at 4 °C at 500× *g* in a fixed-angle centrifuge, and pipette the liquid.
5. Resuspend in 100 μL Wash buffer and transfer to a 1.5 mL LoBind tube. While vortexing gently add washed ConA-coated beads slurry. Place on an end-over-end rotator for 10 min.

6. Quick spin in a tabletop centrifuge to remove the liquid from the lid and place the tubes on a magnetic stand to clear and withdraw the liquid.

3.3 Bind Primary Antibody (Overnight)

1. Resuspend ConA-coated beads coated with nuclei from Subheading 3.2, **step 6** in 50 μL ice-cold Antibody buffer.
2. Add 0.5–1 μL primary antibody (*see* **Note 2**) to each sample while gentle vortexing (speed less than 3).
3. Place on a nutator at 4 °C and incubate overnight (12–16 h). The liquid should remain in the bottom and side of the tube while rocking (*see* **Note 3**).

3.4 Bind Secondary Antibody (1 h)

1. Quick spin to remove liquid from the cap, place each tube on the magnetic stand to clear, and aspirate the liquid.
2. Add 100 μL Dig-wash buffer.
3. Add 1 μL secondary antibody (1:200) and mix gently by pipetting (*see* **Note 4**).
4. Place the tubes on a nutator at room temperature (25 °C) for 1 h.
5. After a quick spin, place the tubes on a magnetic stand to clear and withdraw the liquid.
6. Add 1 mL Dig-wash buffer. Gently vortex (speed less than 3).
7. Repeat **steps 5** and **6** twice.

3.5 Bind pA-Tn5 Adapter Complex (1.5 h)

1. Place the tubes on the magnetic stand to clear and withdraw the liquid.
2. Add 100 μL Dig-300 buffer.
3. Add 1 μL pA-Tn5 mix while gently vortexing (speed less than 3).
4. Place the tubes on a nutator at room temperature (25 °C) for 1 h.
5. After a quick spin, place the tubes on a magnetic stand to clear and pull off the liquid.
6. Add 1 mL Dig-300 buffer. Gently vortex (speed less than 3).
7. Repeat **steps 5** and **6** twice.

3.6 Tagmentation (1 h)

1. After a quick spin, place the tube on the magnetic stand to clear and pull off the liquid.
2. Add 300 μL Tagmentation buffer *while* gently vortexing (speed less than 3).
3. Incubate at 37 °C in a thermoblock with the speed of 500 rpm for 1 h.

3.7 DNA Extraction

1. After a quick spin, add 2 volume (100 μL) Tagment DNA Extract Beads (*see* **Note 5**), mixing by pipetting up and down.
2. After 5 min, place on a magnet and allow to clear before carefully withdrawing liquid.
3. On the magnet and without disturbing the beads, add 200 μL 80% ethanol.
4. Withdraw liquid with a pipette to the bottom of the tube and add 200 μL 80% ethanol.
5. Withdraw liquid and allow to dry for 4–5 min.
6. Remove from magnet stand, add 22 μL 10 mM Tris–HCl pH 8.0, and vortex on full speed.
7. After 5 min, place on a magnetic stand and allow to clear.
8. Transfer 20 μL liquid to a fresh 0.2 mL PCR tube with a pipette.

3.8 PCR (1 h)

1. 20 μL DNA from Subheading 3.7, **step 8** + 2.5 μL 10 μM i5 primer + 2.5 μL 10 μM i7 primers. In a thin-wall 0.2 mL PCR tube, use a different barcode combination for each sample.
2. Add 25 μL NEBNext® Ultra™ II Q5® Master Mix.
3. Mix, quickly spin, and place in Thermocycler and begin the cycling program with a heated lid (105 °C).

   Step 1: 72 °C, 3 min.

   Step 2: 98 °C, 30 s.

   Step 3: 98 °C, 15 s.

   Step 4: 60 °C, 10 s.

   Step 5: 72 °C, 8 s.

   Repeat Steps 3–5 13 times.

   72 °C for 2 min and hold at 10 °C.

3.9 Post-PCR Cleanup and Size Selection (30 min)

1. After the tubes have cooled to room temperature (25 °C), remove them from the PCR cycler and add 25 μL NovoNGS DNA clean Beads (0.5x SPRI cleanup ratio), mixing by pipetting up and down (*see* **Note 5**).
2. Let sit at room temperature (25 °C) for 5 min.
3. Place on a magnet stand for 0.2 mL tubes and allow to clear. Carefully transfer the supernatant to a new tube.
4. Add 40 μL NovoNGS DNA clean Beads (0.8x), mixing by pipetting up and down.
5. Let sit at room temperature (25 °C) for 5 min.
6. Place on a magnet stand and allow to clear. Carefully withdraw the liquid.

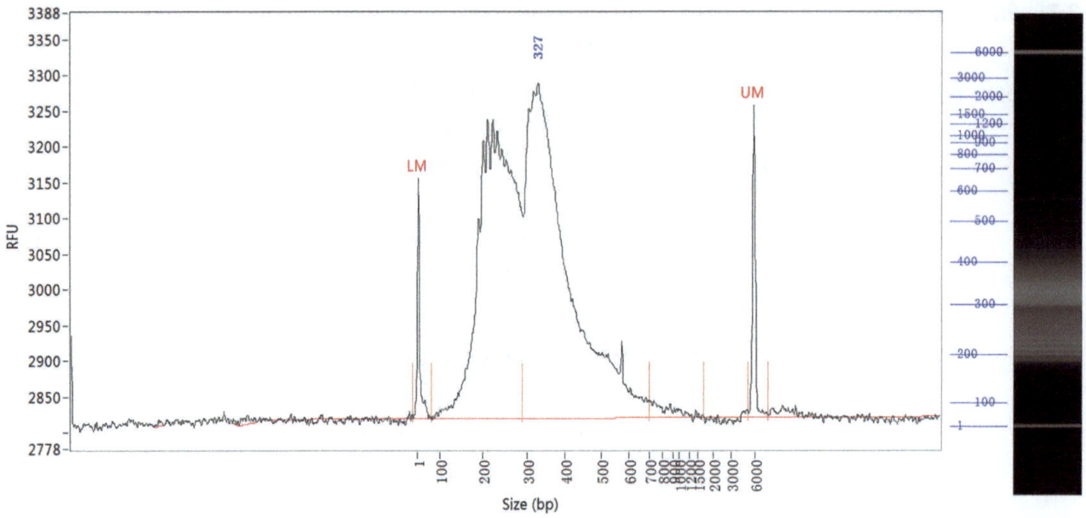

Fig. 2 Bioanalyzer trace of a Cut&Tag library

7. On the magnet and without disturbing the beads, add 200 μL 80% ethanol.
8. Withdraw liquid with a pipette to the bottom of the tube and add 200 μL 80% ethanol.
9. Withdraw liquid and allow to dry for 4–5 min.
10. Remove from the magnetic stand, add 25 μL 10 mM Tris–HCl pH 8.0, and vortex at full speed.
11. After 5 min, place on a magnetic stand and allow to clear.
12. Remove liquid to a fresh 1.5 mL LoBind tube with a pipette.
13. Determine the size distribution with a Bioanalyzer or similar device. The library should look like in Fig. 2. Measure the concentration of libraries using Qubit.
14. Sequencing the library on an Illumina next-generation sequencer using a flow cell with paired-end chemistry (*see* **Note 6**).

4 Notes

1. We do not fix the tissue. The tissues can be snap-frozen in liquid nitrogen and stored at −70 °C. In our hand, the tissues stored for 2 years can be used for Cut&Tag library preparation.
2. We suggest that the primary antibody should be diluted 1:50–100 in antibody buffer. A ChIP-grade antibody is the best choice for Cut&Tag.

3. To evaluate the success of the procedure include in parallel a positive control antibody (e.g., α-H3K4me3 Abcam #ab5800, α-Pol2 CellSignaling #2629) and a negative control antibody (e.g., rabbit α-mouse IgG).
4. Select a secondary antibody against the isotype and species of the primary antibody.
5. These beads can be replaced by AMPure beads.
6. We use PE150 and sequence the library with 10–15 M reads. PE150 in our hand is cost-efficient.

References

1. Kaya-Okur HS, Wu SJ, Codomo CA et al (2019) CUT&Tag for efficient epigenomic profiling of small samples and single cells. Nat Commun 10(1):1930
2. Kaya-Okur HS, Janssens DH, Henikoff JG et al (2020) Efficient low-cost chromatin profiling with CUT&Tag. Nat Protoc 15(10): 3264–3283
3. Henikoff S, Henikoff JG, Kaya-Okur HS et al (2020) Efficient chromatin accessibility mapping in situ by nucleosome-tethered tagmentation. Elife 9:e63274
4. Janssens DH, Meers MP, Wu SJ et al (2021) Automated CUT&Tag profiling of chromatin heterogeneity in mixed-lineage leukemia. Nat Genet 53(11):1586–1596
5. Janssens DH, Otto DJ, Meers MP et al (2022) CUT&Tag2for1: a modified method for simultaneous profiling of the accessible and silenced regulome in single cells. Genome Biol 23(1):81
6. Meers MP, Llagas G, Janssens DH et al (2023) Multifactorial profiling of epigenetic landscapes at single-cell resolution using MulTI-Tag. Nat Biotechnol 41(5):708–716
7. Wu SJ, Furlan SN, Mihalas AB et al (2021) Single-cell CUT&Tag analysis of chromatin modifications in differentiation and tumor progression. Nat Biotechnol 39(7):819–824
8. Yin Q, Li Y, Zhou Z et al (2022) RPA1 controls chromatin architecture and maintains lipid metabolic homeostasis. Cell Rep 40(2):111071

Chapter 13

Integrative Analysis of CUT&Tag and RNA-Seq Data Through Bioinformatics: A Unified Workflow for Enhanced Insights

Niccolò Liorni, Alessandro Napoli, Marta Adinolfi, Manlio Vinciguerra, and Tommaso Mazza

Abstract

Cleavage Under Targets and Tagmentation (CUT&Tag) is a recent methodology used for robust epigenomic profiling that, unlike conventional chromatin immunoprecipitation (ChIP-Seq), requires only a limited amount of cells as starting material. RNA sequencing (RNA-Seq) reveals the presence and quantity of RNA in a biological sample, describing the continuously changing cellular transcriptome. The integrated analysis of transcriptional activity, histone modifications, and chromatin accessibility via CUT&Tag is still in its infancy compared to the well-established ChIP-Seq. This chapter describes a robust bioinformatics methodology and workflow to perform an integrative CUT&Tag/RNA-Seq analysis.

Key words Cleavage under targets and tagmentation (CUT&Tag), RNA sequencing (RNA-Seq), Chromatin immunoprecipitation sequencing (ChIP-Seq), Next-generation sequencing (NGS), Histone marks, Bioinformatics

1 Introduction

Chromatin immunoprecipitation sequencing (ChIP-Seq) is a high-throughput technique used to identify DNA-binding sites of proteins, such as transcription factors, across the genome [1]. This method combines chromatin immunoprecipitation (ChIP) with next-generation sequencing (NGS) to analyze the locations of DNA-protein interactions.

The first step of the ChIP-Seq process involves cross-linking the proteins to the DNA in living cells, which stabilizes the protein-DNA complexes. The protein of interest is then immunoprecipitated using particular antibodies that recognize the protein after the chromatin has been broken up into small pieces via enzymatic digestion or sonication. The resulting DNA-protein complexes are purified, and the cross-links are reversed to release the DNA fragments. The purified DNA fragments are then subjected to NGS,

Franziska Greulich (ed.), *Chromatin Immunoprecipitation: Methods and Protocols*, Methods in Molecular Biology, vol. 2846, https://doi.org/10.1007/978-1-0716-4071-5_13,

which generates millions of short DNA reads. The reads are mapped to the reference genome, and peaks of enrichment are identified, representing regions of the genome where the protein of interest binds. The ChIP-Seq data can be visualized as genome-wide binding maps, which provide insight into the regulatory mechanisms controlling gene expression.

ChIP-Seq has become a widely used method for studying protein-DNA interactions, and it has enabled the identification of novel transcriptional regulators and the characterization of their roles in gene regulation. This technique has also been used to investigate the epigenetic modifications that occur on DNA, such as histone modifications and DNA methylation, providing a deeper understanding of chromatin structure and gene regulation.

However, ChIP-Seq has limitations: reduced resolution, typically in the range of hundreds of base pairs to a few kilobases, due to the DNA fragmentation and immunoprecipitation steps involved in the protocol, which can introduce biases and artifacts [2, 3]. In addition, it requires a relatively high amount of cells or tissue compared to other sequencing-based techniques. This can be a limitation when working with reduced or rare cell populations or when studying dynamic processes involving small numbers of cells. Moreover, ChIP-Seq can suffer from nonspecific binding of antibodies, which can lead to false-positive signals. This is particularly problematic when studying low-abundance proteins or transcription factors with similar DNA-binding motifs and can be challenging to reproduce, especially when using different antibodies or different experimental conditions.

In the last two decades, alternative chromatin profiling strategies have been introduced, such as DamID [4], ChEC-Seq [5], and CUT&RUN [6, 7], which provide base-pair resolution and reduced background levels. Starting from very few cells (~100/1000), CUT&RUN can produce high-quality data, but the DNA end polishing and adapter ligation steps, required for sequencing library preparation, increase the time and cost of the technique [8]. Furthermore, it is not well suited for single-cell platforms [9, 10], limiting its usage to bulk sequencing.

In this scenario, Cleavage Under Targets and Tagmentation (CUT&Tag), a relatively new epigenomic profiling technique, has emerged. It enables targeted sequencing of chromatin-associated proteins and their DNA interactions at specific genomic loci. In CUT&Tag, cells are permeabilized and attached to a surface, usually magnetic beads or a microfluidic device. Then, a fusion protein containing a DNA cleavage enzyme and a chromatin-associated protein is added, and the cells are incubated. The fusion protein binds to the target protein and cleaves the DNA at its recognition site. Subsequently, a DNA sequencing adapter is attached to the cleaved DNA fragments, and the DNA is amplified by PCR and sequenced [8, 11].

CUT&Tag has several advantages over ChIP-Seq, including reduced cell numbers and processing time, higher signal-to-noise ratios, and lower background levels. Additionally, it can be performed on fresh or frozen tissue samples and is compatible with high-throughput sequencing technologies.

The more straightforward CUT&Tag protocol requires fewer steps and uses fewer reagents, reducing the risk of experimental errors. Moreover, the lower cell input requirement is particularly advantageous when working with limited or rare cell populations. Using a more efficient enzyme-based tagging approach reduces nonspecific background signals, and its transposase-based approach, which can accurately map protein-DNA interactions at single-nucleotide resolution, leads to higher resolution compared to ChIP-Seq.

Overall, CUT&Tag mainly offers several advantages over ChIP-Seq, making it a powerful tool for studying protein-DNA interactions. However, it is important to note that each technique has its strengths and weaknesses, and the choice of which one to use should depend on the specific research question being addressed.

2 Materials

In this part, we provide a comprehensive list of links to the principal tools for CUT&Tag analysis, along with information on their dependencies and installation procedures using Conda, a popular package management system and environment manager.

Create a new Conda environment for the CUT&Tag analysis:

```
conda create -n cuTag-env
```

Activate the environment before installing the tools:

```
conda activate cuTag-env
```

Install the tools (e.g., FastQC):

```
conda install -c bioconda fastqc
```

The essential tools required for CUT&Tag analysis include a diverse set of software packages and libraries. These tools encompass a range of functionalities, from quality control to alignment, peak calling, and downstream analysis. The following tools are among the principal ones used in CUT&Tag analysis:

1. FastQC, a tool for assessing the quality of sequencing data, website: https://www.bioinformatics.babraham.ac.uk/projects/fastqc/
   - Bioconda FastQC package: https://anaconda.org/bioconda/fastqc
2. Bowtie2, a fast and efficient aligner for mapping sequencing reads to a reference genome:
   - website: http://bowtie-bio.sourceforge.net/bowtie2/index.shtml
   - Bioconda Bowtie2 package: https://anaconda.org/bioconda/bowtie2
3. Samtools, a suite of tools for manipulating and analyzing SAM/BAM files, which store sequence alignment data:
   - GitHub repository: https://github.com/samtools/samtools
   - Bioconda samtools package: https://anaconda.org/bioconda/samtools
4. Sambamba, a faster alternative to samtools for handling SAM/-BAM files:
   - GitHub repository: https://github.com/biod/sambamba
   - Bioconda sambamba package: https://anaconda.org/bioconda/sambamba
5. SEACR, a software package designed specifically for analyzing CUT&Tag data:
   - GitHub repository: https://github.com/databio/SEACR
   - Bioconda SEACR package: https://anaconda.org/bioconda/r-seacr
6. ChIPQC, an R package for quality control and analysis of data:
   - Bioconductor page: https://bioconductor.org/packages/release/bioc/html/ChIPQC.html
7. bedtools, a collection of utilities for working with genomic interval data:
   - GitHub repository: https://github.com/arq5x/bedtools2
   - Bioconda bedtools package: https://anaconda.org/bioconda/bedtools
8. IDR, a statistical method for assessing the reproducibility of peak calls:
   - GitHub repository: https://github.com/nboley/idr
   - Bioconda IDR package: https://anaconda.org/bioconda/idr

9. ChIPseeker, an R package for annotating and visualizing ChIP-Seq data:
   - Bioconductor page: https://bioconductor.org/packages/release/bioc/html/ChIPseeker.html
10. clusterProfiler, an R library for functional enrichment analysis of gene clusters or sets:
    - Bioconductor page: https://bioconductor.org/packages/release/bioc/html/clusterProfiler.ht
11. BETA, a software package for predicting transcription factor binding sites from ChIP-Seq data:
    - Bioconductor page: https://bioconductor.org/packages/release/bioc/html/BETA.html

All tools mentioned in this work were tested on a workstation equipped with 48 AMD cores and 252 GB of RAM. Some of these software packages support task parallelization thanks to multiple computing cores. Additionally, depending on the experiment being studied, they also require sufficient storage space for the raw data to be used.

3 Methods

3.1 CUT&Tag and RNA-Seq Protocols

CUT&Tag protocols have been previously described in [8, 12] and can usually be performed using commercial kits and following the manufacturer's protocols (for instance, Active Motif, California, USA). Similarly, RNA-Seq protocols conducted on a variety of cell and tissue samples as well as basic bioinformatics procedures to infer differentially expressed genes (DEGs) were largely described in previous works [13–17].

We describe the bioinformatic analysis of CUT&Tag and the integration of differential expression analysis by RNA-seq in the following. *See* Fig. 1 for an overview.

3.2 Quality Control and Mapping

The first step in a CUT&Tag in silico analysis workflow is the quality control of the sequenced reads, typically stored as FASTQ files (Fig. 1, step A).

Quality control is a crucial step to ensure the sequences are high quality and free from errors so that all downstream analyses can be performed on clean data. The most common software to evaluate the quality of FASTQ data is FastQC, a handy quality-control tool widely used in most bioinformatics analyses.

FastQC generates a variety of quality control metrics and plots, including per-base quality scores, per-sequence quality scores, sequence length distribution, GC content, overrepresented sequences, and k-mer content. These metrics can identify potential problems with the sequencing data, such as low-quality base calls, overrepresented sequences, and adapter contamination.

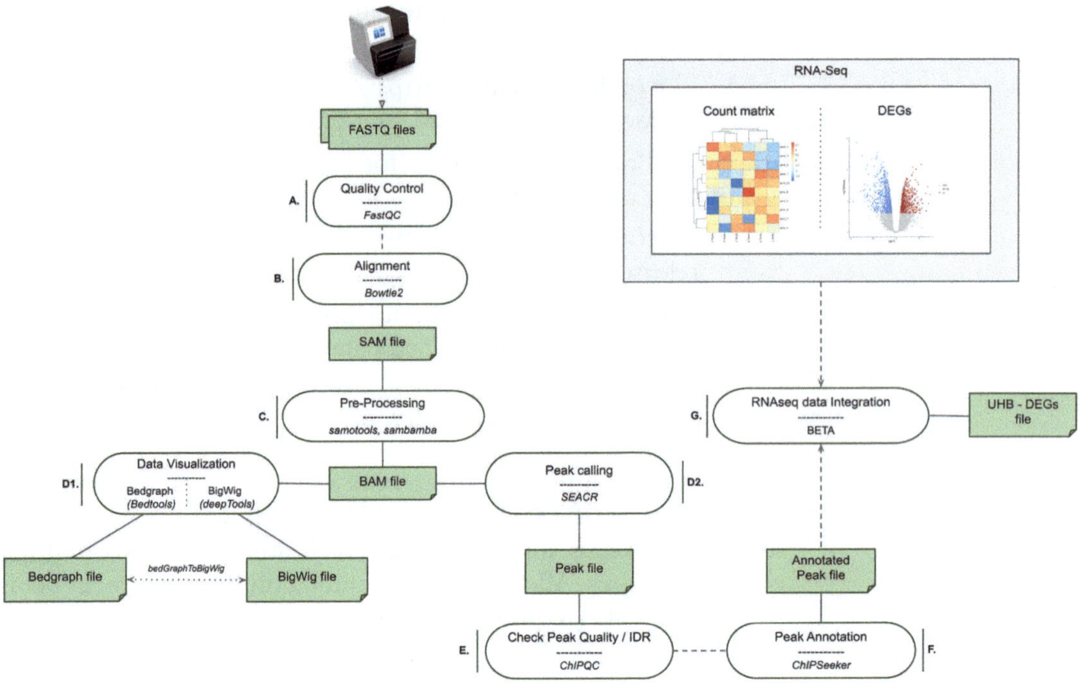

Fig. 1 Graphical abstract of the bioinformatic pipeline for CUT&Tag data and integration with RNA-seq. The green blocks represent input-output files analyzed using methods and tools in the white oval blocks. For steps A–G, refer to the text

FastQC software (https://www.bioinformatics.babraham.ac.uk/projects/fastqc/) is usually executed via its graphical user interface or using the following blueprint command line on paired-end reads:

```
fastqc -t {threads} --outdir {output directory} read1.fastq
read2.fastq
```

A poor output from FastQC (Fig. 2b, *see* **Note 1**) may suggest performing operations like trimming the adapter sequences or re-running the sequencing experiment.

If confident about the quality of the sequences, a user can proceed with the alignment of the sequenced reads on a reference genome (Fig. 1, step B) to identify the genomic coordinates of the DNA fragments obtained through the CUT&Tag experiment. There are several software packages designed to align DNA sequences to a reference genome, with the most popular being Burrows-Wheeler Aligner (BWA) [18], Bowtie 2 [19], and HISAT2 [20]. To align sequencing reads from a CUT&Tag experiment, the most appropriate tool is Bowtie 2—the successor of the original Bowtie version 1 software that comes with several improvements in speed, sensitivity, and accuracy—due to its ability to handle gapped alignments and tolerate mismatches.

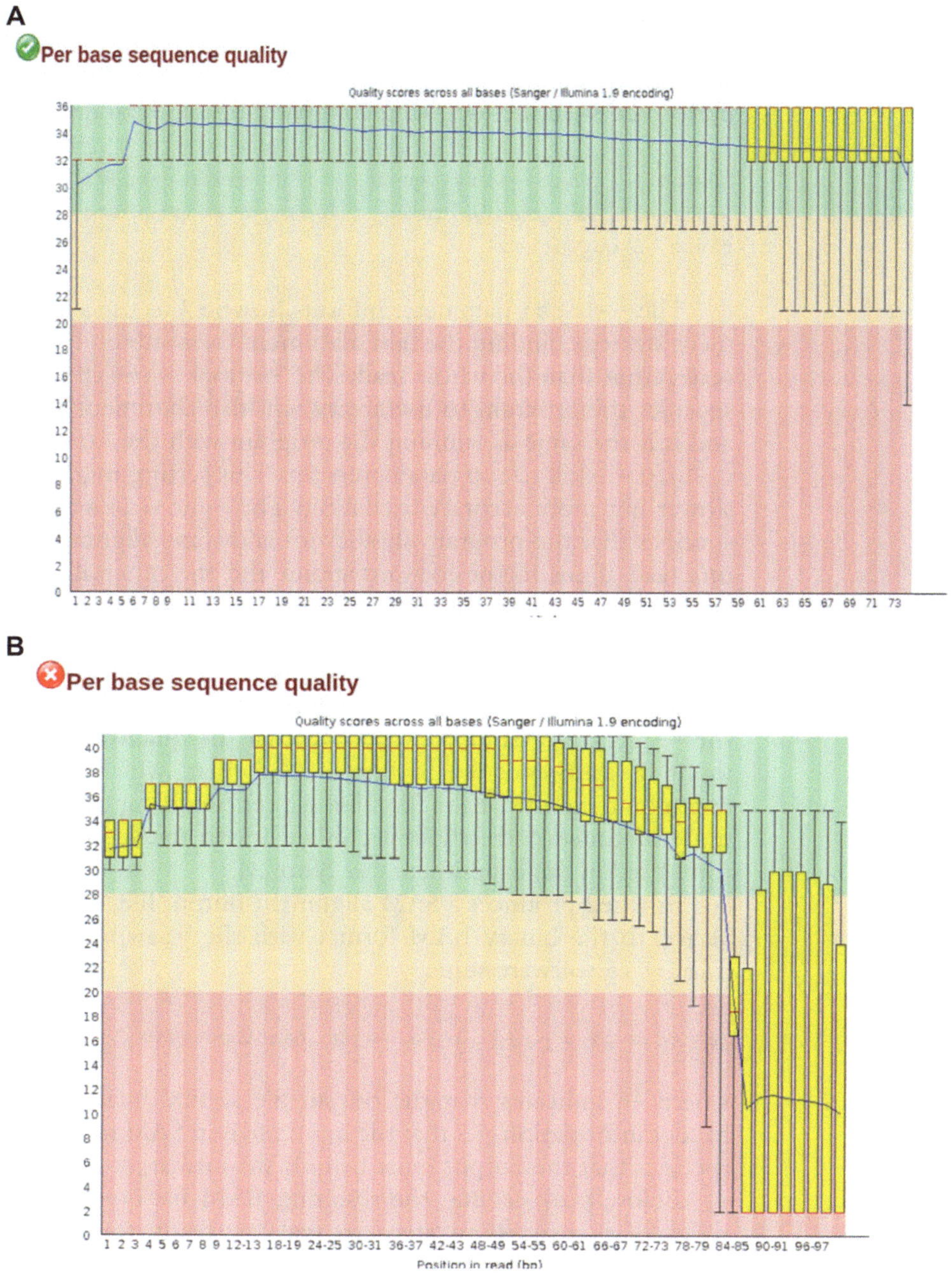

Fig. 2 High (**a**) versus low (**b**) quality profiles as from the FastQC output (*see* **Note 1**)

Bowtie 2 requires building an index of the reference genome, allowing for fast and efficient mapping. After downloading the reference genome in FASTA format (*see* **Note 2**), the Bowtie 2 index is built with the command:

```
bowtie2-build {reference_genome.fa} {index prefix}
```

This produces several *ebwt* and *bt2* index files with the prefix text specified in the command. The indexes are used in the subsequent alignment step that, in a typical CUT&Tag analysis, is performed with the following command line and parameters:

```
bowtie2 --local --very-sensitive --no-mixed --no-discordant -q
--phred33 -x {bowtie index files} -1 read1.fastq -2 read2.
fastq -S output.sam
```

The `--local` parameter informs Bowtie 2 to use a local alignment strategy, finding the best alignment for a subsequence of the read rather than the whole read. The `--very-sensitive` option specifies the program to use a very sensitive alignment algorithm, and it is the same as running the program with the `-D 20 -R 3 -N 0 -L 20 -i S,1,0.50` parameters set. This setting might result in slower but more accurate execution. The `--no-mixed` parameter specifies that the program should not allow any alignments where one read aligns to the forward strand and the other aligns to the reverse strand, while the `--no-discordant` parameter tells Bowtie 2 not to allow alignments where the distance between the two reads is larger than the expected insert size. The option `-q` is needed to specify that the input files are in FASTQ format, and the `--phred33` parameter specifies that the quality scores in the input files are encoded in Phred+33 format, which is the most common data format for Illumina sequencing data; this parameter must be tuned correctly in accordance with the sequencing platform used in the experiment (*see* **Note 3**).

Bowtie2 produces a SAM alignment output file; it can be converted to the binary BAM format with the `view` module of the `samtools` software suite:

```
samtools view -h -S -b -o output.bam input.sam
```

where `-h` indicates that the output file should include the SAM header information, `-S` that the input file is in SAM format, and `-b` specifies that the output must be in BAM format.

Before peak calling, the resulting BAM file is subjected to a series of preprocessing steps to make it ready for downstream analysis (Fig. 1, step C). While there are several possible software tools designed to perform processing operations (*see* **Note 4**) in a CUT&Tag analysis, the tool of choice is *Sambamba* [21], an open-source software used for processing high-throughput sequencing data in the BAM and SAM file formats. It is designed to handle large-scale genomic datasets efficiently and quickly, making it a popular choice.

The first processing steps involve the `sort` module of sambamba, which, by default, coordinate-sorts the input BAM file with the following command-line syntax:

```
sambamba sort -o output.sorted.bam input.bam
```

The resulting sorted BAM is immediately filtered to keep only uniquely mapped reads (*see* **Note 5**) using the `view` module of Sambamba with the following blueprint command line:

```
sambamba view -F "[XS] == null and not unmapped and not
duplicate." input.bam > output.bam
```

The resulting BAM file is ready for downstream analyses, mainly including the generation of bedgraphs and bigWigs, peak calling, and functional enrichment.

3.3 Data Visualization

Bedgraph is a common output format for displaying the coverage or signal strength of a genomic feature, such as the DNA fragment sequences obtained from a CUT&Tag experiment. A bedgraph file can be loaded into well-known visualization programs like the UCSC Genome Browser [22] or the Integrative Genomics Viewer (IGV) [23] to enable the visualization of the signal in particular regions of the genome. This allows researchers to assess the quality of the CUT&Tag data and compare the "signals" between various samples, replicates, or conditions to find differentially enriched regions. Moreover, they can readily and easily be converted into BigWig format (Fig. 1, step D1).

Bedgraph files can be obtained directly from the preprocessed BAM file in a few steps. The first operation is to convert the input BAM file into a BED format file using the module `bamtobed` of the `bedtools` software suite, with the following blueprint command line:

```
bedtools bamtobed -i {input BAM} -bedpe > {output BED}
```

where the `-bedpe` option tells bedtools to write BAM alignments in BEDPE format (*see* **Note 6**). Only one alignment per pair of reads will be reported. Specifically, if each read-mate is aligned to the same chromosome, the BAM alignment reported will be the one with an insert size greater than zero. The resulting BED file should have three columns (chromosome, start position, and end position). Finally, the BED file is converted into a bedgraph using the module `genomecov` of `bedtools`. This program also needs a file containing a list of the sizes of each chromosome in the genome and computes the coverage of each genomic interval, thereby producing the final bedgraph file containing four columns: chromosome, start position, end position, and coverage (*see* **Note 7**). The resulting bedgraph is ready for visualization through the UCSC genome browser or IGV and can be converted to the bigWig format with the UCSC command-line tool "*bedGraphToBigWig*."

Another way to obtain a coverage file in bigWig format directly from the BAM file is through the *deeptools* software suite with the following command:

```
bamCoverage -b {input BAM} -o {output bigWig}
```

3.4 Peak Calling

Peak calling refers to a computational technique employed to detect regions in the genome enriched with aligned reads, which is typically observed following a CUT&Tag or, more generally, a ChIP-Seq experiment (Fig. 1, step D2). The distributions of the reads are assessed using statistical measures and compared against a background to determine if the enrichment site is likely to be a real binding site.

Peak calling algorithms are critical for identifying genome-wide enrichment of aligned reads and distinguishing peaks of modified histones from noise and artifacts. However, these algorithms can be vulnerable to mistaking background signals as peaks, particularly when the background is low.

Sparse Enrichment Analysis (SEACR) [24] is a peak calling algorithm explicitly developed for CUT&RUN data, which, like CUT&Tag data, is characterized by low background noise. Other tools are listed in **Note 8**.

SEACR calculates and uses the global distribution of the background signals to calibrate a basic threshold for calling peaks. It effectively distinguishes between true and false-positive peaks with near-perfect specificity in CUT&RUN datasets and efficiently identifies enriched regions for various protein targets. SEACR takes a file of type bedgraph as input. Bedgraph files should reflect density across read pairs rather than individual reads. To generate a bedgraph from BAM files, we suggest the following steps using bedtools:

- Convert BAM files to paired-end BED files using bedtools `bamtobed` with the `bedpe` flag.
- Select the 5′ and 3′ coordinates of the read pair from the bed file to create a new BED3 file.
- Convert the BED3 file to a bedgraph using bedtools `genomecov`.

Here is an example SEACR (version 1.4) command line:

```
SEACR_1.4.sh -b {input.bedgraph} -c {params.control} -e 5 -n
norm -m stringent -o {params.outdir}
```

To improve the accuracy of peaks calling with CUT&Tag data, researchers often use a blacklist of genomic regions that are known to be problematic (*see* **Note 9**). These regions are typically excluded from the analysis as they can cause false-positive peaks due to technical artifacts or biological reasons. By excluding these problematic regions from the analysis, a CUT&Tag blacklist can improve the accuracy of peak calling and help identify biologically relevant binding sites.

3.5 Handling Replicates

Getting highly reproducible peaks requires combining replicates, which is a crucial step. Due to the significant variability that can occur in any high-throughput experiment, it is recommended to design the experiment with a minimum of 2–3 biological replicates. Even though two replicates measuring the same biological phenomenon should exhibit high consistency, this is not always the case. Therefore, it is essential to use objective metrics to evaluate the reproducibility of high-throughput assays.

A preliminary approach to evaluating the quality of CUT&Tag data and assessing the consistency between replicates is the execution of ChIPQC [25] (Fig. 1, step E). This R software package provides several functions for ChIP-Seq data quality assessment, visualization, and filtering. ChIPQC takes as input a spreadsheet that specifies the samples and the replicates involved in the experiment, along with the paths of both the BAM files and the called peak files for each of them, and generates a complete HTML report of the experiment containing quality control plots and other crucial information such as the number of reads, the peak calling results, and correlation between replicates. ChIPQC can be executed with a few R commands:

```
library(ChIPQC)
samples <- read.delim("/path/to/sample_spreadsheet.csv", sep
= "\t")
ChIP_obj <- ChIPQC(samples, annotation = "hg38")
ChIPQCreport(chipObj, reportName="report name", report-
Folder="/path/to/output_folder/")
```

This will generate the ChIPQC HTML report in the desired output directory, which will contain a summary of filtering and quality metrics generated by the ChIPQC package, such as the percentage of the reads marked as duplicates, the estimated fragment length by the cross-coverage method, and, above all, the percentage of the reads that fall within the called peaks and the percentage of the reads that fall within blacklisted regions. It further provides a heatmap of the correlation between the replicates' peak sets and a PCA analysis, enabling the assessment of the quality of samples and the consistency of replicates.

In cases of poor consistency between replicates, a user should resort to the different approaches available for handling replicates, such as identifying overlapping peak calls across replicates and analyzing differences in binding regions, or more complex methods that involve statistical testing to assess the reproducibility between replicates (*see* **Note 10**). The easiest way to find overlapping peaks is to use bedtools. In the example below, the `intersect` function is used to evaluate the file `file_peaks_replicate1` and find regions of overlap with `file_peaks_replicate2`:

```
bedtools intersect -a {file_peaks_replicate1} -b {file_peaks_-
replicate2} -wo > {replicates_overlaps.bed}
```

Where the `-a` and `-b` command-line options specify the peak files from the first and second replicates, respectively, while the `-wo` parameter tells the program that the output should include the original intervals from both input files along with the number of base pairs overlapping between them. It is also possible to apply this process to more than two replicates. In this case, it suffices to use the intersection file obtained at a previous iteration (i.e., the replicates_overlaps.bed file) as one of the input files and any additional replicates as the second input file (for an example *see* **Note 11**). This procedure returns a file in BED format containing the peaks common to both replicates, featuring eight columns specifying the chromosome, start/end positions of the interval, the name, the score, and the strand (if provided) of the intervals of both the input files.

These reproducible peaks can then be used to conduct further downstream analyses, such as motif discovery, gene annotation, or differential binding analysis.

3.6 Gene Set and Pathway Enrichment Analyses

After peak calling, we obtain a file containing a list of peak coordinates. These can be annotated with information about the nearest genes, considering their distance to the transcription start sites (TSSs) using the annotatePeak function of the ChIPseeker R package [26] (Fig. 1, step F).

```
# Load the ChIPseeker package
library(ChIPseeker)

# Human, ChIPseeker supports wide variety of species
library(TxDb.Hsapiens.UCSC.hg38.knownGene)

peakAnnoList <- lapply(samplefile.bed,
 annotatePeak,
 TxDb=txdb,
 tssRegion=c(-1000, 1000))
```

where `samplefile.bed` is the bed file of peaks, `TxDb` is a database of known genes generated from UCSC relative to the *hg38* human reference genome (imported as `TxDb.Hsapiens.UCSC.hg38.`

knownGene), and tssRegion refers to a genomic region that encompasses the TSS of a gene and its immediate upstream and downstream flanking regions.

Therefore, it is possible to conduct an in silico functional enrichment analysis to find common biological patterns among the genes linked to these peaks. The ability to associate peaks with nearby genes, in particular, enables the identification of biological processes and pathways that are most likely to be influenced by the binding of a transcription factor or histone mark. The application of these techniques can offer insightful information about the regulatory mechanisms underlying gene expression and, for instance, can assist in identifying potential therapeutic targets for illnesses brought on by the dysregulation of these mechanisms. This task can be accomplished by resorting to various bioinformatics tools, including Gene Ontology (GO) analysis, pathway enrichment analysis, and motif analysis. Here, we resort to the enrichGO function of the clusterProfiler [27] package to obtain the GO semantic terms significantly associated with a list of genes.

```
# Load the clusterProfiler package
library(clusterProfiler)

# Genome-wide annotation for Human
library(org.Hs.eg.db)

ego <- enrichGO(gene = entrez,
 keyType = "ENTREZID",
 OrgDb = org.Hs.eg.db,
 ont = "BP",
 pAdjustMethod = "BH",
 qvalueCutoff = 0.05,
 readable = TRUE)
```

where entrez is a list of Entrez Gene identifiers; BP specifies the ontology subtype (Biological Process, in this case); BH (Benjamini-Hochberg) is the chosen procedure to control the multiple testing problem, used to control the false-positive rate when conducting hypothesis tests on multiple comparisons; and qvalueCutoff is a cutoff threshold chosen for the resulting q-value. If readable is set to TRUE, the input gene identifiers will be converted to gene symbols.

The pathway enrichment analysis can be performed with the enrichKEGG function of the clusterProfiler package with the following command:

```
# Interface for querying SQLite-based annotation packages
library(AnnotationDbi)
```

```
ekegg <- enrichKEGG(gene = entrez,
 organism = 'hsa',
 pvalueCutoff = 0.05)
```

where hsa is a parameter that specifies the human organism.

The enrichGO function will retrieve GO terms enriched by a specified set of genes, while enrichKEGG will retrieve KEGG pathways enriched by the genes. Both produce a table containing information about the enriched gene sets, including the names of the genes and the *p*-values and adjusted *p*-values of the statistical tests behind the overall functional enrichment procedure. For enrichGO, additional columns may include the GO subtype (e.g., Biological Process), while enrichKEEG may include the KEGG pathway IDs.

It is possible to save a resulting enrichment table using the following commands in R:

```
summary <- data.frame(ego)
write.csv(summary,"/path/summary.csv")
```

where ego is the result of the enrichGO function, as described above.

3.7 Integrating CUT&Tag with RNA-Seq

Cells employ a vast array of intricate mechanisms to regulate gene expression. The sophisticated machinery regulating gene expression can vary depending on the cellular conditions. This complex system results in the precise increase or decrease of specific gene products, highlighting the nuanced nature of gene regulation. Despite its usefulness, RNA-Seq data analysis alone cannot completely elucidate the intricate mechanisms that govern gene expression regulation. Likewise, ChIP-Seq and its associated techniques are also insufficient for this purpose.

The integrative analysis of RNA-Seq and ChIP-based data has, in fact, many interesting applications. For instance, comparing histone modification states with quantitative gene expression can lead to the discovery of new enhancer regions, and the expression and simultaneous occurrence of different modifications at a gene can reveal gene regulation dynamics along a developmental trajectory.

Efforts to quantitatively integrate available RNA-Seq and CUT&Tag data of histone modifications in various conditions are crucial to improving our understanding of the role of epigenetics in gene regulation. RNA-Seq and CUT&Tag-Seq can be analyzed independently, and then the differentially expressed genes can be compared with the target genes from CUT&Tag using simple overlap metrics.

For these genes, binding evidence is displayed, and a plot of expression changes taken from the RNA-Seq data is placed next to it. Rather than a direct comparison, an alternative approach for integrating gene expression and peak data is to utilize specialized tools such as BETA (Binding and Expression Target Analysis) [28] and IntePareto [29] (Fig. 1, step G). The BETA tool has an advantage over other "nearby gene" (*see* **Note 12**) analyses as it can identify target genes for proteins that bind enhancer regions. The usage command is as follows:

```
BETA -p peaks.bed -e gene_exp.diff -b hg19_CTCF_bound.bed -k R
-g hg38.refseq
```

Where the main arguments are as follows: the peak file in bed format (`-p`), the differential expression file (`-e`), the kind of expression file (`-k`), the conserved CTCF binding sites boundary file in bed format (`-b`), the genome file to search refGenes (`-g`).

4 Notes

1. FASTQC generates a set of reports and graphs that summarize different quality metrics. The main components of the FASTQC output are as follows:
   - *Basic Statistics*: Providing general statistics about the dataset, including the total number of sequences, the length of sequences, GC content, and other summary metrics. It provides an overview of the dataset's composition.
   - *Per-Base Sequence Quality*: This graph displays the quality scores for each base position in the reads. The *y*-axis represents the quality score (Phred score), and the *x*-axis represents the position in the read. Higher scores indicate higher confidence in the base call.
   - *Per-Sequence Quality Scores*: This graph provides a summary of the quality scores for all sequences in the dataset. It shows the distribution of average quality scores across all reads.
   - *Per-Base Sequence Content*: This graph shows the proportion of each nucleotide (A, T, G, and C) at each position in the reads. It helps identify biases such as overrepresented or underrepresented nucleotides at specific positions. Skewed patterns might indicate adapter contamination, biases in PCR amplification, or other technical issues.
   - *Per-Sequence GC Content*: This graph illustrates the distribution of GC content across all sequences in the dataset. It helps detect any systematic biases or contamination issues related to GC content. Unusual or unexpected peaks in the graph may warrant further investigation.
   - *Sequence Duplication Levels*: This graph shows the level of duplication in the dataset. High duplication rates may

indicate PCR artifacts or library preparation issues. Sequencing biases can affect downstream analyses, so it is important to evaluate duplication levels.

- *Overrepresented Sequences*: This section lists any sequences that appear to be overrepresented in the dataset. These sequences may represent contaminants, adapter sequences, or artifacts. Identifying and removing such sequences is crucial for accurate downstream analysis.
- *Adapter Content*: This section analyzes the presence of adapter sequences in the data. Adapters are short DNA sequences used in library preparation and can remain in the sequencing data. The presence of adapter sequences can affect downstream analyses, so it is important to remove them if detected.

2. The reference genome in FASTA format can be obtained from various sources, depending on the organism of interest. The NCBI provides a vast collection of reference genomes for numerous species. You can access them through the NCBI GenBank or NCBI Assembly databases. Visit the NCBI website (https://www.ncbi.nlm.nih.gov/) and search for your organism of interest to find the corresponding reference genome. Here is an example of the hg38 reference genome: https://www.ncbi.nlm.nih.gov/assembly/GCF_000001405.26/.
3. The Phred format represents the quality scores using ASCII characters, where each character corresponds to a numerical quality value. The two most common Phred formats are Phred33 and Phred64.

   The Phred format of FASTQ files can be determined by examining the quality scores in the FASTQ file, and it can be identified following these steps:

   - Open the FASTQ file in a text editor.
   - Look at the quality scores, which are typically represented as a string of characters on the quality score line (usually the fourth line of each read entry in the FASTQ file).
   - Find a representative quality score character within the string.
   - Ascertain the corresponding ASCII value for that quality score character. You can use an ASCII table or an ASCII converter tool to determine the value.
   - Compare the ASCII value with the known values for Phred33 and Phred64 formats: In the Phred33 format, the ASCII values range from 33 to 73 (inclusive) for quality scores of 0 to 40. In the Phred64 format, the ASCII values range from 64 to 104 (inclusive) for quality scores of 0 to 40.

Based on the range of ASCII values observed, you can determine the Phred format of the FASTQ file. If the observed values align with the Phred33 range, the file is likely in Phred33 format. Similarly, if the observed values align with the Phred64 range, the file is likely in Phred64 format. Note that different sequencing platforms and software tools may use different Phred formats. Therefore, it is essential to verify the format of your specific FASTQ files to ensure accurate quality score interpretation for downstream analysis.

4. Alternative tools to perform processing operations in CUT&-Tag analysis are *Samtools* [30] or the *Picard* software suite (https://broadinstitute.github.io/picard/).

5. In CUT&Tag protocols, the standard pipeline recommends keeping uniquely mapped reads for downstream analysis because they provide more reliable and specific information compared to multiply mapped reads. Uniquely mapped reads align to a single location in the reference genome, providing a higher level of confidence that they represent a true signal rather than mapping artifacts or repetitive regions. By focusing on uniquely mapped reads, you can reduce the likelihood of including false-positive signals in your analysis. Including multiple mapped reads in the analysis may introduce ambiguity and noise since they can align to multiple locations in the genome. These reads may be derived from repetitive regions or regions with high sequence similarity, making it challenging to accurately attribute the signal to a specific genomic location. Filtering out multiple mapped reads can help improve the signal-to-noise ratio and enhance the specificity of the analysis.

   Uniquely mapped reads are more likely to represent biologically relevant events, such as specific binding events or chromatin accessibility at particular genomic loci, and can give you more confidence in the biological interpretation of your results. It is worth noting that the decision to keep uniquely mapped reads may depend on the specific analysis goals and the nature of the data. Ultimately, the choice to retain uniquely mapped reads in a CUT&Tag analysis pipeline is aimed at maximizing the accuracy and interpretability of the results.

6. The BED format is used to represent genomic annotations and coordinates. It consists of at least three columns: chromosome, start position, and end position.

   Additional optional columns can include information such as feature name, score, strand, and more. It is widely used for storing and exchanging information about genomic features like genes, transcription factor binding sites, and chromatin modifications.

BEDPE format (BED Paired-End): The BEDPE format is used to represent paired-end genomic alignments or interactions. It is commonly used for techniques like chromatin conformation capture (3C), Hi-C, and ChIA-PET. It consists of multiple columns: chromosome 1, start position 1, end position 1, chromosome 2, start position 2, and end position 2. Additional columns can include information such as interaction score, strand information, and other attributes. It allows for the representation of paired-end information, facilitating the analysis of chromatin interactions and structural variations.

Both formats provide structured representations of genomic data and are widely used in the field of genomics for storing, analyzing, and visualizing genomic features and interactions.

7. To convert a BED file to a BedGraph file, you will need to transform the coordinates and format the data appropriately. The required steps are as follows:

```
bedtools genomecov -bg -i {input BED} -g {chrom sizes list}
> {output bedgraph}
```

where the `-bg` option specifies the program to produce a bedgraph as an output.

Then, to produce the file containing the list of the sizes of each chromosome, a proper solution might be to run the following commands:

```
samtools faidx reference_genome.fa
awk -v OFS='\t' {'print $1,$2'} reference_genome.fa.fai >
reference_genomeFile.txt
```

8. Model-based analysis of ChIP-Seq version 2 (MACS2) [24] is a widely used peak calling algorithm for ChIP-Seq, along with other ChIP-Seq peak calling methods that are designed to address the high rate of "background noise" in ChIP-Seq data.

The MACS2 algorithm captures the influence of genome complexity to evaluate the significance of enriched regions. It was implemented for large binding regions, defined as "broad," but also for the transcription factor binding sites ("narrow" peaks).

The choice between broad and narrow peak calling depends on the characteristics of the biological signal you are interested in and the specific experimental context.

Narrow peak calling is used when you are interested in identifying well-defined, localized binding events as transcription factor binding sites, which tend to have sharp and distinct peaks. Peaks are characterized by a small genomic region with a high signal intensity.

Broad peak calling is used to identify larger, more diffuse regions of enrichment, often associated with histone modifications such as H3K27me3 or H3K9me3, which can spread over larger genomic regions. Broad peaks are characterized by a wider genomic region with a more gradual signal intensity profile.

The choice between narrow and broad peak calling depends on the specific biological question, the type of signal being investigated, and the resolution of the data. In some cases, it may be necessary to analyze both narrow and broad peaks to capture the full range of genomic features and regulatory elements.

GoPeaks [26] is another Peak caller that has been developed for histone modification CUT&Tag data. The tool operates in five general steps. First, the genome is partitioned into small intervals. Second, the number of aligned reads contained within each bin is calculated. Next, GoPeaks utilizes a binomial distribution to assess whether the bin counts significantly differ from the genome-wide distribution of aligned reads in bins containing more than 15 counts. Next, bins with a very high count are selected. Finally, significant bins located within 150 bp of each other are merged into peaks. GoPeaks can effectively identify peaks of histone modification binding from CUT&Tag data in a genome-wide manner.

Currently, only one pipeline is available for processing CUT&RUN and ChIP-Seq data, which includes both the MACS2 and SEACR peak callers. The pipeline, known as CUT&RUNTools [27], provides a feasible alternative for analyzing CUT&TAG data. CUT&RUNTools is a pipeline designed to analyze paired-end sequencing read FASTQ files. To extract useful information, it goes through several analytical processes, including trimming adapter sequences, aligning reads to the reference genome, identifying peaks, estimating the cut matrix at a single-nucleotide resolution, performing de novo motif searching [28], carrying out motif footprinting analysis, identifying direct binding sites, and visualizing the data.

The pipeline generates an aggregate footprint that characterizes chromatin-associated protein binding. It also provides binding log-odds values that can be used to identify direct binding sites at the individual motif level. In version 2.0, CUT&RUNTools include new features specially designed for CUT&Tag experiments. Both bulk and single-cell data can be processed, analyzed, and interpreted.

The tool requires a JSON configuration file that contains a list of paths and parameters for the tools used in the pipeline.

9. In the context of CUT&Tag experiments, a "blacklist" refers to a set of genomic regions that are known to produce artifacts or exhibit nonspecific binding. These regions are often excluded during data analysis to improve the accuracy and specificity of the results. Blacklists can help mitigate potential false-positive signals and reduce noise in the analysis. The specific regions included in a blacklist can vary depending on the organism, cell type, and experimental conditions. Common types of regions found in a blacklist for CUT&Tag experiments may include the following:

   - Repetitive DNA sequences, such as transposable elements or satellite DNA, which can lead to nonspecific binding of the restriction enzyme or antibody used in the CUT&Tag experiment.
   - Highly expressed or active genes, which can produce high levels of background signal and interfere with the identification of actual binding sites.
   - Regions with low mappability, such as those with high GC content or repetitive sequences, which can make it difficult to accurately map sequencing reads and result in false positives.

   Blacklists are often generated and provided by research consortia or bioinformatics resources specific to the organism or experimental context. These resources curate and provide standardized blacklists based on known artifacts and common problematic regions. The commonly used sources where users can find blacklists are as follows:

   - ENCODE Project (https://www.encodeproject.org/)
   - UCSC Genome Browser (https://genome.ucsc.edu/)
   - ChIP-Atlas (https://chip-atlas.org/)
   - Fantom Consortium (https://fantom.gsc.riken.jp/)

10. The tool intervene [25] is an alternative to bedtools, which allows for both the intersection and visualization of multiple gene or genomic region sets. Intervene outputs a BED file containing peaks both replicates share.

    The concordance of peak calls can also be evaluated using the Irreproducibility Discovery Rate (IDR) framework [26], available at https://github.com/nboley/idr. It is a popular method that compares two ranked lists of regions or peaks and assigns values to reflect their reproducibility. The underlying idea behind the IDR framework is that when two replicates measure the same biological phenomenon, the most significant peaks that represent actual signals should show high consistency between replicates. In contrast, peaks with low significance that likely represent noise should show low consistency.

The consistency between a pair of ranked lists that contain both significant and insignificant findings can be plotted to observe a transition in consistency. This transition in consistency is used as an internal indicator of the shift from signal to noise and provides an estimate of the number of peaks that have been accurately identified. The IDR approach relies on ranks and does not require calibrated input signals or a specific fixed scale.

There are three main steps to the IDR:

- Evaluate peak consistency between true replicates.
- Evaluate peak consistency between pooled pseudo-replicates.
- Evaluate self-consistency for each replicate.

The IDR tool typically outputs both a summary and a file of peaks. The summary file includes some statistics about the IDR analysis, including the number of peaks called in each replicate, the number of reproducible peaks, and the IDR threshold. The peaks file, on the other hand, contains the reproducible peaks identified by the IDR analysis and typically includes ten columns: chromosome, start/end position, a unique identifier for the peak, the score assigned to a peak, the strand and the signal value of the peak, the *p*-value and the *q*-value assigned to the peak, and the summit position of the peak; it ranges from 0 to 1, where a lower IDR score indicates higher reproducibility. The IDR score is calculated as the ratio of the number of reproducible peaks (those that appear in both replicates) to the total number of peaks called in either replicate. Typically, an IDR threshold is chosen based on the desired level of reproducibility. In general, a common threshold for IDR is around 0.05 or 0.1, meaning that peaks with an IDR score below this threshold are considered reproducible.

11. Continue the process for more replicates: Repeat the procedure for each additional replicate you want to include in the analysis. Use the previously generated intersection file as one of the input files in each iteration.

```
bedtools intersect -a {replicates_overlaps.bed} -b {file_
peaks_replicate3} -wo > {replicates_overlaps_123.bed}
```

12. The analysis of ChIP-Seq data employs two distinct approaches to identify potential target genes for transcription factors or other DNA-binding proteins: the BETA (Binding and Expression Target Analysis) tool and the "nearby gene" approach.

    The BETA tool is a computational method specifically designed for integrative analysis of ChIP-Seq data and gene expression data. It aims to discover statistically significant relationships between genomic regions (peaks) and changes in gene expression. BETA considers both the proximity of peaks to genes and the correlation between peak intensity and gene expression levels, quantifies the enrichment of peaks near genes using a statistical model, and assesses the significance of the associations. It incorporates multiple statistical tests to identify target genes and provides a ranked list based on the strength of the association.

The "nearby gene" approach is a simpler and more commonly used method to assign potential target genes based on the proximity of ChIP-Seq peaks to nearby genes.

In this approach, the nearest gene(s) to each peak are identified based on genomic distance. Peaks located within a certain distance threshold, typically within a few kilobases upstream or downstream of a gene's transcription start site (TSS), are considered potential target genes. This approach assumes that regulatory elements located in proximity to genes are more likely to influence their expression.

While the "nearby gene" approach is straightforward and computationally efficient, it does not account for the correlation between peak intensity and gene expression. It solely relies on genomic proximity and may result in assigning multiple nearby genes to a single peak or overlooking distal regulatory interactions. In contrast, the BETA tool provides a more sophisticated statistical analysis by considering both peak-gene proximity and the correlation between peak intensity and gene expression, which can help identify target genes with stronger evidence of regulatory interactions.

The choice between these approaches depends on the specific research question, the complexity of the regulatory landscape being studied, and the desired level of statistical rigor and sensitivity in the analysis.

References

1. Zentner GE, Henikoff S (2014) High-resolution digital profiling of the epigenome. Nat Rev Genet 15:814–827
2. Policastro RA, Zentner GE (2018) Enzymatic methods for genome-wide profiling of protein binding sites. Brief Funct Genomics 17:138–145
3. Teytelman L, Thurtle DM, Rine J et al (2013) Highly expressed loci are vulnerable to misleading ChIP localization of multiple unrelated proteins. Proc Natl Acad Sci USA 110:18602–18607
4. van Steensel B, Delrow J, Henikoff S (2001) Chromatin profiling using targeted DNA adenine methyltransferase. Nat Genet 27:304–308
5. Zentner GE, Kasinathan S, Xin B et al (2017) Corrigendum: ChEC-seq kinetics discriminates transcription factor binding sites by DNA sequence and shape in vivo. Nat Commun 8:15723
6. Skene PJ, Henikoff S (2017) An efficient targeted nuclease strategy for high-resolution mapping of DNA binding sites. Elife 6:e21856
7. Skene PJ, Henikoff JG, Henikoff S (2018) Targeted in situ genome-wide profiling with high efficiency for low cell numbers. Nat Protoc 13:1006–1019
8. Kaya-Okur HS, Wu SJ, Codomo CA et al (2019) CUT&Tag for efficient epigenomic profiling of small samples and single cells. Nat Commun 10:1930

9. Mezger A, Klemm S, Mann I et al (2018) High-throughput chromatin accessibility profiling at single-cell resolution. Nat Commun 9:3647
10. Zheng GXY, Terry JM, Belgrader P et al (2017) Massively parallel digital transcriptional profiling of single cells. Nat Commun 8:14049
11. Kaya-Okur HS, Janssens DH, Henikoff JG et al (2020) Efficient low-cost chromatin profiling with CUT&Tag. Nat Protoc 15:3264–3283
12. Janssens DH, Otto DJ, Meers MP et al (2022) CUT&Tag2for1: a modified method for simultaneous profiling of the accessible and silenced regulome in single cells. Genome Biol 23:81
13. Giallongo S, Řeháková D, Biagini T et al (2022) Histone variant macroH2A1.1 enhances nonhomologous end joining-dependent DNA double-strand-break repair and reprogramming efficiency of human iPSCs. Stem Cells 40:35–48
14. Chiodi V, Domenici MR, Biagini T et al (2021) Systemic depletion of histone macroH2A1.1 boosts hippocampal synaptic plasticity and social behavior in mice. FASEB J 35:e21793
15. Frohlich J, Kovacovicova K, Mazza T et al (2020) GDF11 induces mild hepatic fibrosis independent of metabolic health. Aging 12: 20024–20046
16. Borghesan M, Fusilli C, Rappa F et al (2016) DNA hypomethylation and histone variant macroH2A1 synergistically attenuate chemotherapy-induced senescence to promote hepatocellular carcinoma progression. Cancer Res 76:594–606
17. Frohlich J, Mazza T, Sobolewski C et al (2021) GDF11 rapidly increases lipid accumulation in liver cancer cells through ALK5-dependent signaling. Biochim Biophys Acta Mol Cell Biol Lipids 1866:158920
18. Li H, Durbin R (2009) Fast and accurate short read alignment with Burrows-Wheeler transform. Bioinformatics 25:1754–1760
19. Langmead B, Salzberg SL (2012) Fast gapped-read alignment with Bowtie 2. Nat Methods 9: 357–359
20. Kim D, Paggi JM, Park C et al (2019) Graph-based genome alignment and genotyping with HISAT2 and HISAT-genotype. Nat Biotechnol 37:907–915
21. Tarasov A, Vilella AJ, Cuppen E et al (2015) Sambamba: fast processing of NGS alignment formats. Bioinformatics 31:2032–2034
22. Lee BT, Barber GP, Benet-Pagès A et al (2022) The UCSC Genome Browser database: 2022 update. Nucleic Acids Res 50:D1115–D1122
23. Robinson JT, Thorvaldsdóttir H, Winckler W et al (2011) Integrative genomics viewer. Nat Biotechnol 29:24–26
24. Meers MP, Tenenbaum D, Henikoff S (2019) Peak calling by Sparse Enrichment Analysis for CUT&RUN chromatin profiling. Epigenetics Chromatin 12:42
25. Carroll TS, Liang Z, Salama R et al (2014) Impact of artifact removal on ChIP quality metrics in ChIP-seq and ChIP-exo data. Front Genet 5:75
26. Wang Q, Li M, Wu T et al (2022) Exploring epigenomic datasets by ChIPseeker. Curr Protoc 2:e585
27. Yu G, Wang L-G, Han Y et al (2012) clusterProfiler: an R package for comparing biological themes among gene clusters. OMICS 16:284–287
28. Wang S, Sun H, Ma J et al (2013) Target analysis by integration of transcriptome and ChIP-seq data with BETA. Nat Protoc 8: 2502–2515
29. Cao Y, Kitanovski S, Hoffmann D (2020) intePareto: an R package for integrative analyses of RNA-Seq and ChIP-Seq data. BMC Genomics 21:1–9
30. Danecek P, Bonfield JK, Liddle J et al (2021) Twelve years of SAMtools and BCFtools. Gigascience 10:giab008

Chapter 14

Single-Cell Histone Modification Profiling with Cell Enrichment Using sortChIC

Helena Viñas Gaza, Vivek Bhardwaj, and Peter Zeller

Abstract

Histone post-translational modifications (PTMs) influence the overall structure of the chromatin and gene expression. Over the course of cell differentiation, the distribution of histone modifications is remodeled, resulting in cell type–specific patterns. In the past, their study was limited to abundant cell types that could be purified in necessary numbers. However, studying these cell type–specific dynamic changes in heterogeneous in vivo settings requires sensitive single-cell methods. Current advances in single-cell sequencing methods remove these limitations, allowing the study of nonpurifiable cell types. One complicating factor is that some of the most biologically interesting cell types, including stem and progenitor cells that undergo differentiation, only make up a small fraction of cells in a tissue. This makes whole-tissue analysis rather inefficient. In this chapter, we present a sort-assisted single-cell Chromatin ImmunoCleavage sequencing technique (sortChIC) to map histone PTMs in single cells. This technique combines the mapping of histone PTM location in combination with surface staining–based enrichment, to allow the integration of established strategies for rare cell type enrichment. In general terms, this will enable researchers to quantify local and global chromatin changes in dynamic complex biological systems and can provide additional information on their contribution to lineage and cell-type specification in physiological conditions and disease.

Key words Single-cell ChIC-seq technique, Chromatin, Histone modifications, Epigenetics, Sequencing data, FACS sorting

1 Introduction

Epigenetics has been defined as the study of heritable alterations in gene expression that are not caused by changes in the DNA sequence. Epigenetic regulators include DNA methylation, noncoding RNAs, histone variants, and histone post-translational modifications (PTMs). Our understanding of the regulation of gene expression by histone PTMs is rapidly evolving. For a long time, the only technique to assess histone modifications along the genome has been Chromatin Immunoprecipitation (ChIP) [1]. In ChIP experiments, the DNA is captured in its chromatin complex using

Franziska Greulich (ed.), *Chromatin Immunoprecipitation: Methods and Protocols*, Methods in Molecular Biology, vol. 2846, https://doi.org/10.1007/978-1-0716-4071-5_14,

formaldehyde cross-linking. Afterward, cells are lysed and DNA is randomly fragmented into pieces. Chromatin-DNA complexes are then recovered by immunoprecipitation with a PTM specific antibody and can be ligated to adapters for sequencing (ChIP-seq) [2, 3]. The main limitations of this technique are that it requires large amounts of input material, it highly depends on clean enrichment strategies to recover cell type–specific differences, and due to an uneven genome recovery, input controls are necessary for meaningful interpretation.

Recent advancements in the sensitivity of protein-DNA mapping techniques have led to an expansion of several methods to study epigenetics at the single-cell level. Generally speaking, we can distinguish between three types of single-cell chromatin profiling techniques: ChIP based, protein A-micrococcal nuclease (pA-MN) based, and protein A-Tn5 transposase (pA-Tn5) based.

Recent ChIP-based techniques like sc-ChIP-seq [4] and ht-sc-ChIP-seq [5] combine microfluidics with enzymatic fragmentation and single-cell DNA barcoding before pooled cells are subjected to immunoprecipitation. Even though these techniques allow for high numbers of cells to be profiled, they depend on custom-built microfluidic devices. They are also less suitable for rare cell types, as cell enrichment can only be performed at the very beginning of the experiment.

The alternatives to bypass the immunoprecipitation step are antibody tethering of either pA-MN or pA-Tn5, which recover DNA fragments based on in situ chromatin cleavage or tagmentation, respectively. Protocols using pA-MN-based targeted fragmentation are scChIC-seq [6], iscChIC-seq [7], and the protocol described here, sortChIC (Fig. 1) [8]. Although MN-based fragments need further processing, before barcoded adaptors can be ligated, the minimal sequence preference of MN allows for high sensitivity per cell. Thus, it is especially suitable for rare cell populations. PA-Tn5 transposase-based options include ChIL-seq [9], scCUT&Tag [10, 11], CoBATCH [12], and autoCut&Tag [13]. However, Tn5 intrinsic DNA affinity demands more stringent targeting conditions, reducing the sensitivity of pA-Tn5-based methods compared to both other approaches. On the other hand, the ability of Tn5 to directly integrate barcoded adaptors allows for 10X-based adaptations, making them the preferred choice for organism-wide studies.

Before the start of the protocol (Fig. 1), the tissue of interest has to be properly dissociated into single cells. Before fixation or permeabilization, cell surface markers can be stained to allow cell-type enrichment during the experiment. The samples can then be fixed and/or permeabilized before they are incubated with a histone modification antibody of interest overnight (8–12 h). The next day, the unbound antibody is washed away before the fusion protein pA-MN is targeted to the bound antibodies in its inactive

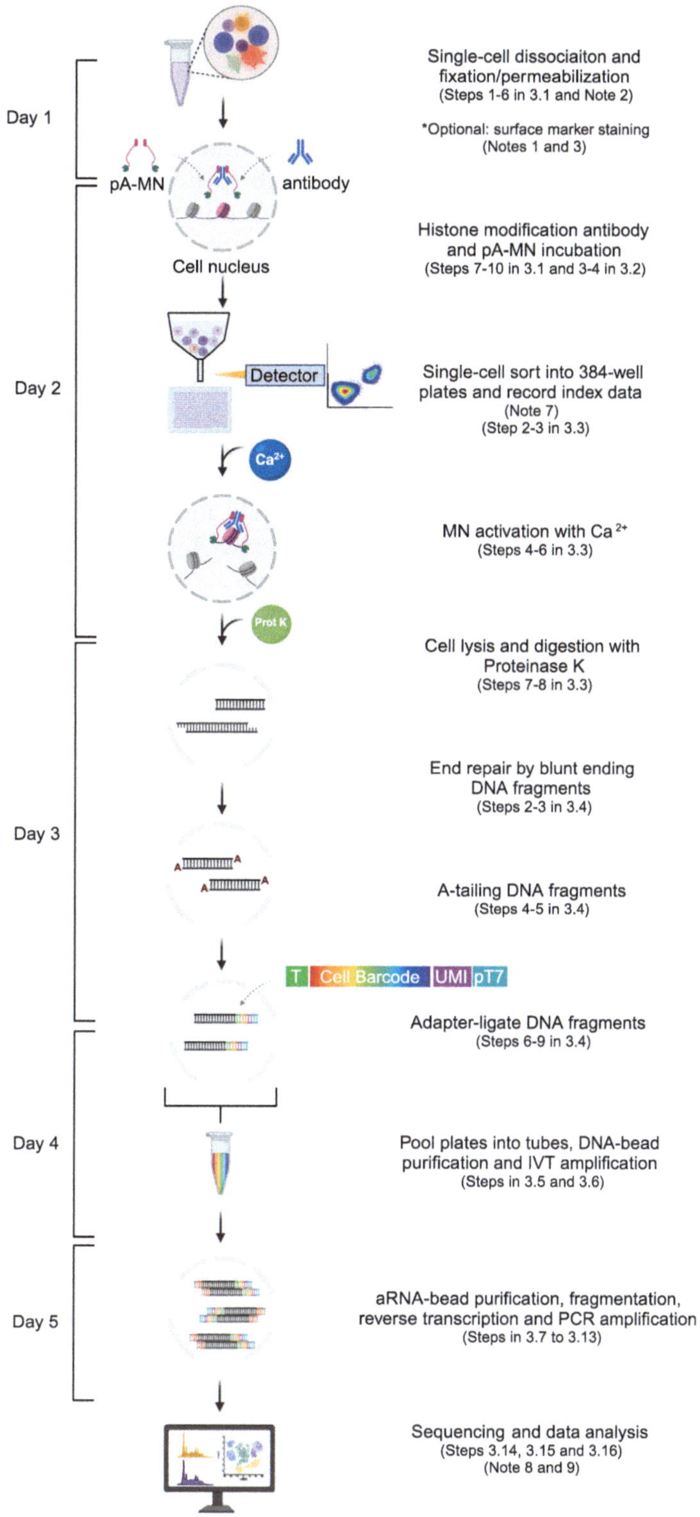

Fig. 1 Overview of the main sortChIC protocol steps. Indicated steps and notes correlate to the relevant sections of the experimental procedure. (Created with BioRender.com)

form and the DNA is stained with Hoechst. After one additional washing step, cells are filtered before fluorescence-activated cell sorting (FACS) into 384-well plates. After sorting, the fusion protein is activated by the addition of calcium and the digestion is performed for 30 min at 4 °C. During this incubation, the tethered MN will digest the genome in the vicinity of the antibody-targeted histone PTM. After 30 min, digestion will be stopped, cells will be lysed, and fragments released. Following digestion, the generated DNA fragments are further processed in three steps at the plate level. First, fragments are end-repaired to generate blunt ends. Second, fragments are A-tailed to prevent religation before forked adapters are ligated overnight. Plates can be now pooled into tubes for easier continuation. The pooled material is purified using DNA beads and linearly amplified overnight during in vitro transcription (IVT). The amplified RNA (aRNA) is bead purified and heat fragmented before yield and quality are controlled by RNA bioanalyzer. The fragmented aRNA is reverse transcribed and amplified by PCR. Final PCR libraries are further purified twice before concentration, size distribution, and quality are assessed by Qubit and high-sensitivity DNA bioanalyzer.

The paired-end FASTQ files obtained from sequencing of the sortChIC libraries are processed to remove adapters (while preserving the cell barcodes and unique molecular identifier (UMIs) and mapped to a suitable reference genome. After mapping, the duplicate fragments (resulting from the IVT and PCR steps) are removed from each cell based on their mapping position and UMI, and the read counts are aggregated into genomic bins/regions of interest, creating the feature x cell table for further downstream analysis. These bioinformatics steps are summarized in Fig. 2 and can be automatically performed using our scChICflow pipeline (https://github.com/bhardwaj-lab/scChICflow).

Finally, the output table can then be used for further downstream analysis such as clustering (or grouping of cells), annotation of cells, and identifying enriched features within each group of cells. Such analysis helps to understand the cell type–specific distribution of histone PTMs.

As mentioned in the introduction, new single-cell chromatin profiling methods greatly expanded our possibilities from the bulk era, allowing the convenient and accurate recovery of cell type–specific chromatin changes in any tissue or organism. From these new targeted fragmentation-based methods, Tn5-based approaches have shown the largest promise when profiling active chromatin states in large cell numbers. SortChIC and other MN-based approaches, on the other hand, proved to benefit from minimal accessibility impact, allowing highly sensitive profiling of also repressive chromatin states in rare cell types.

The FACS integration of sortChIC allows for additional flexibility. By using surface marker stainings, specific cell types can be sorted from one common sample preparation. This allows for the

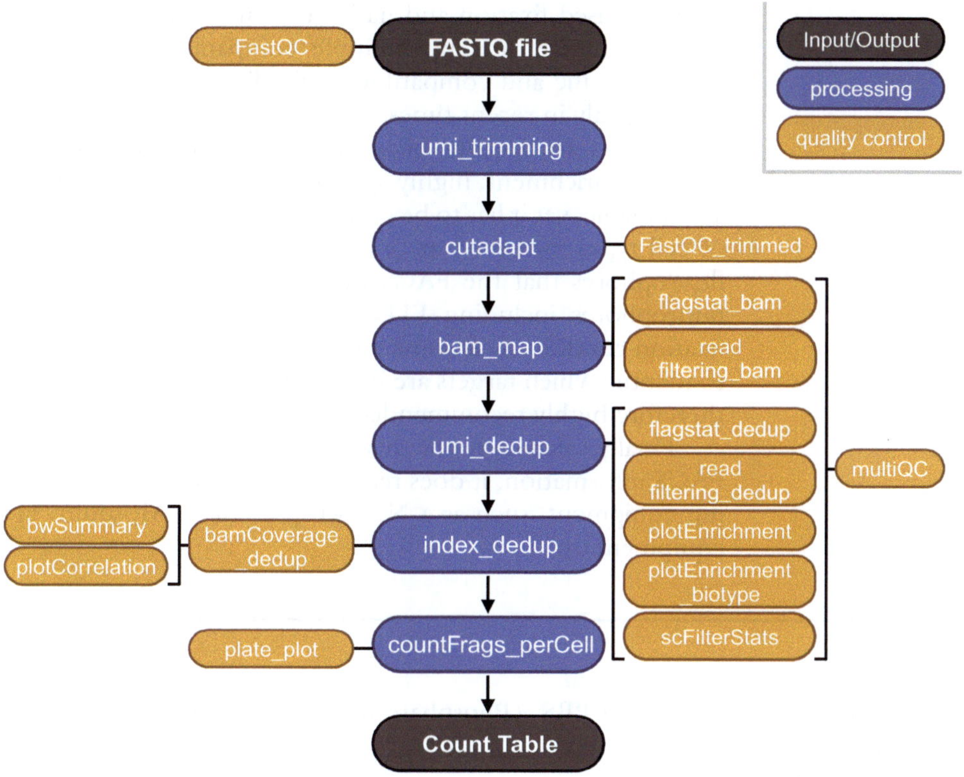

Fig. 2 Processing workflow for sortChIC data. Black: Input and Output file. Violet: Processing steps (a description of the individual steps can be found in the Subheading 3.14). Yellow: Quality Control steps. (Created with BioRender.com)

enrichment and cell type identification in single-cell experiments, but can also be used to prepare highly comparable bulk samples. The recorded single-cell index data can further be used as additional information about the abundance of the selected proteins during data analysis. Using general cell stains such as CFSE (carboxyfluorescein succinimidyl ester), multiple samples of an experiment can be pooled before processing and separated later by FACS, to ensure comparable conditions between samples. Similarly, stained cell line cells can be used as "carrier material" in combination with very small samples, such as patient biopsies. This allows the processing of small samples with little material loss while keeping experimental conditions (cell-to-antibody ratio) constant.

Finally, samples can be maintained at −80 °C which simplifies experimental planning and can minimize experimental variance from the individual animal and cell isolation.

One of the major limitations of all current single-cell chromatin profiling methods is the dependency on high-quality antibodies, which might require testing and further optimization. These optimizations might include different types of fixatives compared to ethanol-based fixation and native conditions presented here. Due to continued efforts by multiple companies, the availability of highly specific and compatible antibodies has been growing even more rapidly in recent times.

Thanks to decades of work into cell populations and their efficient enrichment, highly specific surface antibody combinations are available, yet it has to be pointed out that the maximum number of cell types measured at once will be limited by the number of fluorophores that the FACS can distinguish. Furthermore, some fluorophores including FITC (Fluorescein) are affected by the fixation conditions, as some antigens can be as well. Staining trials to identify which targets are best stained before or after fixation are therefore highly recommended.

Finally, although the plate-based setup is essential to keep FACS information, it does require some specialized liquid dispensing equipment, such as a Nanodrop II and a Mosquito HTP (*see* equipment).

2 Materials

2.1 Single-Cell Isolation, Fixation, and Histone Modification Antibody Incubation

1. 1x D-PBS (Phosphate-buffered Saline without Ca^{2+} and Mg^{2+}).
2. CellTrace available in multiple colors (Invitrogen).
3. EDTA-free protease inhibitor tablets (e.g., Sigma Aldrich).
4. 1 M HEPES-KOH pH 7.5.
5. 5 M NaCl.
6. Spermidine solution (pure).
7. Tween-20.
8. Saponin.
9. 0.5 M EDTA pH 7.5.
10. Ethanol (EtOH).
11. Histone modification antibody (Table 1).
12. Basic wash buffer (WB): 47.5 mL nuclease-free water (water), 1 mL 1 M HEPES-KOH pH 7.5, 1.5 mL 5 M NaCl, 3.6 μL pure Spermidine solution, 0.05% (v/v) Tween20 or Saponin and 1 EDTA-free protease inhibitor cocktail tablet per 50 mL.
13. Wash buffer 1 (WB1): Basic wash buffer supplemented with 4 μL/mL of 0.5 M EDTA pH 7.5.
14. 0.5 mL protein low-binding tubes.
15. 50 mL Falcon tubes.
16. Cooling centrifuge with swing buckets for 0.5 mL tubes.
17. Vortex mixer.

Table 1
Tested and recommended antibodies for single-cell sortChIC experiments

Order number (Manufacturer)	Targeted histone modification	Concentration	Physiological relevance
ab8895 (Abcam)	H3K4me1 (active)	1:200	Gene body, enhancers
MA5-11199 (Thermo Fisher)	H3K4me3 (active)	1:400	Transcription start site (TSS)
MA5-33395 (Thermo Fisher)	H3K9me3 (repressive)	1:100	Constitutive heterochromatin, AT-rich gene-poor regions
9733S (NEB)	H3K27me3 (repressive)	1:100	Facultative heterochromatin, GC-rich gene dense regions

The first column shows the reference number and in () the manufacturer. The second column is which histone modification is targeted and in () if these are active or repressive marks. The third column is the concentration recommended for single-cell sortChIC experiments that yield the best results. The fourth column, explains in which genetic context these marks are usually present

2.2 Protein A-MN Treatment, FACS Sorting, Activation, and Stop of the MNase (MN)

1. Wash buffer 2 (WB2): Basic wash buffer without the EDTA-free protease inhibitor tablet.
2. 0.5 μg/mL Hoechst 34580.
3. pA-MN fusion protein: *see* **Note 1** on pA-MN purification.
4. 70 μm cell strainer.
5. 5 mL FACS tubes with a cell-strainer cap.
6. FACS sorter able to sort into 384-well plates.
7. Cooling centrifuge for 384-well plates.
8. Roller.

2.3 FACS Sorting and Activation of the MNase (MN)

1. 0.5 M EGTA.
2. 1 M $CaCl_2$.
3. 10% (v/v) NP-40: Prepare 10% (v/v) Igepal/NP40 in nuclease-free water.
4. 20 mg/mL proteinase K (prot K, e.g., Invitrogen).
5. Stop solution: Make a master mix with 67 μL nuclease-free water, 8 μL 0.5 M EGTA, 15 μL 10% (v/v) NP40 and 20 mg/mL proteinase K per reaction.
6. Hard-shell 384-well PCR plates, thin wall (has to correspond to the calibration of the liquid handling robots).
7. 0.5 mL DNA low-binding tubes.
8. 384-well cool blocks.
9. Silverseal sealer, aluminum.

10. For single-cell reagent addition: liquid handling robot able to distribute 50–1000 nL per well of a 384-well plate (e.g., Nanodrop II).
11. 384-well plate-compatible thermal cycler (e.g., GeneAmp and/or ProFlex PCR System 9700).
12. BD FACS Software 1.2.0.142.

2.4 End-Repair, A-Tailing and Adapter Ligation of DNA Fragments

1. 5000 units/mL DNA Polymerase I, Large (Klenow) Fragment (e.g., NEB).
2. 10,000 units/mL T4 PNK (e.g., NEB).
3. 10x PNK buffer: 700 mM Tris–HCl, 100 mM $MgCl_2$, 50 mM DTT, pH 7.6 (e.g., NEB).
4. 5000 units/mL none hot start Taq polymerase (e.g., AmpliTaq 360, Thermo Fisher Scientific).
5. 400,000 units/mL T4 DNA ligase (NEB).
6. 0.1 M DTT.
7. 10 mM dNTPs.
8. 25 mM $MgCl_2$.
9. 50% (w/v) PEG8000.
10. 20 mg/mL BSA.
11. 100 mM dATP.
12. 1 M KCl.
13. 1 M $MgCl_2$.
14. 1 M Tris–HCl pH 7.5.
15. Forked-ChIC adapter: *see* **Note 2**.
16. For single-cell adaptor addition: liquid handling robot able to transfer 50 nL per well of a 384-well plate without cross-contamination (e.g., Mosquito HTS robot).

2.5 Pooling Plates

1. V-block collection plates (VBLOCK Reservoir, Click-Bio). Alternatively, the lid of a pipet tip box can be used.
2. 1.5 mL DNA low-binding tubes.

2.6 DNA-Bead Cleanups and In Vitro Transcription

1. IVT kit, MEGAscript™ T7 Transcription Kit (Invitrogen).
2. Bead binding buffer (BBB): 20% (w/v) PEG8000, 2.5 M NaCl, and 0.02% (w/v) NaN_3.
3. DNA-purification beads (Beckman coulter).
4. 80% (v/v) Ethanol.
5. Magnetic stand for 1.5 mL tubes.

2.7 DNAse Treatment and RNA Purification

1. RNA-purification beads (Beckman coulter).
2. Thermal cycler for 0.5 mL tubes (e.g., Mastercycler Nexus Flexlid).
3. Magnetic stand for 0.5 mL tubes.

2.8 Fragmentation

1. Fragmentation buffer: 200 mM Tris-acetate pH 8.1, 500 mM KOAc, 150 mM MgOAc.

2.9 RNA Cleanup and RNA Bioanalyzer

1. Bioanalyzer instrument (e.g., Agilent), with RNA pico assay.

2.10 Reverse Transcription

1. RNaseOUT (Invitrogen).
2. Superscript II (Invitrogen).
3. Random hexamer reverse transcription primer: GCCTTGGC ACCCGAGAATTCCANNNNNN. Prepare a 20 μM working solution.

2.11 RNase Treatment and PCR Amplification

1. 10 mg/mL RNase A (DNase and protease-free).
2. 2x PCR master mix (e.g., NEBNext High-Fidelity).
3. Illumina TrueSeq small RNA Library PCR primers (RP1).
4. 0.2 mL PCR 8 tube-strips.

2.12 Final DNA Cleanup and DNA Bioanalyzer

1. Bioanalyzer instrument (e.g., Agilent), with DNA high-sensitivity assay.
2. Qubit Fluorometer with RNA high-sensitivity assay.

2.13 Data Processing

1. Software for processing and quality control (QC) of sortChIC data are listed in Table 2. We recommend the use of *conda* (with *bioconda*) to install and run these tools (https://bioconda.github.io/).

3 Methods

3.1 Single-Cell Isolation, Fixation (1 h 45 min), and Histone Modification Antibody Incubation

Optional: surface marker antibody staining (*see* **Note 3**) or nuclei isolation (*see* **Note 4**).

1. Before starting: Prepare the basic Wash Buffer (WB). Keep at 4 °C overnight for the next day. Prepare Wash Buffer 1 (WB1) by adding 4 μL 0.5 M EDTA to 1 mL of the WB. Pre-cool absolute EtOH at −20 °C. For bulk samples, follow the same steps as for single cells (*see* **Note 5**).
2. Wash cells two times in 1.5 mL protein low-binding tubes with 500 μL D-PBS and count the cells.
3. Spin at 600× *g* for 3 min at room temperature (20–25 °C).

Table 2
List of software to process and analyze single-cell sortChIC data

Software (version)	Functions	Link
FastQC (≥0.10) [14]	fastqc	https://github.com/s-andrews/FastQC
cutadapt (≥4.0) [15]	cutadapt	https://cutadapt.readthedocs.io/en/stable/guide.html
umi_tools (≥1.1) [16]	extract, dedup	https://umi-tools.readthedocs.io/en/latest/
bwa-mem (≥0.7) [17]	bwa mem	https://bio-bwa.sourceforge.net/bwa.shtml
samtools (≥1.12) [18]	index, flagstat, view	http://www.htslib.org/doc/samtools.html
deepTools (≥3.1) [19]	estimateReadFiltering, plotEnrichment, bamCoverage, multiBigwigSummary, plotCorrelation	https://deeptools.readthedocs.io/en/develop/
sincei (≥0.2) [20]	scCountReads, scFilterStats, scClusterCells, scBulkCoverage	https://sincei.readthedocs.io/en/latest/
MultiQC (≥1.14) [21]	multiqc	https://multiqc.info/docs/

The first column is the name of the software and the version to download. The second column is a list of functions that can be run using that software. The third column is the link to the documentation of each package for more details and information

4. Resuspend the pellet in cold D-PBS at a concentration of 1 million (M) cells per 300 μL before adding 700 μL of absolute EtOH (pre-cooled at −20 °C) while vortexing gently.

   Example: 2 M cells in 600 μL D-PBS before adding 1400 μL of absolute EtOH.

5. Incubate samples at −20 °C for 1 h.

6. After fixation, wash two times with 500 μL WB1 and aliquot 0.5–1 × 10^6 cells (cell numbers should be kept identical between experiments) per 0.5 mL protein low-binding tubes. Spin at 600× *g* for 3 min at 4 °C.

   Optional: cell staining for sample pooling (*see* **Note 5**).

7. Remove the supernatant pipetting carefully without disturbing the pellet and resuspend cells in 400 μL WB1 containing the histone modification antibody in its appropriate concentration (*see* **Note 6** and Table 1) and EDTA to chelate all existing bivalent ions.

8. Tubes are incubated on a rotator overnight (8–12 h) at 4 °C in the dark. Fixed cells can be stored at −80 °C until ready to continue (*see* **Note 7**).

3.2 Protein A-MN Treatment, FACS Sorting, Activation, and Stop of the MNase (MN) (1 h 45 min)

1. Before starting: WB can be stored at 4 °C and be re-used from the previous day. Prepare Wash Buffer 2 (WB2).
2. Wash once with 500 μL of WB containing protease inhibitor to remove unbound antibody. Spin at 600× *g* for 3 min at 4 °C.
3. Remove the supernatant carefully without disturbing the pellet and resuspend samples in 500 μL of WB containing 300 ng/mL pA-MN and 0.5 μg/mL Hoechst 34580 to stain the DNA.
4. Mix for 1 h at 4 °C in the dark on a roller.
5. Spin down and resuspend in 500 μL of WB two times. Spin at 600× *g* for 3 min at 4 °C.
6. Filter samples through a 70 μm cell strainer into 5 mL FACS tubes.

3.3 FACS Sorting and Activation of the MNase (MN) (2–3 h, Depending on the Number of Plates)

1. Before starting: Have 384-well plates containing 5 μL of sterile-filtered mineral oil to prevent evaporation of the droplets ready. For bulk samples (*see* **Note 8**), prepare 0.5 mL DNA-low binding tubes with 5 μL of WB2, without protease inhibitor to allow proper digestion by Proteinase K (Prot K) later on.
2. Sort single cells in 384-well plates containing sterile-filtered mineral oil. Gate for single cells in the desired cell cycle stage based on the Hoeschst profile (Fig. 3 and *see* **Note 9**).
3. Seal the plates with aluminum foil and spin at 2000× *g* for 2 min at 4 °C. Keep plates at 4 °C until all cells are sorted.
4. After the sorting, prepare the activation solution by adding $CaCl_2$ to a final concentration of 2 mM to WB2. The subsequent small volumes are distributed using a liquid

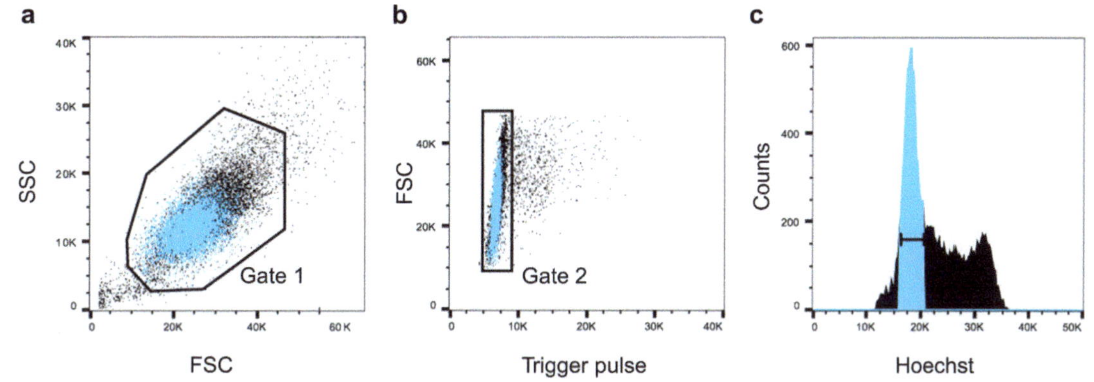

Fig. 3 FACS gating strategy for single cells in G1 phase of the cell cycle. (**a**) Dot plot of all the events in the sample. Cells are selected based on FSC (Forward Scatter) and SSC (Side Scatter) to remove debris. The desired population is shown in light blue. (**b**) Dot plot of cells passing Gate 1. Trigger pulse versus FSC is used to remove cell doublets. (**c**) DNA content histogram (Hoechst) of cells passing Gate 2. Cells are selected in the G1 phase of the cell cycle (light blue area)

handling robot such as the Nanodrop II system, and plates are sealed and spun down at 2000× *g* for 2 min at 4 °C after each dispensation.

5. Dispense 100 nL per well of Activation Solution to activate the MN, if available using 384-well cool blocks. If not, make sure to keep plates at 4 °C right after dispensation.
6. Place the plates in a 384-PCR machine set at 4 °C and let the MN digest the chromatin for 30 min.
7. Stop the reaction by adding 100 nL per well of Stop Solution (containing EGTA which chelates Ca^{2+} thus stopping the MN digestion. NP40 as detergent, together with Prot K, permanently destroys the MN and lyse the cells, leading to the release of DNA fragments). Seal and spin the plates.
8. Incubate plates on a 384-PCR machine for 20 min at 4 °C followed by 6 h at 65 °C. Heat inactivate the Prot K for 20 min at 80 °C, and hold plates at 10 °C overnight (8–12 h).
9. Pause point: Samples can be stored at −80 °C indefinitely.

3.4 End-Repair, A-Tailing, and Adapter Ligation of DNA Fragments (3 h per 2 Plates)

1. Before starting: Thaw the forked-ChIC adapter plate (5 μM). Thaw all the reagents needed for the master mix preparation.
2. Prepare end-repair master mix as in Table 3 and dispense 150 nL per well. Seal and spin the plates.
3. Incubate on a 384-well PCR machine for 30 min at 37 °C to allow blunt ending of the fragments followed by 20 min at 75 °C to heat inactivate Klenow and hold at 4 °C. This step repairs the 5′ and 3′ overhangs of the fragments as well as phosphorylates the 5′ ends for the adapter ligation later on.

Table 3
End-repair master mix

End-repair mix	Amount per well (nL)	Amount for one 384-well plate (μL)
Klenow large	2.5	3.75
10 mM dNTPs	6	9.00
T4 PNK	2.5	3.75
100 mM ATP	3.5	5.25
10x PNK buffer	35	52.50
25 mM $MgCl_2$	10	15.00
50% PEG8000	7.5	11.25
20 mg/mL BSA	1.8	2.63
Nuclease-free water	81.3	121.88

Table 4
A-tailing master mix

A-tailing mix	Amount per well (nL)	Amount for one 384-well plate (μL)
AmpliTaq 360	1	1.50
100 mM dATP	1	1.50
1 M KCl	25	37.50
50% PEG8000	7.5	11.25
20 mg/mL BSA	0.8	1.13
Nuclease-free water	114.8	172.13

Table 5
Master mix for adapter ligation

Adapter ligation mix	Amount per well (nL)	Amount for one 384-well plate (μL)
T4 DNA ligase	25	37.50
1 M Tris–HCl pH 7.5	10.5	15.75
1 M $MgCl_2$	3.5	5.25
0.1 M DTT	52.5	78.75
100 mM ATP	3.5	5.25
50% PEG8000	10	15.00
20 mg/mL BSA	1	1.50
Nuclease-free water	44	66.00

4. Prepare a master mix for A-tailing as in Table 4 and dispense 150 nL per well. Seal and spin the plates.
5. Incubate on a 384-well PCR machine for 15 min at 72 °C and hold at 4 °C (*see* **Note 10**).
6. Prepare a master mix for adapter ligation (*see* Table 5). Keep the master mix at 4 °C while dispensing the adapters.
7. Dispense two plates at a time with 50 nL of 5 μM T-tail containing forked-ChIC adapters (*see* **Note 2**) using a liquid handling robot (e.g., Mosquito HTS). Seal and spin the plates. After adapter ligation, each fragment will be distinguishable from each other containing a cell-specific barcode (BC) and a Unique Molecular Identifier (UMI).
8. Dispense 150 nL of the adapter ligation mix. Seal and spin the plates.

9. Incubate on a 384-well PCR machine for 20 min at 4 °C, followed by 16 h at 16 °C for the ligation of the adapters to the fragments. Continue with the inactivation of the ligase for 10 min at 75 °C and hold at 4 °C overnight (8–12 h).
10. Pause point: Plates can be stored at −80 °C. However, we recommend to directly proceed with the first amplification.

3.5 Pooling Plates (30 min per 2 Plates)

1. Before starting, coat the V-block collection plates with sterile-filtered mineral oil. If collection plates are not available, the lids of pipette tip boxes can be used instead.
2. Take the plates from the 384-well PCR machines or the freezer and dispense 1 μL of water per well to reduce material loss. Seal the plates and spin at 2000× *g* for 2 min at 4 °C.
3. Invert the plates onto V-block collection plates (or clean lid of a pipette tip box) and spin at 250× *g* for 1 min at 4 °C.
4. Transfer all the collected material from the V-block (liquid and oil phase) into 5 mL DNA-low binding tubes and spin at 13,000× *g* for 1 min at room temperature (20–25 °C).
5. Transfer only the liquid phase into a 1.5 mL DNA-low binding tube and spin at 13,000× *g* for 1 min at room temperature (20–25 °C).
6. Repeat **step 5** once.

3.6 DNA-Bead Cleanups and In Vitro Transcription (IVT) (1 h)

1. Before starting: Equilibrate the Ampure XP DNA beads to room temperature (20–25 °C) (DNA beads) and vortex them well until homogenous. Dilute DNA beads 1 in 8 with BBB. Freshly prepare 80% EtOH for the cleanups, as it can evaporate over time. Thereby, the 80% EtOH gets more diluted, which would wash away more DNA/RNA fragments. Equilibrate all the reagents of the MEGAscript T7 Transcription Kit at room temperature (20–25 °C) for 30 min, except for the enzyme. Vortex them well and prepare the IVT mix as described in Table 6.
2. Determine how much liquid phase has been collected. It should be between 400 and 500 μL (*see* **Note 11**).
3. Purify and size-select DNA fragments by adding diluted DNA beads at a bead-to-sample ratio of 0.8 (0.8X) to get rid of primer dimers but keep fragments larger than 200 bp. Mix well by repeated pipetting with a P1000.

   For example: If 480 μL of liquid phase has been recovered, add 384 μL of diluted DNA beads (*see* **Note 12**).
4. Incubate the samples with the beads for 15 min at room temperature (20–25 °C) to allow the DNA to bind to the beads.

Table 6
In vitro transcription (IVT) master mix

IVT mix	Amount per one reaction (μL)
ATP	2
GTP	2
CTP	2
UTP	2
T7 Buffer	2
T7 Enzyme	2

Reagents from MEGAscript T7 Transcription Kit

5. Place in a magnetic stand and allow DNA beads to accumulate. Remove the supernatant pipetting carefully without disturbing the beads.
6. Take the tubes from the magnetic stand and add 1 mL of fresh 80% EtOH. Resuspend the beads fully. Let the EtOH wash for 1 min.
7. Place the samples back in the magnetic stand and wait for the beads to accumulate again before removing the EtOH.
8. Wash once more with 1 mL of fresh 80% EtOH. This time, keep the samples in the magnetic stand and do not resuspend the beads. Let the EtOH wash for 1 min and remove it carefully.
9. Air-dry beads for 10 min in the magnetic stand with the tube lids open.

   This step is critical. Beads should appear "matte" in color, not shiny when looked under the light. Start the elution with water the latest when cracks appear on the pellet (*see* **Note 13**).
10. Resuspend in 8 μL of nuclease-free water.
11. Transfer the 8 μL of the sample with the beads (to increase the yield) into a new 0.5 mL DNA low-binding tube.
12. Add 12 μL of the IVT mix prepared as in Table 6.
13. The "clean" DNA is linearly amplified overnight on a thermal cycler as follows: 13 h at 37 °C (with lid at 70 °C) and hold at 4 °C.

3.7 DNase Treatment and RNA Purification

1. Before starting: Equilibrate DNA and RNA Clean XP beads to room temperature (20–25 °C) and vortex them well until homogenous. RNA beads should be RNAse-free. Get the fresh 80% EtOH for the cleanups.
2. Put the samples in a magnetic stand and transfer the 20 μL of material into new 0.5 mL DNA-low binding tubes.

3. Add 2 μL of Turbo DNase I (from IVT kit), to remove template DNA. Resuspend well and incubate on a thermal cycler at 37 °C for 30 min.
4. Purify and size-select the IVT product (aRNA) by adding a 0.8 sample-to-bead ratio of RNA beads and mix well. Incubate for 10 min at room temperature (20–25 °C) to allow the material to bind to the beads.
5. Put the tubes in a magnetic stand and allow the beads to accumulate.
6. Remove the supernatant by pipetting carefully without disturbing the beads.
7. Add 500 μL of fresh 80% EtOH and incubate for 1 min.
8. Remove the EtOH carefully without disturbing the beads.
9. Wash again repeating **steps** 7 and **8**.
10. Air-dry beads for 15 min in the magnet stand with the tube lids open.

    This step is critical. Beads should appear "matte" in color, not shiny when looked under the light. Start the elution with water the latest when cracks appear on the pellet.
11. Resuspend samples in 22 μL of nuclease-free water and incubate them at room temperature (20–25 °C) for 2 min.
12. Place the tubes in the magnetic stand.
13. Transfer the supernatant into new 0.5 mL DNA-low binding tubes and keep on ice.

3.8 Fragmentation

1. Add 4.4 μL of Fragmentation buffer to the supernatant and mix well by pipetting up and down.
2. Put samples on a pre-heated thermal cycler at 94 °C for 2 min to allow the fragmentation, before moving them back on ice. Do not leave samples at 94 °C longer than 2 min to prevent over-fragmentation.
3. Stop the fragmentation by transferring samples to ice and adding 2.64 μL 0.5 M EDTA.

3.9 RNA Cleanup and RNA Bioanalyzer

1. Add 23 μL of undiluted RNA beads and resuspend by pipetting up and down. Incubate for 10 min at room temperature (20–25 °C) to allow the material to bind to the beads.
2. Wash twice for 1 min with 500 μL of fresh 80% EtOH (just like in Subheading 3.7, **steps 5–9**).
3. Air-dry beads for 15 min in the magnetic stand with the tube lids open. Beads should appear "matte" in color, not shiny when looked under the light. Start the elution with water the latest when cracks appear on the pellet.

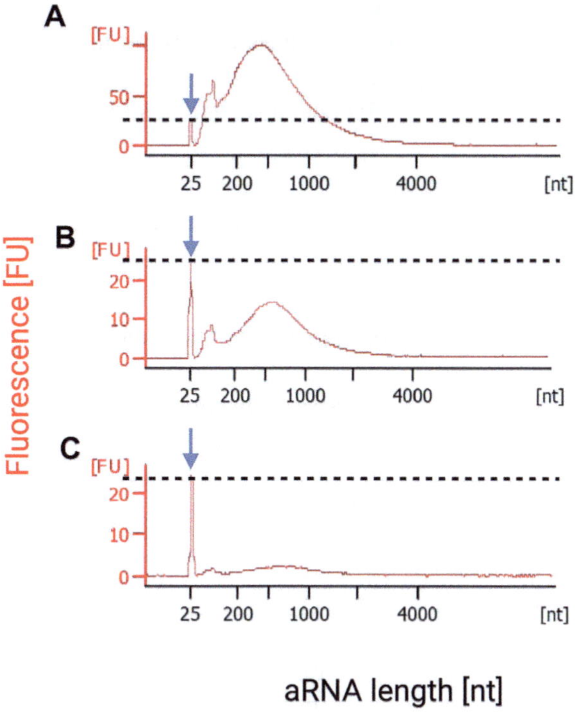

Fig. 4 Examples of aRNA bioanalyzer profiles of the fragmented aRNA. It is always advised to do a minimum of 8 PCR cycles to ensure that the final DNA library mainly consists of fully indexed material. (**a**–**c**) Show positive examples with correct size distribution (between 300 and 700 nt). Because the adaptor-to-library ratio can result in misleading Qubit quantifications, yield at this stage is estimated on the bioanalyzer. (**a**) If the material goes above the marker peak (dashed line), 8 PCR cycles are sufficient. (**b**) If the yield is around half of the marker peak, 10 PCR cycles are required. (**c**) Example with very low yield. Here, we recommend the maximum number of PCR cycles, which is 15. The gray arrow indicates the reference marker; nt: nucleotides; FU fluorescence; dashed line: reference concentration given by marker

4. Resuspend the material with 12 μL of nuclease-free water and let the samples elute for 2 min at room temperature (20–25 °C).
5. Place the tubes on a magnetic rack and transfer supernatant to new 0.5 mL DNA-low binding tubes.
6. Continue to reverse transcription with 5 μL of the sample. Keep the rest of the aRNA at −80 °C in case more PCR cycles are needed or the final steps have to be repeated.
7. Run 1 μL of the aRNA on a Bioanalyzer RNA pico Chip to assess the quality and size distribution of the fragments. Examples of RNA libraries and the appropriate amount of PCR cycles needed for each are shown in Fig. 4.

Table 7
Random hexamer master mix

Random hexamer master mix	Amount per one reaction (μL)
10 mM dNTPs	0.5
20 μM Random hexamer	1

Table 8
Reverse Transcription mix

Reverse Transcription mix	Amount per one reaction (μL)
First Strand buffer	2
0.1 M DTT	1
RNaseOUT	0.5
SuperscriptII	0.5

3.10 Reverse Transcription

1. Transfer 5 μL of fragmented aRNA to new 0.5 mL DNA-low binding tubes and add the dNTPs and the random hexamer as described in Table 7 (*see* Subheading 2.10, **item 3**) to prime the fragments for reverse transcription.
2. Mix samples well and allow the hybridization of the primer by incubating the sample at 65 °C for 5 min on a thermal cycler, before moving it to ice.
3. Add 4 μL of the Reverse Transcription mix as described in Table 8.
4. Incubate the samples on a thermal cycler, with the lid at 50 °C, for 10 min at 25 °C and 1 h at 42 °C for the production of the reverse transcript (cDNA) and hold at 4 °C.

3.11 RNase Treatment and PCR Amplification

1. Single-stranded cDNA is purified by the addition of 0.5 μL of RNase A at 37 °C for 30 min on a thermal cycler.
2. Prepare the PCR master mix according to Table 9.
3. Add 38 μL of the PCR master mix and 2 μL of the selected RPIx to the samples.
4. Mix well by pipetting and incubate as follows on a thermal cycler: Activation for 30 seconds (s) at 98 °C, amplification for 8–15 cycles (*see* **Note 14**) of (10 s at 98 °C, 30 s at 60 °C, 30 s at 72 °C), final amplification for 10 min at 72 °C and hold at 4 °C.
5. Pause point: Samples can be stored at −20 °C indefinitely or until ready to continue.

Table 9
PCR mix

PCR mix	Amount per one reaction (μL)
NEBNext High-Fidelity 2x PCR master mix	25
RNA PCR primer (RP1)	2
Nuclease-free water	11

3.12 Final DNA Cleanup and DNA Bioanalyzer

1. Add 40 μL of DNA beads to the PCR product. Mix well and incubate for 15 min at room temperature (20–25 °C).
2. Wash twice for 1 min with 500 μL of fresh 80% EtOH (just like in Subheading 3.7, **steps 5–9**).
3. Air-dry beads for 10 min in the magnetic stand with the tube lids open. Beads should appear "matte" in color, not shiny when looked under the light. Start the elution with water the latest when cracks appear on the pellet.
4. Resuspend in 30 μL of nuclease-free water and incubate at room temperature (20–25 °C) for 2 min.
5. Place tubes on a magnetic rack and transfer the supernatant to a new 0.5 mL DNA-low binding tube.
6. Add 24 μL of fresh DNA beads. Mix well and incubate at room temperature (20–25 °C) for 15 min.
7. Repeat **steps 2** and **3**.
8. Resuspend in a final volume of 7 μL nuclease-free water and elute at room temperature (20–25 °C) for 2 min.
9. Place the tubes on a magnetic rack and transfer supernatant to new 0.5 mL DNA-low binding tubes.
10. Pause point: The finished libraries can be kept at −20 °C until submission for sequencing.

3.13 Final Library Assessment and Sequencing

1. Measure the concentration of the final library by following the manual for the Qubit dsDNA HS Assay kit with 1 μL of each sample.
2. Quality and size distribution of each sample is assessed on a Bioanalyzer High Sensitivity DNA chip. Examples of successful and less successful libraries are shown in Fig. 5.
3. Dilute samples to correct molarity for sequencing. We dilute to 4 nM (*see* **Note 15**). Pool according to the desired number of reads per sample, which will depend on the library complexity which correlates with the genomic abundance of the mark used.
4. If mainly single-cell sortChIC libraries are sequenced, add around 5% PhiX control.

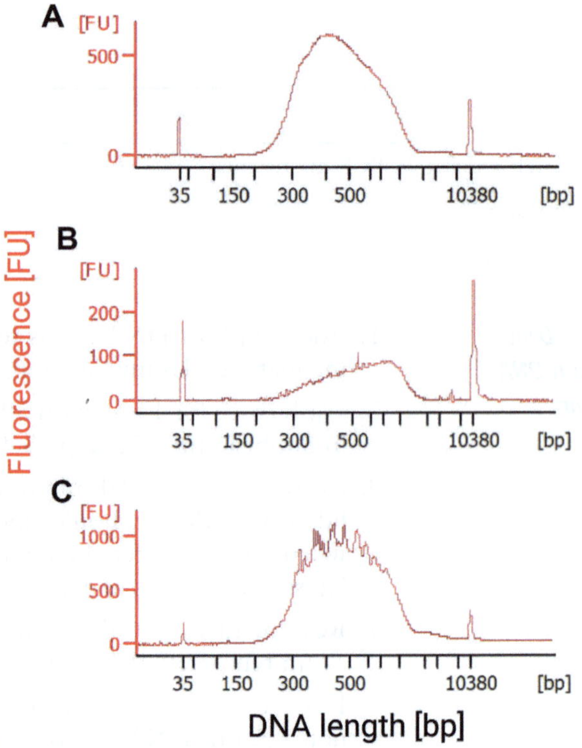

Fig. 5 Example DNA bioanalyzer plots of the final sequencing libraries. (**a**) and (**b**) show positive examples. (**a**) The final DNA library has the correct size distribution (between 300 and 800 bp) and is ready to be send for sequencing. (**b**) The final DNA library shows a slightly increased size distribution (300–1000 bp). This can be the result of targeting a rare chromatin component (ready for sequencing), or inefficient digestion of an abundant target (optimization recommended). (**c**) The final DNA library has a spiky regular profile indicating adapter multimerization, which might be caused by high concentrations of adaptors at the aRNA stage. Depending on the prominence of the peaks, an additional RNA cleanup for size selection should be performed on the frozen aRNA aliquots. bp: base-pairs; FU: fluorescence

3.14 Data Processing

Sc-sortChIC data can be processed automatically using the scChICflow pipeline available from https://github.com/bhardwaj-lab/scChICflow. Users are recommended to install the required software tools for the pipeline through conda, using the `env.yaml` file provided with the pipeline: `conda env create -n chicflow -f env.yaml`. Additionally, Table 2 contains download and usage instructions for the software tools used in the pipeline, for manual installation.

After installation, users can configure the workflow, by providing the genome-specific files, and parameters within the config.yaml file. The workflow can be executed as follows:

```
scChICflow -i <input_directory> --outdir <output_directory> -j
30 --config config.yaml --countRegions windows -binSize 50000
```

Here, `<input_directory>` contains paired-end .fastq formatted files, and `-j` specifies the number of parallel jobs (or CPU cores). The last two options specify that reads must be counted in 50 Kb windows after deduplication (*see* **Note 16**). The data processing steps for sortChIC are summarized in Fig. 2. Please visit https://github.com/bhardwaj-lab/scChICflow for an explanation of the underlying steps and the instructions to run a test workflow on published sc-sortChIC data.

3.15 Assessment of sortChIC Data Quality

Several single-cell QC metrics created by scChICflow can inform us about the quality and complexity of sortChIC signal and help us filter out poor-quality cells (Fig. 6). As MN has a strong preference to cut in an AT context, overall specificity can be seen through the enrichment of the "A" base at the start of read1 (Fig. 6a). At the single-cell level, the minimal filtering we recommend is the exclusion of cells with a low number of unique cuts, as well as a low fraction of reads starting with "A" (Fig. 6b). Depending upon the mark of interest, enrichment scores can be determined by counting reads in "expected regions" and comparing them with the total number of reads obtained from this cell.

3.16 Downstream Analysis

After removing the poor-quality cells, a typical downstream analysis step is to perform unsupervised clustering of the cells based on the variance in the ChIC signal between cells. A visual exploration of our cell clusters and a 2D embedding of the cells (using methods such as tSNE and UMAP) provide a good estimate of the number of cell types present in the experiment (Fig. 6c). The tool `scClusterCells` from the sensei toolkit can be used for this purpose (command: `scClusterCells -i <counts_in_bins>.loom -o <output>.loom -op <output_umap>.png`).

To visualize cell-type specific signals, users can group cells based on the determined clusters and create coverage files, merging the signal from each group (also called in silico bulk coverage) and loading them on a genome browser (Fig. 6d). Quality of the resulting coverage tracks is a result of the total number of cells in each group, but also the specificity of the signal. These grouped coverage files can be obtained using the tool `scBulkCoverage` (command: `scBulkCoverage --binSize <desired_binsize> -b <bamfiles> -i <cell_groupings>.tsv -o <output_prefix>`).

These downstream steps provide a good insight into the cell type complexity and difference of signal enrichment between cell types in the data. However, depending upon the target histone mark, tissue, and initial quality assessment, users would have to tweak the filtering, clustering, and aggregation of ChIC signals from cells.

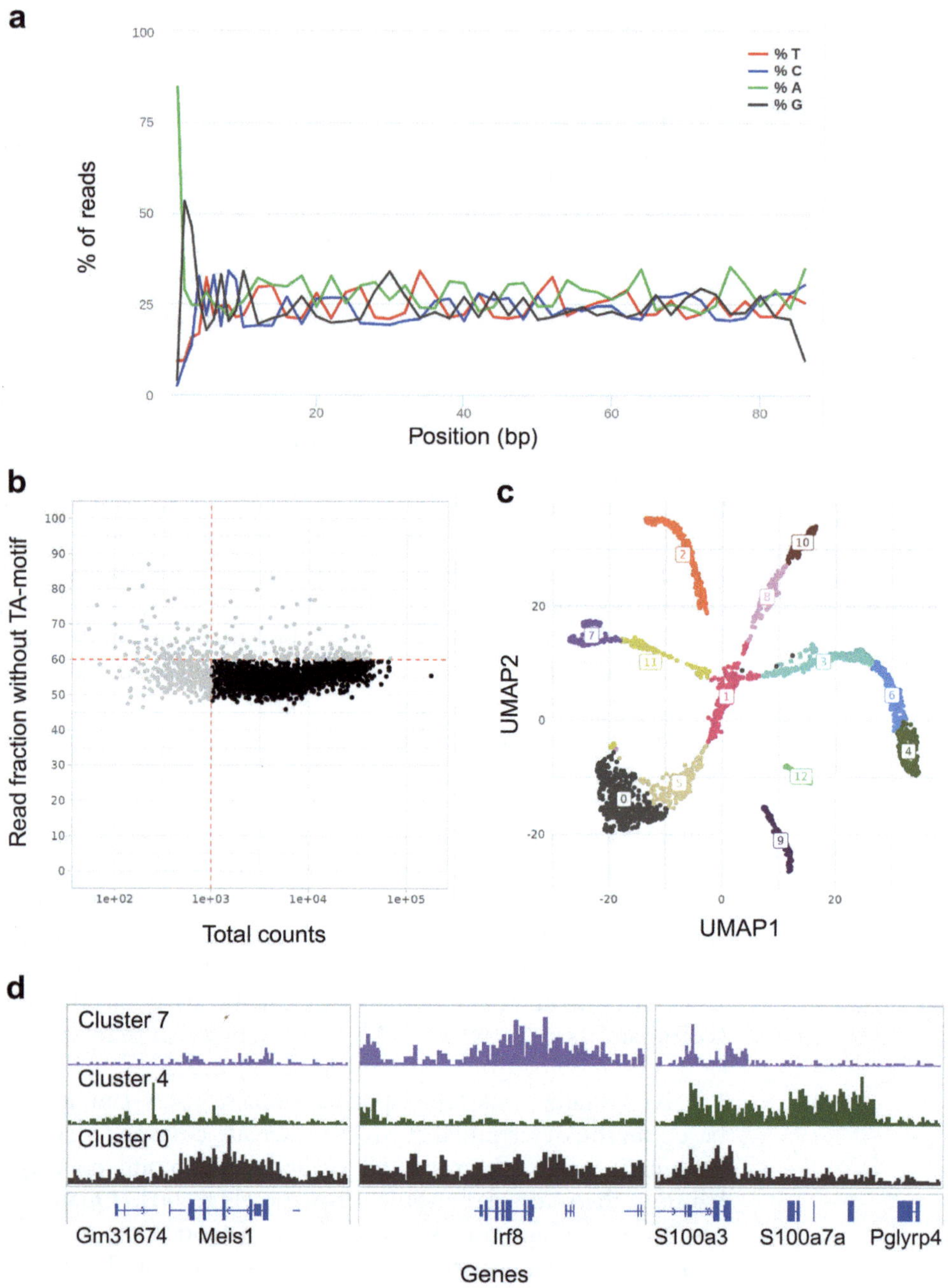

Fig. 6 Quality control (QC) and downstream analysis based on processed data from scChICflow. (**a**) FastQC output after trimming the adapters (found in `QC/multiqc.html` file from the workflow). It shows the enrichment of an "A"-base at the MN cut site (in green), an indication of successful MN activity. (**b**) Main QC plot used for high-quality cell filtering (plotted from the columns *Total_sampled* vs. *Wrong_motif* from `QC/scFilterStats.txt`). Cells with a low number of recovered reads or a too-high fraction of first-in-pair reads starting with the MN restriction bias are removed. (**c**) 2D UMAP embedding of cells. This is obtained by running `scClusterCells` (from the sincei toolkit) on the reads counted in 50 kb bins on sortChIC data (*see* "Downstream Analysis"). (**d**) IGV profile of H3K4me1 read coverage of three clusters shown in C, obtained by running `scBulkCoverage` (*see* "Downstream Analysis"). The enrichment is observed on promoter and enhancer regions, as expected

4 Notes

1. JM101 bacteria were transformed with pK19pA-MN (Addgene plasmid # 86973). Bacteria are grown in LB media containing 50 mg/L kanamycin and protein production is activated at OD = 0.4 by adding IPTG to 2 mM final concentration and harvested after 2 h. For each 200 mL starting culture, bacteria are centrifuged and resuspended in 10 mL TEN (10 mM Tris–HCl, pH 7.5, 1 mM EDTA, and 150 mM NaCl) containing 5 mM DTT. Bacteria were lysed by adding 1 mg chicken egg white lysozyme (Sigma) and incubation for 10 min at 4 °C and sonication using a Branson SFX250. Debris was removed by centrifugation at 17,000× *g* for 20 min at 4 °C. Protein was recovered from the supernatant using IgG Sepharose® 6 Fast Flow (Sigma Aldrich) following the manufacturer's instructions, with the addition of 0.03% Empigen to the TEN-based washing steps. After elution with 0.5 mL 0.5 M HAc/NH_4Ac, pH 3.4 and neutralized by the addition of 5 M NaOH, the protein concentration is determined and Glycerol is added to a final concentration of 50%. Aliquots are stored at −80 °C. For more details, check the method section in [22].
2. Forked-ChIC adapter design:

   Top strand:

GGTGATGCCGGTAATACGACTCACTATAGGGAGTTCTACAGTCCGACGATC*NNN*ACACACTAT

Bottom strand:

/5Phos/TAGTGTGT*NNN*GATCGTCGGACTGTAGAACTCCCTATAGTGAGTCGTATTACCGGC**GAGCTT**

Sequence features from left to right on the top strand: Bases written in bold form a fork to prevent adaptor dimer- or multimerization. Bases in green represent T7 polymerase binding sites for IVT-based amplification. Bases in blue are the binding sites (RA5) for the Illumina small RNA PCR primer indexes (RPIx). The three random nucleotides (Ns) in italics are the unique molecular identifier used for read deduplication, and the eight underlined bases afterward represent the cell barcode, which is different for each of the 384 wells.

These oligonucleotides are ordered from Integrated DNA Technologies (IDT) in two separate plates at a concentration of 100 μM. With a liquid handling robot (or by hand) mix 7.5 μL of the two oligonucleotide strands with 15 μL of 2x Annealing buffer (20 mM Tris–HCl pH 7.5, 100 mM NaCl, 2 mM EDTA) into a new 384-well plate. The new stock plate (25 μM) is heated at 94 °C for 2 min and gradually cooled down to 4 °C at 1.5 °C/min. Dilute the stock plate further in Tris–HCl pH 7.5 to a final working concentration of 5 μM. Keep stock and working plates at the −20 °C until ready to use. Alternatively, pre-annealed adaptors can be directly ordered from IDT, so only the working dilution has to be prepared.

3. Surface marker antibodies can be used before fixation to allow the enrichment of desired populations during the cell sorting step. If no optimal dilution is known, we recommend titrating the antibodies, to ensure specificity and fixation compatibility. To ensure clean separation of the fluorescence signal, the number and selection of the antibody-coupled fluorophores have to be adjusted to the specifications of the FACS machine.

   Stainings are performed for 1 M cells per 100 μL in the tissue-specific buffer containing 5% of serum of the primary antibody host for blocking for 30 min at 4 °C in the dark. As all our surface antibodies used within this protocol were rat, we used rat serum for blocking. Cells are washed once with 500 μL of D-PBS before proceeding to EtOH fixation. As a hint, PE (phycoerythrin) and Brilliant Blue–based Fluorophores show no signal reduction through the Ethanol fixation and GFP (Green Fluorescent Protein) and APC (Allophycocyanin) variances only show some signal reduction. FITC, on the other hand, proved to be highly sensitive and completely lost its fluorescent capacity.

4. Nuclei can be isolated by omitting the fixation step and replacing the 0.05% Tween20 with 0.05% Saponin (prepared as 10% stock fresh in D-PBS) in all wash buffers (WB, WB1, and WB2). Saponin is a detergent-like molecule that interacts with the cholesterol in the membrane, selectively removing it and leaving holes without destroying it. Therefore, single-cell suspensions are washed once with the wash buffer 1 (WB1), before direct incubation with the histone modification antibody overnight (8–12 h). The next day, keep using the buffers containing Saponin instead of Tween20 for the rest of the steps until sorting. The Gate 1 (Fig. 3a) for nuclei will be slightly different because nuclei are smaller than cells.

5. To allow optimal comparability between samples, cell-to-reagent ratios should be kept constant. One possibility to ensure this is by processing samples in the same tube. This can be performed by staining cells after fixation with general protein dyes such as CFSE or CellTrace. Following two washes, cells from multiple samples can then be pooled for the antibody and pA-MN targeting and can be separated during the FACS step. Similarly, small samples such as biopsies can be adjusted in number by combining them with stained cell line cells.

6. To ensure a sensitive signal from histone modification antibodies, we suggest titrating them 0.5x and 2x of the concentration recommended by the manufacturer/publication. For the titration, we recommend the usage of a cell line (e.g., K562), where if possible the target was already characterized and start with the same amount of cells in each tube. Sequence the samples shallowly and compare them to each other (e.g., using IGV tracks) and if available public data to decide which concentration produces best results.
7. To store cells after EtOH fixation, samples are washed twice with 500 μL WB1 to remove as much EtOH as possible. Count cells and resuspend with WB1 to a maximum of 1 M cells per 100 μL. Add DMSO to a final concentration of 10% and keep at −80 °C indefinitely.
8. Bulk samples can be treated just like single-cell preparations up until the sorting step (**step 2** in Subheading 3.3). Instead of sorting into 384-well plates, prepare 0.5 μL DNA-low binding tubes with 5 μL of WB2. Cells can be sorted in bulk (any amount up to 100,000 does not need volume adjustments) in the tubes directly. Depending on the volume added during the sort, spin cells down and reduce the volume back to 5 μL. Continue with the addition of 5 μL of 4 mM $CaCl_2$ in WB2 to activate the MN and incubate for 30 min at 4 °C on a thermal cycler. Add 10 μL of stop solution to chelate the calcium and lyse the cells. Incubate following the same program on a thermal cycler overnight. The next day, proceed to the fragment treatment in the tubes with the addition of 15 μL of end-repair mix, 15 μL of A-tailing mix, 5 μL of forked-ChIC adapters (5 μM), and 15 μL of adapter ligation mix. There will be no need for pooling so you can directly add 56 μL of DNA beads (diluted 1 in 2 with BBB). From here onward, the volumes for each sample are the same as for the single-cell plate processing.
9. To ensure the sorting of single cells as well as avoiding cell cycle–related biases, set the gates as follows: FSC and SSC to discard debris. Exclude doublet cells in Hoechst versus Hoechst areas. With a DNA content histogram (Hoechst staining), you can gate according to the cell cycle phase. Select those that fall into the first peak of Hoechst resembling the G1 population (Fig. 3). Also, during the sort, index information can be recorded from every single cell that can be used for cell type identification or surface marker quantification.
10. AmpliTaq360 has a strong preference to incorporate dATP as a single base overhang; therefore, no general dNTP removal is needed. The addition of an A to the 3′ end will prevent the concatemerization of temples, thereby improving the ligation of the adapters which are T-tailed.

11. Larger volumes can result from incomplete sealing of the plates allowing condensate to accumulate in the plate. The resulting suboptimal reaction conditions will lead to reduced yield.
12. A higher DNA-bead ratio (e.g., 1.5X, 2X) will select for shorter fragments while a smaller ratio (e.g., 0.5X) will recover longer fragments. Fragments larger than 1 kb will not cluster efficiently during sequencing.
13. Make sure that all EtOH is removed as it could inhibit the RNA polymerase during the IVT reaction later on.
14. The number of cycles will depend on the aRNA amount (Fig. 4). In turn, the amount of material recovered will depend on the histone mark targeted as well as the organism and potentially cell type. More abundant histone marks (e.g., H3K27me3) will produce more material than less abundant histone marks (e.g., H3K4me3). Overamplification during PCR might lead to a stronger bias toward certain fragments.
15. To prevent over- or underrepresentation of any individual molecules in the same sequencing run, samples need to be pooled at equal molarity rather than mass. We use 4 nM. The sequencing depth will depend on the amount of samples and the budget available. For example, if 12 PCR cycles were sufficient for amplification, we recommend starting with 3 Million reads per plate. If 8 PCR cycles are enough, 10 Million reads are a recommended starting point to check the quality and complexity. Samples can be resequenced later to adjust the depth to their actual complexity. Generally speaking, paired-end and longer-read sequencing provides the most accurate results and is recommended especially for areas of the genomes with low complexity or many repetitive elements.
16. By default, we aggregate read counts per single cell in 50 kb windows, which provides a reasonable signal for clustering of single cells. But depending on the genome size and the density of genes, users may benefit from decreasing this bin size. Also, sometimes the users might expect specific regions of enrichment (such as promoters for H3K4me3), in this case they can aggregate signals based on a custom bed file containing the expected regions. Since this depends on the target of interest (histone mark/transcription factor), we recommend users to start with genome-wide 50 kb bins, and later explore different counting options with `scCountReads`.

Acknowledgments

This protocol was developed in the lab of Alexander van Oudenaarden.

References

1. Solomon MJ, Larsen PL, Varshavsky A (1988) Mapping protein DNA interactions in vivo with formaldehyde – evidence that histone-H4 is retained on a highly transcribed gene. Cell 53(6):937–947
2. Johnson DS, Mortazavi A, Myers RM et al (2007) Genome-wide mapping of in vivo protein-DNA interactions. Science 316(5830):1497–1502
3. Mikkelsen TS, Ku M, Jaffe DB et al (2007) Genome-wide maps of chromatin state in pluripotent and lineage-committed cells. Nature 448(7153):553–560
4. Rotem A, Ram O, Shoresh N et al (2015) Single-cell ChIP-seq reveals cell subpopulations defined by chromatin state. Nat Biotechnol 33(11):1165–1172
5. Grosselin K, Durand A, Marsolier J et al (2019) High-throughput single-cell ChIP-seq identifies heterogeneity of chromatin states in breast cancer. Nat Genet 51(6):1060–1066
6. Ku WL, Nakamura K, Gao W et al (2019) Single-cell chromatin immunocleavage sequencing (scChIC-seq) to profile histone modification. Nat Methods 16(4):323–325
7. Ku WL, Pan L, Cao Y et al (2021) Profiling single-cell histone modifications using indexing chromatin immunocleavage sequencing. Genome Res 31(10):1831–1842
8. Zeller P, Yeung J, Vinas Gaza H et al (2023) Single-cell sortChIC identifies hierarchical chromatin dynamics during hematopoiesis. Nat Genet 55(2):333–345
9. Harada A, Maehara K, Handa T et al (2019) A chromatin integration labelling method enables epigenomic profiling with lower input. Nat Cell Biol 21(2):287–296
10. Kaya-Okur HS, Wu SJ, Codomo CA et al (2019) CUT&Tag for efficient epigenomic profiling of small samples and single cells. Nat Commun 10(1):1930
11. Wu SJ, Furlan SN, Mihalas AB et al (2021) Single-cell CUT&Tag analysis of chromatin modifications in differentiation and tumor progression. Nat Biotechnol 39(7):819–824
12. Wang Q, Xiong H, Ai S et al (2019) CoBATCH for high-throughput single-cell epigenomic profiling. Mol Cell 76(1):206–216.e207
13. Janssens DH, Meers MP, Wu SJ et al (2021) Automated CUT&Tag profiling of chromatin heterogeneity in mixed-lineage leukemia. Nat Genet 53(11):1586–1596
14. Andrews S (2010) FastQC: a quality control tool for high throughput sequence data. Babraham Bioinformatics
15. Martin M (2011) Cutadapt removes adapter sequences from high-throughput sequencing reads. EMBnet J 17:10
16. Smith T, Heger A, Sudbery I (2017) UMI-tools: modeling sequencing errors in Unique Molecular Identifiers to improve quantification accuracy. Genome Res 27(3):491–499
17. Li H (2013) Aligning sequence reads, clone sequences and assembly contigs with BWA-MEM. arXiv:1303.3997. https://doi.org/10.48550/arXiv.1303.3997
18. Li H, Handsaker B, Wysoker A et al (2009) The sequence alignment/map format and SAMtools. Bioinformatics 25(16):2078–2079
19. Ramirez F, Ryan DP, Gruning B et al (2016) deepTools2: a next generation web server for deep-sequencing data analysis. Nucleic Acids Res 44(W1):W160–W165
20. Bhardwaj V, Mourragui S (2023) sincei: A user-friendly toolkit for QC, counting, clustering and plotting of single-cell (epi)genomics data. (Version 0.3.1). https://doi.org/10.5281/zenodo.8105536
21. Ewels P, Magnusson M, Lundin S et al (2016) MultiQC: summarize analysis results for multiple tools and samples in a single report. Bioinformatics 32(19):3047–3048
22. Schmid M, Durussel T, Laemmli UK (2004) ChIC and ChEC; genomic mapping of chromatin proteins. Mol Cell 16(1):147–157

Chapter 15

XL-DNase-Seq: Footprinting Analysis of Dynamic Transcription Factors

Kyu-Seon Oh, Mohammad Aqdas, and Myong-Hee Sung

Abstract

We have developed a novel method for genomic footprinting of transcription factors (TFs) that detects potential gene regulatory relationships from DNase-seq data at the nucleotide level. We introduce an assay termed cross-link (XL)-DNase-seq, designed to capture chromatin interactions of dynamic TFs. A mild cross-linking step in XL-DNase-seq improves the detection of DNase-based footprints of dynamic TFs. The footprint strengths and detectability depend on an optimal cross-linking procedure. This method may help extract novel gene regulatory circuits involving previously undetectable TFs. The XL-DNase-seq method is illustrated here for activated mouse macrophage-like cells, which share several features with inflammatory macrophages.

Key words Transcription factor footprinting, Chromatin accessibility, High-throughput sequencing, Genomic footprinting, DNase-seq

1 Introduction

The chromatinscape is an important physicochemical structure that impacts genome regulation and cellular phenotypes of eukaryotes. All the DNA-templated processes in chromatin occur at the level of individual nucleosomes [1]. Generally, accessible regions of chromatin such as enhancers, promoters, and insulators are bound by chromatin-interacting factors, which may affect gene expression [2]. Therefore, it is necessary to investigate the chromatin characteristics that correspond to discrete cell state-specific genome regulation. With the enormous data coming from high-throughput sequencing methods, it is still challenging to delineate the mechanisms that regulate factor binding and thus their genomic targets.

In the 1970s, treatment of a protein-DNA complex with a nonspecific nuclease was found to produce contiguous nucleotide stretches that were protected from nucleolytic attack, which were

Franziska Greulich (ed.), *Chromatin Immunoprecipitation: Methods and Protocols*, Methods in Molecular Biology, vol. 2846,
https://doi.org/10.1007/978-1-0716-4071-5_15,

called protein "footprints" [3]. Genomic footprinting extends this further to cellular nuclei and whole cells [4–6]. Open chromatin regions are hypersensitive to nuclease digestion [7]. These hypersensitive sites can be captured in DNase I hypersensitivity (DHS) assays, where the isolated nuclei are subjected to limited digestion with the enzyme DNase I. DHSs can be identified by DNase I digestion, followed by high-throughput DNA sequencing (DNase-seq). Transcription factor (TF)-bound regions within DNase I-hypersensitive sites may be protected from DNase I cleavage, leaving detectable footprints. Differential TF footprinting can be integrated with gene expression data to generate gene regulatory networks.

DNase-seq has become a widely accepted technique for studying chromatin accessibility, often related to gene expression analysis in eukaryotes. DNase-seq generates genome-wide DHS maps that classify functional regulatory sites, and the assay has also been shown to reveal some single-nucleotide resolution TF footprints from local DNA protection due to TF–DNA binding [8, 9]. In genomic TF footprinting, the presence of DNase I protection at a binding motif was associated with TF occupancy at that site. However, many TFs have highly dynamic interactions with target chromatin, and TFs with short DNA residence times result in no footprints at bound motif elements [10]. Additionally, the enzymes used to probe chromatin have non-negligible DNA sequence preferences for their reactions, thereby compromising attempts to identify TF occupancy in an unbiased manner [11, 12]. Such sequence bias has been accounted for through a combination of different approaches using computational algorithms, molecular biology, and deeper sequencing [13–15]. Although genomic TF footprinting aims to identify diverse TF occupancy regardless of their specific DNA sequence content, short-lived TFs with high exchange rates or short chromatin residence times represent a major hurdle [10, 12].

In this chapter, we present a method that improves upon existing TF footprinting protocols by introducing a mild cross-linking step to capture highly dynamic interactions of TFs with target DNA elements in chromatin [16]. This cross-linking step is intended to preserve and further enhance the footprinting depths of most TFs. The cross-linking conditions need to be optimized for each cell type. We show how to use various concentrations of formaldehyde for different cross-linking durations that support the ideal parameters and enhance the footprinting depths of TFs. In summary, the footprinting protocol, termed "XL-DNase-seq," with an optimal cross-linking step can be a straightforward modification to conventional DNase-seq protocols.

2 Materials

RAW264.7 cells were grown in a 150 mm dish in 20 mL Dulbecco's Modified Eagle medium (DMEM), 10% (v/v) fetal bovine serum (FBS) at 37 °C, and 5% CO_2. Cells were treated with 10 ng/mL LPS from *Salmonella minnesota* R595 for 8 h.

2.1 Formaldehyde Cross-Linking

1. Methanol-free 16% (w/v) Formaldehyde.
2. 20x Glycine (2.5 M).
3. 1x PBS.
4. PBS + PI: Transfer 10 mL of PBS into a 15 mL tube and add 10 tablets of PI (1 tablet/50 mL PBS). Warm at 37 °C and vortex until dissolved. Mix 490 mL of PBS with 10 mL of PBS with PI to make the final 500 mL solution, which can be stored at 4 °C.
5. Phase-Contrast Hemacytometer or Automated Cell Counter.

2.2 Collection of Cells After Cross-Linking

1. 6.4 M Spermidine.
2. 1 M Tris–HCl pH 8.0.
3. 5 M NaCl.
4. 1 M KCl.
5. 0.5 M EDTA pH 8.0.
6. 0.5 M EGTA pH 8.0.
7. Pefabloc SC Plus.
8. 1,4-Dithiothreitol (DTT).
9. Complete EDTA-free Protease Inhibitor (PI).
10. Sterile dH_2O or Molecular Biology Grade Water.
11. Buffer A: Add 7.5 mL of 1 M Tris–HCl pH 8.0, 1.5 mL of 5 M NaCl, 30 mL of 1 M KCl, 1 mL of 0.5 M EDTA pH 8.0, 0.5 mL of 0.5 M EGTA pH 8.0, 40 μL of 6.4 M Spermidine, and add 459.5 mL sterile dH_2O, then mix thoroughly by inverting. All buffers and reagents should be filtered (0.22 μM) and prechilled on ice before use. Store the buffer at 4 °C. Immediately before use, add Pefabloc SC Plus, DTT, and PI.
12. 50 mL Conical Centrifuge Tubes.
13. Cell Lifter.
14. Cooling centrifuge for 50 mL tubes.
15. Microscope for inspection of cells.

2.3 Nuclei Isolation from Cross-Linked Cells

1. IGEPAL®CA-630.
2. 10% (v/v) IGEPAL®CA-630: 100% IGEPAL is extremely viscous at room temperature (20–25 °C), so warm stock bottle of IGEPAL®CA-630 to 37 °C. Add 1 mL of 100% IGEPAL to 9 mL of nuclease-free, sterile dH_2O. It is important to carefully and accurately draw up 100% IGEPAL due to its high viscosity. 10% IGEPAL should be warmed in a 55 °C water bath for 30 min before use, to ensure proper dissolution. Store at 4 °C.
3. 2x IGEPAL Lysis Buffer (Buffer A + PI + 1 mM Spermidine + 0.1% (v/v) IGEPAL): Aliquot 490 mL of buffer A with 0.1% IGEPAL into a beaker. Aliquot 10 mL of buffer A into a 15 mL tube and add 10 tablets of PI. Warm at 37 °C and vortex until dissolved. Mix the 490 mL solution with the 10 mL of the PI solution to make the final 500 mL 2x IGEPAL lysis buffer. Finally, add 80 μL of spermidine and store at 4 °C. Prepare this buffer freshly before use (Subheading 3.3, **step 1**).
4. Buffer A + PI + 0.5 mM Spermidine: Prepare freshly on the day of usage. Aliquot 490 mL of buffer A into a beaker, and another aliquot 10 mL of buffer A into a 15 mL tube and add 10 tablets of PI tablet (1 tablet/50 mL buffer). Warm at 37 °C and vortex until dissolved. Mix the 490 mL solution with the 10 mL of the PI solution to make the final 500 mL 2x IGEPAL lysis buffer. Finally, add 40 μL of spermidine and store at 4 °C until needed in Subheading 3.3, **step 5**.
5. Trypan Blue.
6. 15 mL Conical Centrifuge Tube.

2.4 DNase I Digestion of Nuclei

1. Deoxyribonuclease I (DNase I).
2. 2 M $CaCl_2$.
3. 10% (w/v) SDS Solution.
4. 10x DNase I Digestion Buffer: Add 3 mL of 2 M $CaCl_2$, 15 mL of 5 M NaCl, and 82 mL nuclease-free, sterile dH_2O. Mix thoroughly by inverting, then store at 4 °C.
5. 1x DNase I digestion buffer in 50 mL conical tube: Prepare this buffer freshly on the day of use. Add 5 mL of 10x DNase I digestion buffer, 45 mL of buffer A, and 8 μL spermidine (1 mM). Then aliquot 5 mL into 15 mL conical tubes (1 tube per DNase I concentration, 1.5 mL per digestion). Store at 4 °C. Before use, equilibrate in a 37 °C water bath for 1 h (*see* Subheading 3.4, **step 6**).
6. Stop buffer: Add 5 mL of 1 M Tris–HCl pH 8.0, 2 mL of 5 M NaCl, 1 mL of 10% (w/v) SDS, 20 mL of 0.5 M EDTA pH 8.0, and 72 mL sterile dH_2O. Store at 4 °C until ready to use. On the day of usage, add 25 mL of stop buffer into a 50 mL conical

tube. Storing the stop buffer at 4 °C can cause it to precipitate. If there is precipitation, then warm the stop buffer to 37 °C for 5 min and mix thoroughly until SDS goes back into solution. Finally, add 4 μL of spermidine (1 mM). Equilibrate in a 37 °C water bath for 1 h before use.

7. Water Bath or Thermomixer for 15 mL tubes.

2.5 Formaldehyde Reverse Cross-Linking

1. 20 mg/mL Proteinase K.
2. 30 mg/mL Ribonuclease A (RNase A).

2.6 Quality Control

1. 10,000x SYBR™ Green I Nucleic Acid Gel Stain.
2. TAE buffer.
3. TaqI Restriction Enzyme and Buffer 4.
4. 1 kb DNA Ladder.
5. Ultrapure Agarose.
6. 1 mg/mL Bovine Serum Albumin (BSA).
7. iQ™ SYBR® Green Supermix or alternative qPCR master mix.
8. Primers for qPCR.
   - DNase I-sensitive region primers located on Chr 1:
     5′-GACCTAGCCCCGGAACTAAC-3′ and 5′-CAGC CCTTCTCGAGACACTT -3′, Weak sensitivity, 66 bp amplicon length.
   - DNase I-sensitive region primers located on Chr 2:
     5′-GGATTCCTTTCTGCTCGAAG-3′ and 5′-CGGT TCGGTTTTTCCTTGTA -3′, High sensitivity, 116 bp amplicon length.
   - DNase I-sensitive region primers located on Chr 7:
     5′-GCAGGACGAAGAAAACAAGC-3′ and 5′-TTCC CTTTTCACCATTTCCA-3′, Medium sensitivity, 169 bp amplicon length.
   - DNase I-insensitive negative control primers located on Chr 3 (NEG):
     5′-GCAGCAGATGGCAAGTAATACTAAGAT-3′ and 5′- CCCTTATTCTCTGAGCATTAGACAGTTATA -3′, corresponding to a 58 bp amplicon length.
9. Microfuge Tubes 1.7 mL.
10. Benchtop Centrifuge.
11. Thermomixer.
12. qPCR machine.
13. Agarose gel equipment and documentation device.

2.7 Phenol: Chloroform:Isoamyl Alcohol Extraction

1. Phenol:Chloroform:Isoamyl Alcohol (PCI, 25:24:1).
2. Biosafety Cabinet.

2.8 Size Fractionation of DHS over Sucrose Gradient Centrifuge

1. Sucrose Bio Ultra Molecular Biology Grade.
2. 9% (w/v) sucrose solution: Add 90 g sucrose, 200 mL 5 M NaCl, 20 mL 1 M Tris–HCl pH 8.0, 10 mL 0.5 M EDTA to sterile dH_2O. Adjust final volume to 1 L. Once completely dissolved, store at 4 °C until use.
3. 20 mg/mL Glycogen.
4. 3 M Sodium Acetate (NaAcetate).
5. Isopropanol.
6. 70% (v/v) Ethanol (EtOH).
7. Qubit dsDNA BR/HS Assay Kit and Qubit fluorometer.
8. Open-top, thick-walled polyallomer ultracentrifuge tube.
9. SW41 Swinging Bucket Rotor.
10. Beckman LE-80 Ultracentrifuge.

3 Methods

Many TFs have highly dynamic interactions with target chromatin, and TFs with short DNA residence times result in no footprints at bound motif elements [10]. We used formaldehyde cross-linking to stabilize the TF-DNA interactions that are short-lived. Thus, we developed a new XL-DNase-seq protocol to capture the short-lived binding of genomic TFs and preserve chromatin accessibility profiles.

To determine the optimal parameters that enhance footprinting depths of TFs that have previously been difficult to detect from the standard DNase-seq protocol, we tested a range of formaldehyde concentrations and cross-linking durations. This protocol includes mild formaldehyde cross-linking that can easily be incorporated before the steps in the conventional DNase-seq protocols [17, 18]. In this approach, cells are cross-linked with formaldehyde solution to preserve chromatin accessibility profiles. Briefly, cells were cross-linked with either 0.1% or 1% formaldehyde at room temperature (20–25 °C) for various times (30 s to 10 min), then quenched with glycine. Cross-linked cells were pelleted, followed by nuclei isolation. Chromatin from five million cells was digested with 20–100 U/mL of DNase I for 3 min at 37 °C followed by termination of the reaction. After proteinase K treatment, digested cross-linked chromatin was then incubated overnight (16–24 h) at 65 °C for formaldehyde reverse cross-linking. On the following day, additional proteinase K and RNase were

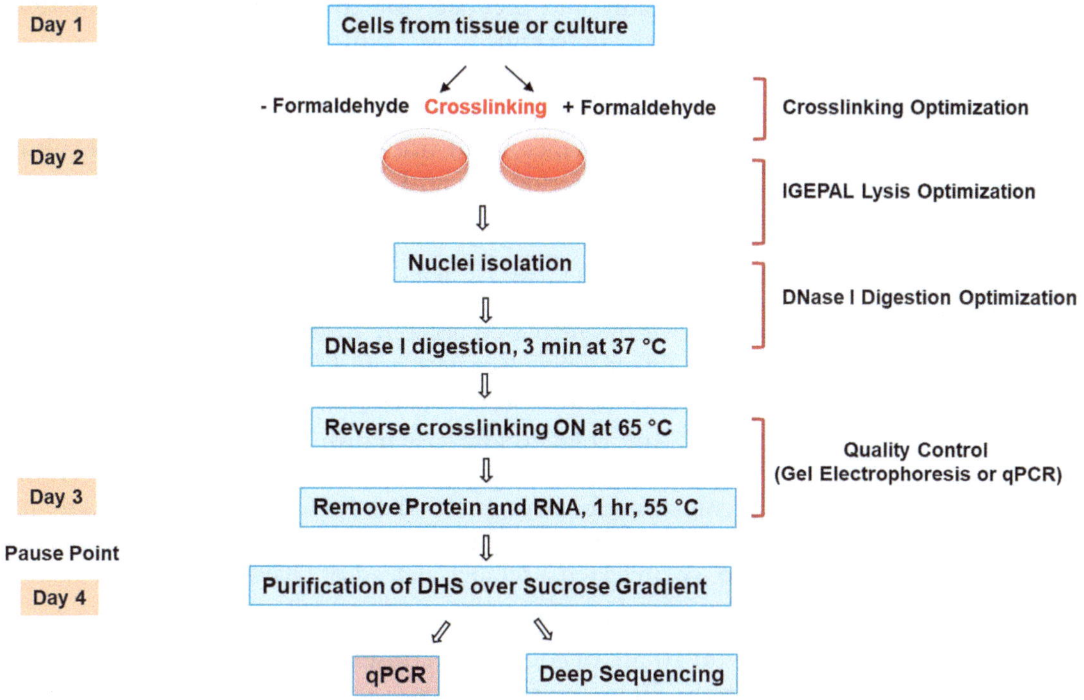

Fig. 1 Overview of the DNase-seq protocol after cross-linking. Briefly, cells are cross-linked with various formaldehyde concentrations and incubation times followed by nuclei isolation. DNase I-digested nuclei are reverse cross-linked and purified over sucrose gradient centrifugation, followed by deep sequencing and analysis

added and incubated at 55 °C for 1 h. DNA fragments were isolated by phenol:chloroform:isoamyl alcohol extraction and were fractionated over a sucrose gradient to isolate 100–1000 bp fragments. These fragments were pooled, precipitated, and then assembled into libraries. Two biological replicate samples per treatment group were prepared separately for high-throughput sequencing (Fig. 1).

3.1 Formaldehyde Cross-Linking

To preserve chromatin accessibility profiles, cells are cross-linked with formaldehyde solution. Importantly, cross-linking conditions need to be optimized for different cell types.

1. Plate 20×10^6 cells in a 150 mm culture dish in 20 mL of media per condition. Check cell cultures daily for morphology, cell density, and the absence of contamination (*see* **Note 1**).
2. Next day, cross-link cells with 0.1% or 1% formaldehyde (*see* **Note 2**), as empirically determined (*see* **Note 3**). Add 0.125 mL (0.1%) or 1.25 mL (1%) of 16% methanol-free formaldehyde into each 150 mm cell culture dish in 20 mL of media.

3. Mix rapidly and incubate for the optimal incubation time, previously determined (*see* **Note 3**) at room temperature (20–25 °C).
4. Once the treatment times are completed, quench formaldehyde with glycine. Add 1 mL of 20x glycine to the cell culture (final concentration 125 mM) and incubate for 10 min at room temperature (20–25 °C).

3.2 Collection of Cells After Cross-Linking

Centrifuge should be precooled to 4 °C before sample preparation.

1. Remove the media completely from the dishes and wash the cross-linked cells twice with cold PBS. At this stage, all procedures can be done at room temperature (20–25 °C).
2. Add 5 mL cold PBS into each dish and gently scrape the cells off the bottom of the dishes with a cell scraper. Apply consistent scraping force in all dishes. Of note, trypsinization is assumed the most aggressive method of enzymatic reactions, so we preferred a scraper instead of trypsin.
3. Transfer the scraped cells into 50 mL conical tubes.
4. Add 10 mL of cold PBS to rinse dishes, and transfer to cell mix in the 50 mL conical tubes.
5. Ensure the cells are completely detached from the dish by inspecting the bottom of the dishes under a microscope.
6. Spin the tubes for 5 min at 500× *g* and 4 °C.
7. Aspirate the supernatant and gently flick the cell pellet to break up the pellet.
8. Resuspend each cell pellet with 10 mL of cold PBS per 50 mL conical tube.
9. Spin the cell suspension for 5 min at 500× *g* and 4 °C.
10. Resuspend the cell pellet in a total of 10 mL of buffer A with spermidine and proteinase inhibitor (PI).
11. Count the cells manually using a hemocytometer or an automated cell counter. Approximately 30×10^6 cells are expected after cross-linking from each 150 mm culture dish.
12. Ensure that all samples have the same cell density. Add buffer A to bring all samples to 2×10^6 cells/mL. For example, if tube #1 has 2×10^6 cells/mL in 10 mL and tube #2 has 3×10^6 cells/mL in 10 mL, then add 5 mL buffer A to tube #2.
13. Mark the total volume of cell suspension on the cap of tubes (e.g., 10 mL on tube #1 and 15 mL on tube #2).

3.3 Nuclei Isolation from Cross-Linked Cells

The amount of detergent in the lysis buffer should be optimized for each cell type since some cell lines are very sensitive to IGEPAL (*see* **Note 4**). The optimal IGEPAL concentration for RAW264.7 cells is 0.05%, which is used in **step 1** and beyond.

1. Slowly add 2x IGEPAL lysis buffer equal to the volume of cell suspension in each of the samples. For example, add 10 mL 2x IGEPAL buffer to tube #1 and 15 mL to tube #2 (*see* Subheading 3.2, **step 13**).
2. Mix gently by inverting five times, followed by a 10 min incubation on ice. Work quickly without vortexing and pipetting. Make sure no precipitates are formed at this step.
3. During the 10 min incubation, check the efficiency of lysis under the microscope. Intact nuclei should be observed without visible clumps. Some clumping during the incubation of cells with lysis buffer is an indication of excessive IGEPAL that leads to nuclear lysis.
4. After the 10 min incubation, spin down immediately for 5 min at 500× *g* and 4 °C to pellet the nuclei. Carefully remove the supernatant and do not leave any solution on the pellet. Any remaining IGEPAL detergent will lyse the nuclei with time, so removing all of it promptly is very important.
5. Resuspend nuclei in 15 mL buffer A with PI and spermidine. Transfer to a new tube to avoid any IGEPAL contamination.
6. Check the nuclei integrity with trypan blue under a microscope and count the nuclei using either an automatic cell counter or a hemocytometer. Trypan blue–stained nuclei should account for more than 95% of the nuclei with no visible clumping. Clumping indicates excessive cellular lysis. Nuclei counts should always be slightly lower than the initial cell count.
7. Transfer 5×10^6 nuclei for each experiment/sample into an individual 15 mL conical tube. We used three DNase I concentrations (0, 60, and 80 U/mL) per sample (Fig. 2 in **Note 4**). Therefore, we divided the nuclei into three new 15 mL conical tubes per DNase I digestion (5×10^6 nuclei each).
8. Spin the nuclei for 5 min at 500× *g* and 4 °C.
9. Discard the supernatant and keep the nuclei on ice until DNase I digestion.

3.4 DNase I Digestion of Nuclei

The optimal DNase I concentration should be determined empirically for each cell type (*see* **Note 5**, Figs. 2 and 3). The following protocol illustrates the procedure for three different cross-linked samples with three DNase I concentrations (0, 60, and 80 U/mL). Adjust according to the number of samples, as necessary.

1. Place the stop buffer and 1x DNase I digestion buffer in a 37 ° C water bath for at least 20 min.
2. Label a new 15 mL conical tube with the appropriate DNase I concentrations for each DNase I concentration, e.g., 0, 60, and 80 U/mL and keep them on ice (*see* Subheading 3.3, **step 7**).

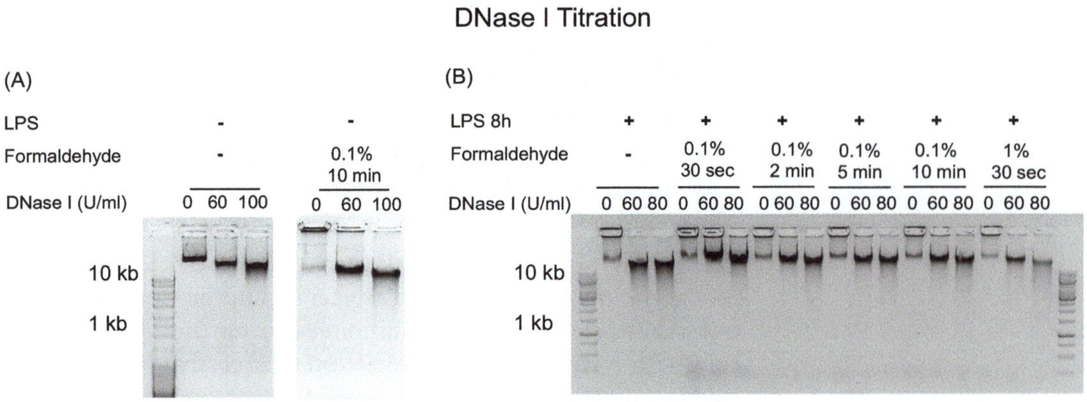

Fig. 2 Titration of DNase I concentration for each cross-linking condition. The light smear is observed with 60 and 80 U/mL digestion in both cross-linked and non-cross-linked chromatin. Cross-linking with 0.1% formaldehyde does not affect DNase activity. (**a**) Cells were either non-cross-linked or cross-linked with 0.1% formaldehyde for 10 min at room temperature (20–25 °C), and then quenched with glycine followed by DNase I digestion with 60 or 100 U/mL. A light smear was observed with 60 U-digests in both non-cross-linked and cross-linked chromatin. (**b**) The cells were activated either with LPS or w/o LPS for 8 h followed by various cross-linking conditions with 0.1% or 1% formaldehyde and cross-linking durations. The gel image shows gradual and consistent changes in high molecular weight DNA fragment sizes as DNase I concentrations increase

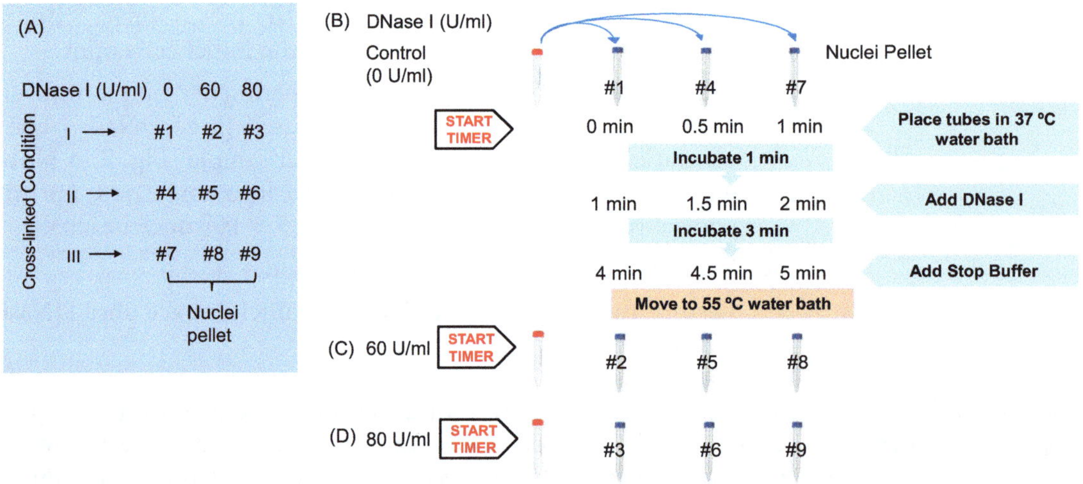

Fig. 3 Schematic representation of three concentrations of DNase I digestion with 0, 60, and 80 U/mL for three different cross-linked nuclei. The DNase I digestion steps must be handled carefully and promptly. (**a**) Nuclei from different samples are divided into three individual 15 mL conical tubes. (**b**) Place tubes with nuclei #1, #4, and #7 in a 37 °C water bath for 1 min. Carefully resuspend nuclei pellet with prewarmed 1.5 mL of 1x DNase buffer without DNase I (for 0 U/mL control samples), followed by incubation for 3 min at 37 °C. After 3 min incubation, add equal volumes of stop buffer to each reaction tube and move tubes to 55 °C water bath. (**c**) Repeat steps in (**a**) for digestion with 60 U/mL (tubes #2, #5, #8). (**d**) Repeat steps in (**a**) for digestion with 80 U/mL (tubes #3, #6, #9)

3. Add 5 mL of 1x DNase I digestion buffer into each labeled 15 mL tube. Then add the proper amount of DNase I to the tubes immediately before starting the digestion reaction. For the control sample (0 U/mL), add 5 mL of 1x DNase I digestion buffer only; for 60 U/mL digestion, add 30 μL of enzyme; and for 80 U/mL digestion, add 40 μL of DNase I (10 units/μL) to 5 mL 1x DNase digestion buffer (*see* Fig. 3a). Make sure each dilution is mixed thoroughly by gentle pipetting. Do not vortex as the DNase I enzyme denatures easily with aeration.
4. Set a timer for the first enzymatic reaction (e.g., the control).
5. Place tubes with nuclei #1, #4, and #7 in a 37 °C water bath and allow the nuclear pellet to equilibrate for 1 min (Fig. 3b). This and the following steps must be handled carefully with precise timing.
6. Carefully resuspend nuclei pellet with 1.5 mL warm (37 °C) 1x DNase buffer without DNase I (Fig. 3b). This process needs to be done quickly without vortex and pipetting.
7. Incubate for 3 min at 37 °C. Keep the nuclei in suspension by briefly tapping the tube during the incubation.
8. After 3 min, add an equal volume of stop buffer to each reaction tube and place the tubes in a 55 °C water bath for 2 h.
9. Repeat **steps 4–8** for digestion with 60 U/mL (tubes #2, #5, #8) and 80 U/mL (tubes #3, #6, #9). The DNase I reaction from the samples with different conditions results in reproducible digestion patterns (Fig. 2).

3.5 Formaldehyde Reverse Cross-Linking

1. Add 7.5 μL proteinase K per 3 mL of digested nuclei (final concentration 50 μg/mL). The tubes can be left at 55 °C for 2 h before treating with proteinase K (*see* Subheading 3.4, **step 8**).
2. Incubate samples overnight (16–24 h) at 65 °C for formaldehyde reverse cross-linking.
3. Next day, add an additional 8 μL of Proteinase K per 3 mL of digested nuclei and incubate for 1 h at 55 °C. The aim is to break the interactions between protein and DNA to help DNA purification.
4. Add 4.5 μL of RNase A (final concentration 45 μg/mL) and incubate the samples at 37 °C for 30 min to digest the RNAs.
5. Store samples at 4 °C until phenol-chloroform extraction (Subheading 3.7).

3.6 Quality Control

The DNase I digestion efficiency is evaluated by qPCR.

1. Take 50 μL from each DNase I-digested sample for qPCR and gel electrophoresis validation. Keep the remaining sample at 4 ° C for Subheading 3.7.
2. These aliquots need to be sheared by pipetting vigorously until the samples are no longer viscous.
3. Incubate aliquots for 10 min at 95 °C.
4. Mix 4 μL of each denatured digested aliquot with 2 μL buffer 4, 0.2 μL BSA, 0.2 μL 100 U/μL TaqI enzyme, and 13.6 μL nuclease-free water in 1.5 mL reaction tubes.
5. Incubate for 30 min at 65 °C followed by 10 min incubation at 95 °C in a thermomixer without shaking.
6. Perform qPCR assays using SYBR Green Supermix or any alternative qPCR master mix. We used three qPCR primer pairs as a positive control for validation of DNase I digestion (*see* **Note 6** and Fig. 4).

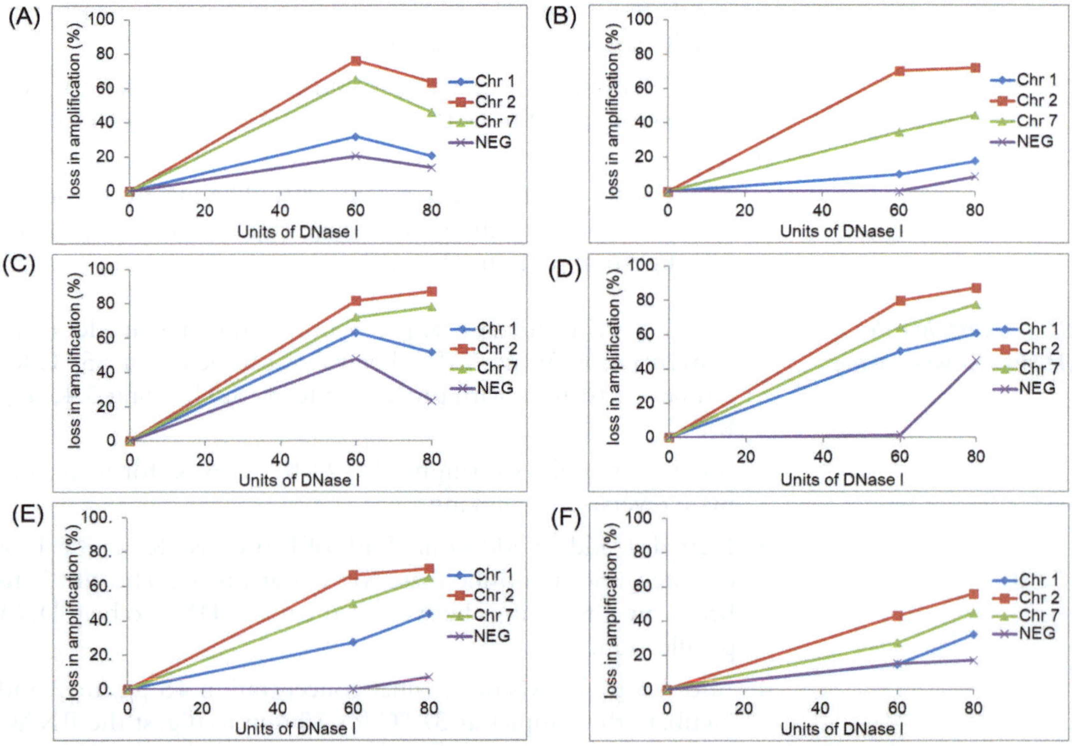

Fig. 4 Validation of DNase I digestion by qPCR. A DNase I sensitivity index is obtained by measuring the loss in amplification at the various regions analyzed across a range of DNase I concentrations (*see* **Note 6**). Primers are designed for strong DNase I sites (Chr 2), intermediate sites (Chr 7), and weak sites (Chr 1) as well as to control DNase I-insensitive regions (NEG). Loss in amplification patterns is similar across all different conditions. (**a**) LPS 8 h/No cross-linking. (**b**) LPS 8 h/0.1% Formaldehyde, 30 s. (**c**) LPS 8 h/0.1% Formaldehyde, 2 min. (**d**) LPS 8 h/0.1% Formaldehyde, 5 min. (**e**) LPS 8 h/0.1% Formaldehyde, 10 min. (**f**) LPS 8 h/1% Formaldehyde, 30 s

7. Run 5 μL of the denatured digested samples on a 1% (w/v) agarose gel for 1 h at 75 V. In our example, the gel image shows identical DNase I patterns in cross-linked and non-cross-linked chromatin by DNase I (*see* Fig. 2).

3.7 Phenol: Chloroform:Isoamyl Alcohol Extraction

1. Retrieve the samples stored in Subheading 3.5, **step 5**. If any precipitation is observed, heat samples in a 55 °C water bath for 15 min.
2. Add 3 mL phenol:chloroform:isoamyl alcohol solution to the 3 mL of DNase I-digested chromatin + stop buffer and mix gently. The phenol:chloroform:isoamyl alcohol mixture may be stored under 100 mM Tris–HCl pH 8.0 in a light-tight bottle at 4 °C for periods of up to 1 month.

   Important: Wear safety glasses when performing phenol: chloroform:isoamyl alcohol extraction, even when working in a fume hood. It is, therefore, required that the user wear appropriate laboratory clothing including gloves and protective eyewear. Dispose of phenol waste according to local regulations.
3. Spin for 10 min at 3000× *g* at 20–25 °C to remove the cellular debris.
4. Carefully remove the top layer and transfer the extracted DNA to a new tube. Do not process to ethanol precipitation.
5. Store the extracted DNA at 4 °C until further purification of the DNA fragments using a sucrose gradient.

3.8 Size Fractionation of DHS over Sucrose Gradient Centrifuge

1. Place samples in a 55 °C water bath for 15 min to dissolve any precipitates.
2. Layer 9 mL of a 9% (w/v) sucrose solution in an open-top, thick-walled polyallomer ultracentrifuge tube.
3. Weigh tubes before spinning in an ultracentrifuge. Adjust weight imbalances with sucrose buffer. Place the bucket-tube assembly on a scale, and remove the sucrose gradient from the top of the tube as necessary, to ensure paired tubes are within 0.05 g of each other.
4. Gently layer 2 mL of each purified DNA sample on top of the sucrose. An interface should be visible between the sample and the sucrose solution.
5. Centrifuge samples in a SW41 swinging bucket rotor using a Beckman LE-80 Ultracentrifuge at 106,979× *g* (25,000 rpm) for 24 h at 20 °C. Use slow acceleration and no brake during the deceleration (*see* **Note 7**).
6. Next day, collect 0.5 mL fractions from the top of each of the gradients using a 1000 μL pipette. Make sure to collect about 15 fractions very slowly and carefully. Each fraction needs to be

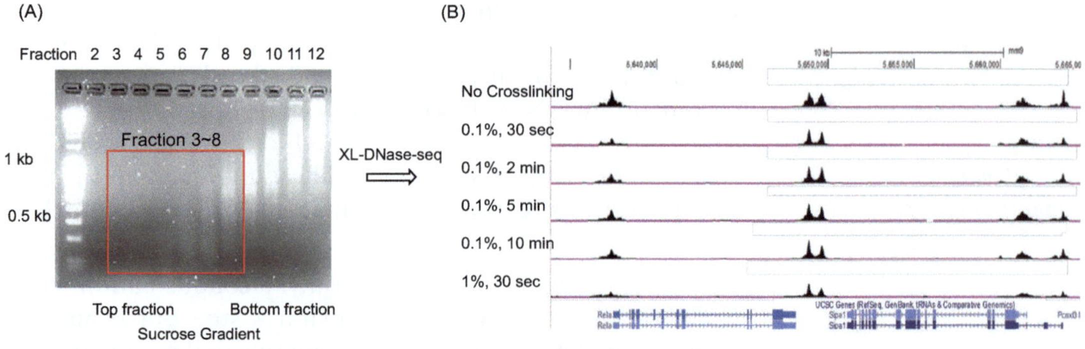

Fig. 5 (**a**) Sucrose gradient profiles of DNase I-hypersensitive sites. The short DNase I fragments are separated from larger fragments by sucrose gradient ultracentrifugation. Pool fractions (3–8) containing fragments smaller than 1 kb. (**b**) The genome browser shot of the *Rela* locus from DNase-seq. Well-preserved similar chromatin accessibility profiles of all different conditions. Near-identical DNase-seq density profiles across cross-linking conditions. XL-DNase-seq preserves the chromatin accessibility profile as observed by the native DNase-seq

collected very gently with as little perturbation to the gradient as possible. Be mindful of the pipette tip position to avoid disturbing the interface between the layers.

7. Load 10 μL of each fraction on a 2% (w/v) agarose gel in 1x TAE buffer without dye.
8. Run the gel for 1 h at 100 V.
9. Stain the gel with 1x SYBR Green in 150 mL TAE buffer with 15 μL SYBR Green (1:10,000) for at least 1 h. Cover the staining container with aluminum foil to protect it from light or place it in the dark.
10. Check DNA fragments with an imaging system. The first two fractions should be blank (Fig. 5a).
11. Pool fractions containing fragments smaller than 1 kb into a 15 mL conical tube.
12. Precipitate DNA from fractions using 0.7 volume isopropanol, 0.1 volume 3 M NaAcetate, and 20 μg glycogen.
13. Incubate the mixture at −80 °C for at least 30 min.
14. Spin by centrifuge at full speed for 20 min.
15. Wash in 5 mL cold 70% EtOH. Spin at full speed for 10 min.
16. Remove the supernatant carefully and air-dry completely.
17. Resuspend the pellet in 30 μL nuclease-free water.
18. Determine the DNA concentration using a Qubit fluorometer.
19. Store the purified DNA samples at −20 °C, or at −80 °C for long-term storage.

Sequencing libraries were generated using Illumina TruSeq protocol and subjected to 76 bp paired-end sequencing on Illumina NextSeq. Footprinting analysis requires a larger sequencing depth than chromatin accessibility analysis, and a minimum of 300 million uniquely mapped quality-passed reads (after filtering chrM-aligned reads) is recommended for each sample. We obtain approximately 700 million reads per sample in our analysis. Given the cost, two biological replicates are used for each condition. Sample quality and library complexity were checked using an established pipeline at the NCI Center for Cancer Research Sequencing Facility. Exemplary results are shown in Fig. 5b.

4 Notes

1. We plate 20×10^6 immortalized mouse macrophage-like RAW264.7 cells into 150 mm dishes per condition and culture until cells are 80–90% confluent. We activate cells with bacterial products, such as lipopolysaccharide (LPS) to induce efficient occupancy of many dynamic TFs in the chromatin, thereby providing an ideal platform for assessing TF footprint characteristics.
2. Formaldehyde is a carcinogen/flammable solution. Hence, it should be tightly sealed and stored at room temperature (20–25 °C). Use it in a chemical fume hood only. Wear appropriate personal protective equipment. We use a higher grade, methanol-free 16% (v/v) formaldehyde and fresh formaldehyde for every chromatin preparation to get more reproducible results.
3. Formaldehyde cross-linking preserves chromatin accessibility profiles while enhancing footprint detection. The ChIP signal increases by formaldehyde in an incubation-time-dependent manner [19], and formaldehyde concentrations for cross-linking may be tissue-specific [20]. The challenge is to optimize the cross-linking conditions since different cell types may require different cross-linking conditions. If the cross-linking is too strong, the DNase I reaction may not be efficient on the chromatin. On the other hand, insufficient cross-linking may result in the dissociation of the dynamic TFs from the target DNA during subsequent steps of the experiment. We recommend performing a small-scale pilot experiment to see if DNase I can digest formaldehyde cross-linked chromatin. Either cells were treated without or with 0.1% formaldehyde, followed by 60 or 100 U/mL DNase I digestion. Note the gradual and consistent changes in high molecular weight DNA fragment sizes with increasing DNase I concentrations. We first verified that the 0.1% mild formaldehyde cross-linking does not affect

DNase I activity (Fig. 2a). We also tested various concentrations of formaldehyde and cross-linking durations to capture the interaction between chromatin and short-lived TFs (Fig. 2b). In our system, the lower 0.1% formaldehyde cross-linking concentration shows better DNase I digestion patterns than 1% formaldehyde followed by 10 min DNase I incubation (Fig. 2b). As previous reports have indicated a dominant effect of cross-linking duration over formaldehyde concentration, we focused on varying the duration of formaldehyde cross-linking. A 0.1% formaldehyde concentration was applied with various cross-linking durations as a cross-linking kinetics study [20], and our pilot study indicated that 1% formaldehyde for 10 min cross-linking may interfere with DNase I reactions. In this experiment, we used two formaldehyde concentrations (0.1% or 1%) and different cross-linking times (30 s, 2 min, 5 min, 10 min) with RAW264.7 cells that have been treated with LPS for 8 h. This experiment aims to determine the optimal cross-linking parameters for a designated cell line, which may affect the footprinting characteristics of dynamic TFs. The DNase I concentrations of 60 and 80 U/mL showed better digestion patterns with either the cross-linked or non-cross-linked chromatin (Fig. 2b).

4. A high-quality nuclei preparation is critical for the generation of high-quality sequencing data from the DNase-seq experiment, which provides an accurate location of regulatory proteins in the genome. During chromatin preparation, the amount of detergent in the lysis buffer needs to be optimized for each cell type to maximize nuclear recovery with minimal damage to nuclear integrity. We performed a pilot experiment to find the optimal IGEPAL concentration. Cells were lysed with 0.02%, 0.05%, and 0.1% IGEPAL lysis buffer and nuclei were harvested followed by 60 or 80 U/mL DNase I treatment. We examined the efficiency of different detergent conditions under a phase-contrast light microscope and found that 1% IGEPAL is excessive for RAW264.7 cells. The excessive lysis inhibits DNase I and gives rise to a less concentrated chromatin preparation. A lysis condition of 0.05% IGEPAL with 10 min incubation on ice is optimal for RAW264.7 cells. IGEPAL is a nonionic, nondenaturing detergent and chemically indistinguishable from Nonidet P-40 (NP-40). 10x IGEPAL lysis buffer stored at 4 °C often shows precipitation. Therefore, the buffer should be prewarmed in a 55 °C water bath for 30 min before performing the lysis step to ensure proper dissolution.

5. The ideal DNase I concentration that results in the release of the most accessible compartments can vary significantly between different cell and tissue types. The optimal DNase I

concentration should be determined empirically for each cell type. Excessive digestion with DNase I can result in extensive release of DNA fragments from genomic regions that show little to no true DHS sites. Conversely, insufficient digestion results in low yield and a greater portion of the released DNA will be derived from nonspecific DNA fragmentation of the nuclei. We treated cells with LPS for 8 h followed by various cross-linking conditions. Different DNase I concentrations were used to digest either non-cross-linked or cross-linked chromatin, and we have found the optimal DNase I concentration to be between 40 and 100 U/mL for RAW264.7 cells. DNase I concentrations of 60 and 80 U/mL showed better digestion patterns with either the cross-linked or non-cross-linked chromatin (Fig. 2). 20×10^6 cells from 150 mm plates were sufficient for three conditions of DNase I digestion. Each DNase I dilution should be mixed by gentle inversion as the enzyme is easily denatured by shaking. DNase I may show lot-to-lot variation; therefore, it is important to calibrate the activity of each lot before use. Store in small aliquots at −20 °C to avoid repeated freezing and thawing. Diluted DNase I should be discarded after use.

6. Before DNA purification by phenol:chloroform:isoamyl extraction, save 50 μL of each of the DNase I-digested samples for gel electrophoresis and qPCR validation. Pipette until samples are no longer viscous. The qPCR amplicons of the DNase I-hypersensitive and the DNase I-insensitive genomic regions should be of similar length and extend for several hundred base pairs. The amplicon should cover the entire hypersensitive region to encompass as many DNase-cutting events as possible. The qPCR primers were designed using Primer3 based on DHS sites in mouse ENCODE DNase-Seq data that have been previously published [21]. These positive control sites are prominent and abundant in various tissues and cell types [22–24]. We tested multiple primer pairs with genomic DNA to identify primer pairs that give the best amplification efficiency and the highest specificity [16]. We used negative control primers for DNase I-insensitive regions that were previously described [18]. The DNase I digestion efficiency was assessed by measuring the loss in amplification at the various sites analyzed across a range of DNase I concentrations. DNase I-digested hypersensitivity sites (DHS) are amplified less efficiently (higher Ct) than identical sites in the undigested control DNA (lower Ct). Delta (Δ) Ct values (the difference in Ct values) display the relative amount of DNase I sensitivity between DNase I-digested and -undigested DNA. Higher ΔCt values represent higher sensitivity to DNase digestion than

lower ΔCt values. Strong DHS sites (Chr 2) showed an approximate 70–80% loss in amplification in mild cross-linked chromatin (Fig. 4).

7. Hook all buckets, loaded or empty, to the rotor. If fewer than six tubes are being run, they must be arranged symmetrically in the rotor. Opposing tubes must be filled to the same level with liquid of the same density.

Acknowledgments

We thank Margaret Lim and Preeyaporn Songkiatisak for their critical reading of the chapter. We also thank Ido Goldstein, Louis Schiltz, and Lyuba Varticovski for technical assistance with qPCR primer information.

References

1. Klein DC, Hainer SJ (2020) Genomic methods in profiling DNA accessibility and factor localization. Chromosom Res 28(1):69–85
2. Song L, Crawford GE (2010) DNase-seq: a high-resolution technique for mapping active gene regulatory elements across the genome from mammalian cells. Cold Spring Harb Protoc 2010(2):pdb.prot5384
3. Galas DJ, Schmitz A (1978) DNAse footprinting: a simple method for the detection of protein-DNA binding specificity. Nucleic Acids Res 5(9):3157–3170
4. Jackson PD, Felsenfeld G (1985) A method for mapping intranuclear protein-DNA interactions and its application to a nuclease hypersensitive site. Proc Natl Acad Sci USA 82(8): 2296–2300
5. Becker PB, Ruppert S, Schutz G (1987) Genomic footprinting reveals cell type-specific DNA binding of ubiquitous factors. Cell 51(3): 435–443
6. Vierstra J, Stamatoyannopoulos JA (2016) Genomic footprinting. Nat Methods 13(3): 213–221
7. Cockerill PN (2011) Structure and function of active chromatin and DNase I hypersensitive sites. FEBS J 278(13):2182–2210
8. Boyle AP, Davis S, Shulha HP et al (2008) High-resolution mapping and characterization of open chromatin across the genome. Cell 132(2):311–322
9. Alvarez JM, Moyano TC, Zhang T et al (2019) Local changes in chromatin accessibility and transcriptional networks underlying the nitrate response in Arabidopsis roots. Mol Plant 12(12):1545–1560
10. Sung MH, Guertin MJ, Baek S et al (2014) DNase footprint signatures are dictated by factor dynamics and DNA sequence. Mol Cell 56(2):275–285
11. Lazarovici A, Zhou T, Shafer A et al (2013) Probing DNA shape and methylation state on a genomic scale with DNase I. Proc Natl Acad Sci USA 110(16):6376–6381
12. He HH, Meyer CA, Hu SS et al (2014) Refined DNase-seq protocol and data analysis reveals intrinsic bias in transcription factor footprint identification. Nat Methods 11(1):73–78
13. Neph S, Vierstra J, Stergachis AB et al (2012) An expansive human regulatory lexicon encoded in transcription factor footprints. Nature 489(7414):83–90
14. Pique-Regi R, Degner JF, Pai AA et al (2011) Accurate inference of transcription factor binding from DNA sequence and chromatin accessibility data. Genome Res 21(3):447–455
15. Gusmao EG, Allhoff M, Zenke M et al (2016) Analysis of computational footprinting methods for DNase sequencing experiments. Nat Methods 13(4):303–309
16. Oh KS, Ha J, Baek S et al (2019) XL-DNase-seq: improved footprinting of dynamic transcription factors. Epigenetics Chromatin 12(1):30
17. John S, Sabo PJ, Canfield TK et al (2013) Genome-scale mapping of DNase I hypersensitivity. Curr Protoc Mol Biol Chapter 27:Unit 21.27

18. Ling G, Waxman DJ (2013) DNase I digestion of isolated nulcei for genome-wide mapping of DNase hypersensitivity sites in chromatin. Methods Mol Biol 977:21–33
19. Deng X, Ma W, Ramani V et al (2015) Bipartite structure of the inactive mouse X chromosome. Genome Biol 16(1):152
20. Poorey K, Viswanathan R, Carver MN et al (2013) Measuring chromatin interaction dynamics on the second time scale at single-copy genes. Science 342(6156):369–372
21. Jin W, Tang Q, Wan M et al (2015) Genome-wide detection of DNase I hypersensitive sites in single cells and FFPE tissue samples. Nature 528(7580):142–146
22. Crawford GE, Holt IE, Whittle J et al (2006) Genome-wide mapping of DNase hypersensitive sites using massively parallel signature sequencing (MPSS). Genome Res 16(1): 123–131
23. John S, Sabo PJ, Johnson TA et al (2008) Interaction of the glucocorticoid receptor with the chromatin landscape. Mol Cell 29(5):611–624
24. Grontved L, Bandle R, John S et al (2012) Rapid genome-scale mapping of chromatin accessibility in tissue. Epigenetics Chromatin 5(1):10

Chapter 16

ChEC-Seq: A Comprehensive Guide for Scalable and Cost-Efficient Genome-Wide Profiling in *Saccharomyces cerevisiae*

Tamar Gera, Divya Krishna Kumar, Gilad Yaakov, Naama Barkai, and Felix Jonas

Abstract

Chromatin endogenous cleavage coupled with high-throughput sequencing (ChEC-seq) is a profiling method for protein-DNA interactions that can detect binding locations in vivo, does not require antibodies or fixation, and provides genome-wide coverage at near nucleotide resolution.

The core of this method is an MNase fusion of the target protein, which allows it, when triggered by calcium exposure, to cut DNA at its binding sites and to generate small DNA fragments that can be readily separated from the rest of the genome and sequenced.

Improvements since the original protocol have increased the ease, lowered the costs, and multiplied the throughput of this method to enable a scale and resolution of experiments not available with traditional methods such as ChIP-seq. This method describes each step from the initial creation and verification of the MNase-tagged yeast strains, over the ChEC MNase activation and small fragment purification procedure to the sequencing library preparation. It also briefly touches on the bioinformatic steps necessary to create meaningful genome-wide binding profiles.

Key words ChEC-seq, Transcription factors, Genome-wide, In vivo binding, *S. cerevisiae*, Short-read sequencing, High-throughput

1 Introduction

Protein binding to DNA is a central step in many cellular processes ranging from DNA replication to transcriptional regulation. Often the exact binding loci are critical and indicative of function. Profiling these binding sites has therefore been of great interest to the biological community. With the emergence of high-throughput sequencing techniques, pre-existing methods such as chromatin immunoprecipitation (ChIP) have been adopted to give a genome-wide view of the specific binding sites of individual pro-

Franziska Greulich (ed.), *Chromatin Immunoprecipitation: Methods and Protocols*, Methods in Molecular Biology, vol. 2846,
https://doi.org/10.1007/978-1-0716-4071-5_16,

teins. However, ChIP-seq and its derivatives, e.g., ChIP-exo [1], ChIP-Nexus [2], still rely on two critical steps: (1) fixation of cells to cross-link the protein to DNA and (2) target binding with antibodies, to precipitate the protein-DNA complexes. Both of these steps can miss binding locations, e.g., transient or indirect protein-DNA interactions not stable enough to be fixated, and introduce additional noise from nonspecific binding of the antibody to DNA or other DNA-binding proteins. Furthermore, most immunoprecipitation procedures involve expensive consumables and equipment and take several days.

Chromatin endogenous cleavage (ChEC), introduced by Schmidt et al. [3], circumvents these steps by directly fusing MNase, a calcium-dependent endonuclease, to the protein of interest. While the intracellular/intranuclear calcium concentration is not sufficient to activate MNase under physiological conditions, permeabilizing the cells and adding calcium activates the MNase and therefore leads to DNA cleavage around the binding site of the protein of interest. While the cutting of DNA was initially quantified with qPCR and Southern blot, MNase cleavage also generates small fragments around the binding site. These can be selectively purified and used to prepare next-generation sequencing libraries, i.e., ChEC-seq, providing a genome-wide view of binding sites [4]. As the MNase cuts very close to the actual binding site, which itself is protected by the bound protein, the resolution of this method usually approaches nucleotide resolution. This high resolution also distinguishes it from other in vivo methods such as Calling Cards [5] and DamID [6], which fuse constitutively active enzymes to the protein of interest. Avoiding immunoprecipitation removes the associated costs and reduces the required sample size. The main disadvantage of ChEC-seq is that the protein of interest must be tagged with MNase, a relatively small, globular protein ~18 kDa, which can potentially interfere with the function of the protein (as can attaching an additional Epitope tag for ChIP-seq). As MNase is active on either the N- or C- termini, and possibly also the middle of a protein, this risk can be reduced, which might be important for larger DNA-binding complexes, such as chromatin remodelers, histone modifiers, or RNA polymerases. Another source of noise might be nonspecific binding or sequence-preference of MNase; however, it is possible to estimate the extent of this effect by comparing between different DNA-binding proteins or to free MNase.

Despite its advantages, the ease and throughput of the original ChEC-seq procedure [7, 8] was limited by at least two factors that we alleviate in our protocol [9–11]: strain generation and library preparation. Moreover, we also increased the throughput of the actual ChEC experiment by performing it in a scalable 96-well plate setup.

While endogenous MNase tagging of yeast proteins using classical homologous recombination is possible, it requires specific materials for each protein and, in the case of N-terminal tagging, might introduce new promoters and interfere with protein expression. To overcome these hurdles, we utilized existing libraries for genome-scale tagging of the N- or C-terminus, as well as CRSIPR-based gene editing to enable the generation of dozens of yeast strains with a small set of compounds and a streamlined protocol [12, 13].

Another limiting factor in the ChEC-seq procedure is library preparation where the purification of small DNA fragments (<100 bp) involves delicate and time-consuming Phenol-Chloroform extraction and DNA precipitation. To facilitate this procedure and increase its throughput, we (1) reduced the number of cleanup steps using a one-tube library preparation protocol [14] and (2) also substituted Phenol-Chloroform extractions, when possible, with a SPRI-bead-based procedure optimized for small DNA fragments.

This protocol guides the reader through all necessary steps of a ChEC-seq experiment from the initial high-throughput tagging of proteins using available yeast libraries, over the actual ChEC procedure to sequencing library preparation (Fig. 1).

In the end, it will also briefly touch on bioinformatic tools that can be used to process the data, but we would refer the reader to our GitHub page (https://github.com/barkailab/) where we publish a version of our pipeline to generate the genome-wide binding profiles. While this protocol focuses on the standard ChEC-seq procedure, there are at least two possible variations: (1) using fixed cells for time-sensitive experiments and (2) adding an external spike-in control for absolute quantification of DNA-binding strength. We focus here on the model organism *S. cerevisiae* but note that similar protocols have been used in other organisms [15] or cell lines [16].

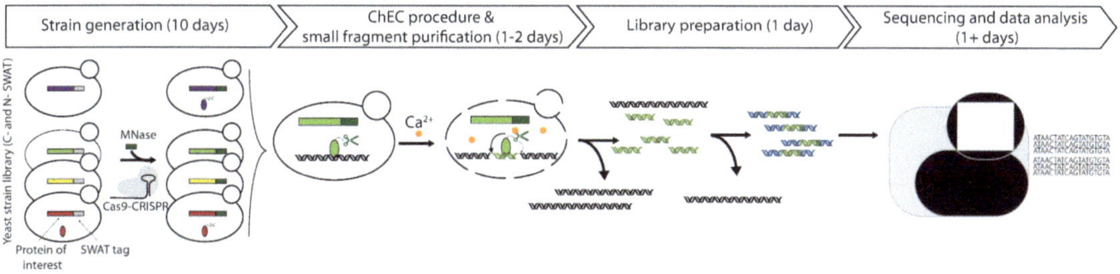

Fig. 1 Flowchart summarizing the main steps of the protocol. After preparing MNase-tagged yeast strains from C/N-SWAT strain libraries, they can be used for the ChEC procedure. Here, Calcium (Ca^{2+})-triggered endogenous MNase cleavage releases small fragments that are further extracted and selectively purified. An optimized protocol is used to prepare libraries from small DNA fragments. The estimated duration of each step is indicated at the top

2 Materials

2.1 High-Throughput MNase Tagging of Proteins in S. cerevisiae

2.1.1 Generating URA3 CRISPR Plasmid

1. bRA89 CRISPR vector (from Addgene #100950).
2. LB medium with Ampicillin for *Escherichia coli* growth: For 1 L liquid LB medium mix 10 g tryptone, 5 g yeast extract, and 10 g NaCl into 1 L ddH_2O and autoclave. Let the medium cool down and store at room temperature (18–22 °C). Before use add Ampicillin to 50 μg/mL.
3. LB agar plates with Ampicillin: For 1 L LB agar mix 10 g tryptone, 5 g yeast extract, 10 g NaCl, and 20 g of agar agar into 1 L ddH_2O and autoclave. Let the medium cool down to ~65 °C, add Ampicillin to 50 μg/mL, pour into 10 cm plates (~ 20–25 mL per plate) and let solidify. Store at 4 °C.
4. URA3 guide RNA insert (targeting URA3 gene in the swap module, the same guide RNA can be used for every parental strain from both N- and C-SWAT libraries) (*see* **Note 1**).

   >URA3 guide RNA (gRNA) insert with bRA89 overhang fwd:

   5′-GGTGTGGGTTTAGATGACAA gtttt

   >URA3 guide RNA (gRNA) insert with bRA89 overhang rev:

   5′-TTGTCATCTAAACCCACACC gatca
5. Competent *E. coli* for CRISPR plasmid amplification, e.g., DH5 alpha.
6. BplI restriction enzyme with Tango Buffer and SAM (Thermo Fisher, #ER1311).
7. Column-based reaction cleanup kit (e.g., RBC Bioscience, #YDF100).
8. T4 ligase and buffer for ligation (e.g., NEB #M0202S).
9. M13 rev primer (5′-CAGGAAACAGCTATGAC) for checking BRA89 plasmid integration.
10. Plasmid MiniPrep Kit (e.g., RBC Bioscience, #YPD100).
11. Shaking incubator for yeast and bacteria growth.
12. Thermocycler.
13. UV-Vis Spectrophotometer for measuring plasmid DNA concentration (e.g., NanoDrop ThermoFisher Scientific).
14. Supplies and equipment to run and document agarose gels.

2.1.2 CRISPR-Based Gene Editing of Library Strains

1. MNase DNA template (e.g., pGZ108 from Addgene #70231).
2. Parental strain (e.g., strain with N- or C-terminal integration module from [13]).
3. Primers with homologous regions (L1, L2, L3, and L4) to amplify the MNase repair replacing the swap module after the

CRIPSR cutting (*see* **Note 2**). The N- and C-SWAT libraries allow using the same set of primers for all N- or C-terminal integrations.

N-terminal:

>L1-ATG-MNase for:

5′-cgtacgctgcaggtcgacggtggcggttctggcggtggcggatc*c* ATG GCCACCAGTACCAAGAAGCT-3′

>MNase-L2 rev:

5′-catcgatgaattctctgtcgcaccaccaccaccagaggaaccgcc CTGGCCGCTATCGGCGTTAT

C-terminal:

>L3-MNase for

5′-cgtacgctgcaggtcgacggtggcggttctggcggtggcggatcc GCCACCAGTACCAAGAAGCT

>MNase-Stop-L4 rev

5′- atcgatgaattcgagctcgttcaatcgatatctcaaagaagaact TTA CTGGCCGCTATCGGCGTTAT

4. HiFi Polymerase Master Mix (e.g., Kapa HiFi, Roche #07958935001).
5. 10 mg/mL salmon sperm DNA for yeast transformation.
6. 50% (w/v) PEG-3350 for yeast transformation (sterile) (e.g., Sigma P4338-500G).
7. 1 M LiAc (Lithium Acetate) for yeast transformation (sterile).
8. YPD medium for yeast growth: For 1 L liquid YPD medium mix 10 g yeast extract, 20 g peptone, and 20 g glucose into 1 L ddH_2O and autoclave. Let the medium cool down and store at room temperature (18–22 °C).
9. YPD agar plates with or without Hygromycin: For 1 L YPD agar mix 10 g yeast extract, 20 g peptone, 20 g glucose, and 20 g of agar agar into 1 L ddH_2O and autoclave. Let the medium cool down to ~65 °C, add Hygromycin to 200 μg/mL if desired, pour into 10 cm plates (~ 20–25 mL per plate), and let solidify. Store at 4 °C.
10. Gene-specific forward primers for verifying MNase integration in the 5'-UTR or ORF for N- and C-terminal MNase, respectively. These can be designed with Primers4Yeast [17].
11. MNase reverse primer for verifying MNase integration [5′-GGCGGAAGGTCATGGGCTG].
12. 80% (v/v) glycerol (autoclaved) for glycerol stock (e.g., from Sigma #G7893-500ML).

2.2 Chromatin Endogenous Cleavage (ChEC) Experiment

2.2.1 ChEC Procedure

1. Yeast strain with MNase-tagged protein of interest (generated in Subheading 3.1).
2. 5 M NaCl (Sodium Chloride).
3. 0.5 M EDTA pH 8.0.

4. 0.2 M EGTA.
5. 10% (w/v) SDS (Sodium-dodecyl sulfate) in ddH_2O.
6. ddH_2O (DNAse-free).
7. 1 M Tris–HCl pH 7.5.
8. 1 M KCl (Potassium Chloride).
9. 0.2 M Spermine in ddH_2O.
10. 1 M Spermidine in ddH_2O.
11. Complete Protease Inhibitor EDTA-Free tablets (e.g., Roche #4693132001 one tablet for 50 mL).
12. 25 mM $CaCl_2$ (Calcium Chloride) in ddH_2O: from 1 M stock solution.
13. 2% (w/v) digitonin in DMSO.
14. 2× stop buffer: For 3.3 mL (16 reactions), add 240 μL 5 M NaCl, 120 μL 0.5 M EDTA, 60 μL 0.2 M EGTA, 2580 μL ddH_2O, 300 μL 10% (w/v) SDS. 0.15 mL per sample is required. Prepare buffer on the day of the experiment (*see* **Note 3**).
15. ChEC Buffer A1: For 50 mL (16 samples) add 0.75 mL 1 M Tris–HCl pH 7.5, 4 mL 1 M KCl, 45.2 mL ddH_2O, 25 μL 0.2 M EGTA, 50 μL 0.2 M Spermine, 25 μL 1 M Spermidine, 1 tablet EDTA-free Protease Inhibitor. 3 mL per sample is needed. Prepare on the day of the experiment and keep it at room temperature (18–22 °C) (*see* **Note 4**).
16. ChEC Buffer A2: For 3.0 mL (16 samples) mix 2850 μL Buffer A1 and 150 μL 2% (w/v) digitonin. 0.15 mL per sample is required. Prepare this buffer on the day of the experiment.
17. 96 deep-well plates for cell washing.
18. 96-well PCR Plate for the ChEC procedure.
19. Centrifuge for 50 mL tubes and deep-well plates.
20. Suitable yeast growth medium (e.g. SD or YPD medium).

2.2.2 Cell Lysis, DNA Extraction

1. 20 mg/mL Proteinase K in ddH_2O.
2. Buffered Phenol:Chloroform:Isoamyl alcohol (P:C:I; 25:24:1, saturated with 10 mM Tris–HCl pH 8.0 and 1 mM EDTA, e.g. Sigma #P3803).
3. Absolute ethanol (EtOH) (precooled to −20 °C for DNA precipitation).
4. 20 mg/mL Glycoblue (or other DNA precipitation agent, e.g. Thermo Scientific #AM9516).
5. 3 M Sodium Acetate.
6. 75% (v/v) EtOH: Mix 7.5 mL absolute EtOH with 2.5 mL ddH_2O.

7. DNA low-binding Microcentrifuge tubes (2 per sample, e.g. Eppendorf #0030108078).
8. Low-retention filter tips in all sizes.
9. Thermoblock for 1.8 mL micro tubes.
10. Tabletop cooling centrifuge for 1.8 mL reaction tubes.
11. Freezer −80 °C.

2.2.3 RNA Digest and Reverse SPRI Cleanup to Enrich Small DNA Fragments

1. 10 mg/mL RNase A in 1× TE Buffer (*see* below).
2. Isopropanol.
3. 1 M Tris–HCl pH 8.0.
4. Ultrapure DNase-free ddH_2O.
5. SPRI beads (e.g., Beckman Coulter #A63881).
6. 85% (v/v) EtOH: Mix 8.5 mL absolute EtOH with 1.5 mL ddH_2O.
7. 1× TE Buffer: 10 mM Tris–HCl pH 8.0, 1 mM EDTA pH 8.0. For 50 mL mix 500 μL 1 M Tris–HCl pH 8.0, 100 μL 0.5 M EDTA pH 8.0, and fill to 50 mL with ultrapure H_2O. Keep at room temperature (18–22 °C).
8. 0.1× TE Buffer: 1 mM Tris–HCl pH 8.0, 0.1 mM EDTA pH 8.0. For 1 mL, mix 100 uL 1× TE Buffer with 900 uL ddH_2O. Keep at room temperature (18–22 °C).
9. 96-well-skirted PCR plate for SPRI cleanup (e.g., Eppendorf #0030129504).
10. Magnet stand for Microcentrifuge tubes.
11. Magnet stand for 96-well plates.

2.3 ChEC-Seq Library Preparation for NGS Short-Read Sequencing

2.3.1 End-Repair and A-Tailing

1. 10× T4 DNA ligase Buffer (e.g., NEB #B0202S). Aliquots can be stored at −20 °C.
2. 10 mM ATP. Aliquots can be stored at −80 °C.
3. 10 mM (each) dNTPs.
4. 50% (w/v) PEG 4000.
5. 5 U/μL T4 DNA polymerase (e.g., Thermo Scientific #EP0062).
6. 10 U/μL T4 Polynucleotide Kinase (PNK) (e.g., NEB #M0201L).
7. 5 U /μL Taq Polymerase (e.g., Bioline #BIO-21040).

2.3.2 Cleanup and Size Selection

1. SPRI beads (e.g., Beckman Coulter #A63881).

2.3.3 Barcoded Adapter Ligation and Reaction Cleanup

1. Quick Ligase Kit (NEB #M2200S or other highly concentrated T4 DNA Ligase).
2. 40% (w/v) PEG 8000 in ddH_2O.
3. Y-shaped adapters with barcodes (as described in [18]). The concentration should be 0.18 μM (*see* **Note 5**).
   >Fwd: 5′ACACTCTTTCCCTACACGACGCTCTTCC GATC*T-3′ (*: phosphorothioate)
   >Rev: 5′P-GATCGGAAGAGCACACGTCTGAACTC CAGTCACnnnnnnnnATCTCG-TATGCCGTCTTCTGCTTG-3′ (nnnnnnnn: index1 sample barcode sequence) (*see* **Note 6**)
4. HXP buffer: For 16 samples mix 500 μL 5 M NaCl and 500 μL 40% (w/v) PEG 8000. Prepare freshly.

2.3.4 Enrichment PCR and Final SPRI Cleanup

1. 2× HIFI Polymerase Master Mix (e.g., Roche #07958927001),
2. P5 and P7 primer (prepare stocks at 100 μM and store at −20 °C).
   >P5: AATGATACGGCGACCACCGAGATCTACACnn nnnnnnACACTCTTTCCCTACACGAC (nnnnnnnn: index 2 barcode) (*see* **Note 7**).
   >P7: CAAGCAGAAGACGGCATACGAGAT
3. Fluorometer to measure library DNA concentration, e.g., Qubit (ThermoFisher Scientific),
4. Fragment analyzer, e.g., TapeStation (Agilent).

2.4 Bioinformatic Processing Pipeline to Generate Genome-Wide Binding Profiles

1. Cutadapt [19].
2. Bowtie2 [20].
3. Bedtools [21].

3 Methods

***3.1 High-Throughput MNase Tagging of Proteins in* S. cerevisiae**

There are at least three ways to generate MNase-tagged yeast strains for ChEC-seq: (1) classical homologous recombination with selection markers as performed in the initial ChEC and ChEC-seq paper, (2) seamless-genetic engineering using CRISPR-Cas9 and homologous repair, or (3) a high-throughput method building on recently generated yeast strain libraries for N- and C-terminal tagging [22]. In this protocol, we will focus on the latter method, as it enables the rapid tagging of dozens of strains with one set of materials. As this protocol, in contrast to the initially published LoxP-based method, utilizes CRIPSR-based gene editing to switch the swap tag for MNase, it can easily be adapted for gene editing of non-library strains. To choose the terminus, N- or C-, for MNase integration, it is helpful to verify that the similarly tagged GFP variant is correctly localized, as can be done on YeastRGB [13, 23]. While there are many different methods for CRISPR-based gene editing, in this protocol, we decided to use a previously published method [12], as it only requires a single plasmid and offers a streamlined cloning procedure.

3.1.1 Generating URA3 CRISPR Plasmid (3 Days, Only Done Once)

1. To cut the bRA89 CRIPSR plasmid, incubate 1 μg of the plasmid with 2 μL of BplI in 1× Tango buffer and 1× SAM for at least 16 h at 37 °C.
2. Purify the cut CRISPR plasmid using any column-based cleanup kit. Measure plasmid concentration with a spectrophotometer. The cut plasmid can be stored in aliquots (50 ng/reaction) and at −20 °C for several months.
3. Generate the gRNA insert duplex of "URA3 gRNA insert fwd" and "URA3 gRNA insert rev" by incubating 5 μL of each oligo stock (100 uM) with 8 μL ddH_2O and 2 μL 10× T4 Ligase Buffer, for 5 min at 99 °C followed by a slow cool down to room temperature (18–22 °C, ~0.2 °C /sec temperature rate). This can be done in any thermocycler with a heated lid at 105 ° C.
4. Dilute insert 1:100 and ligate 2 μL of diluted oligo duplex with 50 ng of cut BRA89 plasmid from **step 2** following the T4 DNA ligase protocol for sticky ends. It is recommended to also include a control ligation and transformation without the insert to assess the cutting efficiency of the BplI enzyme.
5. Subsequently, transform the ligation mix into competent *E. coli* and plate them on LB + Amp plates for positive selection.
6. Screen positive transformants for URA3 gRNA integration using colony PCR with primers "M13 rev" and "URA3 gRNA insert fwd." The expected amplicon size is ~300 bp.
7. Amplify and perform mini plasmid preparation of positive Bra89-URA3 plasmid-containing *E. coli.*

3.1.2 CRISPR-Based Gene Editing of Library Strains (7–10 Days, Can Be Done for Many Strains in Parallel)

1. Take a single colony from the N- or C-SWAT library parental strain and grow it overnight (10–24 h) in liquid YPD medium at 30 °C and 200 rpm until saturation ($OD_{600nm} = 20$).
2. Dilute the saturated overnight culture 1:100 in 5 mL of fresh YPD medium and grow for an additional 6 h at 30 °C and 200 rpm to reach the mid-exponential phase.
3. In the meantime, amplify the MNase repair template from the pGZ108 template using the primer set for N- or C-terminal tagging in a 20 μL PCR reaction using a HiFi Polymerase.
4. After the 6 h growth in **step 2**, transform yeast cells with 1 ug of plasmid and 20 μL of repair PCR, around 2–5 ug of DNA, using standard LiAc-PEG transformation [24]. It is

recommended to include a negative control transformation using the same amount of plasmid but substituting the repair PCR product with H_2O to verify a functional CRISPR system (*see* **Note 8**).

5. Recover cells in YPD medium for 1.5 h at 30 °C and 200 rpm.
6. After recovery, plate the transformation mix on YPD + Hygromycin plates for selection.
7. After 2 days of growth at 30 °C, emerging positive colonies will all contain the CRISPR plasmid, but not all of them will necessarily have integrated the MNase repair fragment. The negative control transformation will indicate the probability of false positives arising from a nonfunctional CRISPR plasmid or background mutation in the target sequence (*see* **Note 9**).
8. Confirm MNase integration using PCR with gene-specific forward primer and MNase reverse primer, and isolate positive colonies for the subsequent verification with Sanger sequencing.
9. To lose the CRIPSR plasmid (and thus the hygromycin resistance), strongly dilute a single yeast colony in 1–2 mL liquid YPD medium and grow it for 36 h at 30 °C in a shaking incubator (200 rpm) until it reaches full saturation ($OD_{600nm} = 20$).
10. Dilute the fully saturated yeast culture 1:100,000 in fresh YPD medium, e.g., 2 sequential dilutions of 3 μL in 1000 mL fresh YPD medium, and plate 100 μL of the final diluted culture onto a fresh YPD plate.
11. After 2 days at 30 °C, select single colonies and confirm plasmid loss by testing their growth on a hygromycin plate (*see* **Note 10**).
12. After plasmid loss is confirmed, glycerol stocks of MNase-tagged yeast strains can be stored at −80 °C until the ChEC-seq experiment.

3.2 Chromatin Endogenous Cleavage (ChEC) Experiment

This part of the protocol describes the improved ChEC procedure where MNase-tagged yeast cells are first permeabilized and then exposed to calcium to activate MNase and trigger DNA cleavage. This protocol can be performed on yeast cells in any growth stage, but we found transcription factor profiling convenient and reproducible from 5 mL of yeast cells at an OD_{600} of 4 in a synthetic complete medium with glucose. For other conditions, e.g., different media and growth phases, the total number of yeast cells should be between 1×10^8 and 3×10^8 cells, e.g., 15 mL of OD_{600} 0.4. After the ChEC procedure, the DNA is extracted, and the small fragments are selectively purified with a reverse SPRI cleanup. Reverse SPRI cleanup relies on the fact that, in the presence of

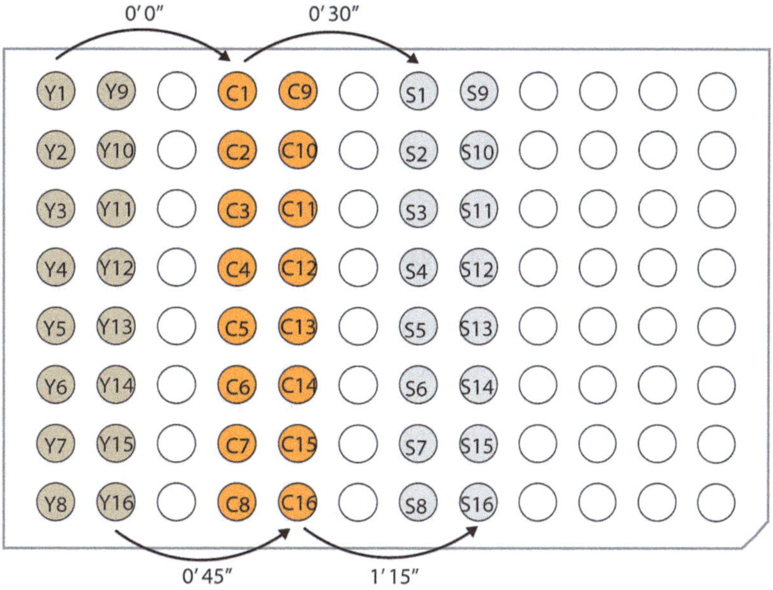

Fig. 2 Recommended ChEC plate layout for a 16-sample experiment. The plate is organized in columns and blocks to enable efficient processing with an 8-channel pipette. The first block (columns 1–2) is reserved for yeast cells (Y1–16), while 13 μL of 25 mM calcium chloride (C1–16) and 120 μL of 2× stop solution (S1–16) are prepared in the second (columns 4–5) and third block (columns 7–8), respectively. The layout can easily be adapted to more or less samples by changing the number of columns in each block accordingly. Also indicated are recommended times to transfer the solutions between the wells, assuming a 30 s MNase cleavage duration

PEG and salt, DNA can bind to paramagnetic beads and the minimal size of the binding DNA can be controlled by adjusting the PEG and salt concentration. Finally, adding isopropanol to the SPRI buffer enables the binding of very small DNA fragments. This part of the protocol can be completed in a single day, but there is a possible pausing step during the DNA precipitation at −70 °C (Subheading 3.2.2, **step 6**).

*3.2.1 ChEC Procedure (1–2 h, see **Note 11**)*

1. Prepare the ChEC plate from the 96-well plate (Fig. 2):
   (a) Leave one well per sample empty for adding yeast cells (labeled Y in Fig. 2).
   (b) Prepare 13 μL of 25 mM $CaCl_2$ for each sample in a different well (labeled C in Fig. 2).
   (c) Prepare 120 μL of 2× stop buffer for each sample in a different well (labeled S in Fig. 2).
   (d) Cover wells until use.

2. When yeast strains reach the desired OD (e.g., 4), pellet cells by centrifugation at 3250× *g* for 2 min, discard the supernatant, and resuspend cells in 0.5 mL Buffer A1 (*see* **Notes 12** and **13**).
3. Transfer each sample to one well in a 96-deep-well plate, and pellet again at 3250× *g* for 2 min. Discard the supernatant by flipping the plate. From here on, use a multichannel pipette to facilitate and accelerate the protocol.
4. Wash the pellet, by resuspending it in 1 mL Buffer A1, centrifuging at 3250× *g* for 2 min, and discarding the supernatant by flipping.
5. Repeat **step 4** once more.
6. For permeabilization of the cells, resuspend each pellet in 150 μL Buffer A2 and transfer to the reserved empty wells in the ChEC plate. Try to avoid bubbles when resuspending in Buffer A2. Incubate at 30 °C for 5 min.
7. After 5 min, transfer 150 μL of permeabilized cells to wells containing 25 mM $CaCl_2$ in the ChEC plate (from **step 1**, Fig. 2) to activate MNase. Mix thoroughly by pipetting up and down.
8. After 30 s in $CaCl_2$, transfer 120 μL of calcium-treated cells to wells containing 120 μL 2× stop buffer in the ChEC plate (from **step 1**, Fig. 2). Mix thoroughly by pipetting up and down. The incubation time with $CaCl_2$ can range from as little as 5 s to 1 min. Longer incubation times lead to nonspecific DNA fragments. In our hands, 30 s gave a good signal for all factors tested (*see* **Note 14**).

3.2.2 Cell Lysis, DNA Extraction (1–2 h, see ***Note 11****)*

1. Add 6 μL of 20 mg/mL Proteinase K to each sample and transfer samples to 1.8 mL DNA-low-binding tubes. Vortex the tubes vigorously.
2. For cell lysis, incubate the cells for 30 min at 55 °C in a thermoblock.
3. For DNA extraction, add an equal amount (240 μL) of buffered Phenol:Chloroform (*see* **Note 15**) to each sample, vortex vigorously, and centrifuge at full speed (17,100× *g*) for 10 min at room temperature (18–22 °C).
4. In the meantime, prepare 1.5 μL of Glycoblue in a fresh DNA-low-binding microcentrifuge tube for each sample.
5. After centrifugation, transfer the aqueous upper phase (~200–240 μL), which contains the DNA, to the Glycoblue-containing tubes. Avoid disturbing the white interphase or the lower phase. Discard the remaining lower phase to the appropriate waste disposal (Phenol:Chloroform is a halogenated organic solvent).

6. To precipitate DNA, add 2.5× Volumes of ice-cold absolute Ethanol (~ 500–600 μL) and 0.1× Volumes of 3 M Sodium Acetate (~20–24 μL) to each sample. Vortex the tubes briefly and transfer them to −80 °C for at least 1 h (can be up to 16 h).
7. After precipitation, pellet the DNA by centrifugation at full speed (17,100× *g*) for 10 min at 4 °C. Aspirate the supernatant carefully.
8. Wash the DNA pellets once with 1 mL 75% EtOH, aspirate EtOH, and air-dry the pellets for 5 min.

*3.2.3 RNA Digest and Reverse SPRI Cleanup to Enrich Small DNA Fragments (2–3 h, see **Note 11**)*

1. During the air-dry **step 8** in Subheading 3.2.2, mix 29 μL of 1× TE Buffer with 1 μL of 10 mg/mL RNase A for each sample.
2. For RNA digestion, resuspend the dried DNA pellet in 30 μL of TE-RNase A mix and incubate for 20 min at 37 °C in a thermoblock for 1.8 mL tubes.
3. After the digestion, add 0.8× volumes of SPRI beads (e.g., 24 μL) to each sample, mix thoroughly by pipetting and incubate for 3 min at room temperature (18–22 °C). In this step, large fragments of uncut genomic DNA are bound to the SPRI beads.
4. Put tubes onto a magnet stand and wait for the beads to settle.
5. Transfer the supernatant, which contains the small DNA fragments, to a clean 96-well skirted PCR plate (Cleanup plate).
6. Add 1× of the original sample volume (30 μL) of SPRI beads to the supernatant, increasing the total concentration to 1.8×. Then carefully add 5.4× of the original sample volume (~162 μL) isopropanol (*see* **Note 16**) to each well, ensuring that the sample does not overflow. The total well volume is ~246 μL. Mix thoroughly, but carefully.
7. Incubate for 3 min at room temperature (18–22 °C) to bind all remaining DNA fragments to the beads.
8. Move the Cleanup plate to a magnet stand and wait 3 min for the beads (bound to small DNA fragments) to settle. Discard supernatant using a pipette.
9. Wash the beads by adding 180 μL 85% EtOH to each well and carefully remove it with a vacuum aspirator.
10. Repeat wash **step 9** once.
11. After the second wash, remove the remaining EtOH carefully and air-dry beads for 5 min.
12. Remove the plate from the magnet stand and resuspend beads in 30 μL 0.1× TE buffer.
13. After 3 min, move the plate back to the magnet stand and wait 3 min for the beads to settle.

14. Collect the supernatant containing the small DNA fragments, i.e. eluate, into wells of a new 96-well PCR plate. The small DNA fragments can be stored at −20 °C until library preparation.

3.3 ChEC-Seq Library Preparation for NGS Short-Read Sequencing

Most standard library preparation methods are not optimized for the short DNA fragments generated by ChEC. The following protocol is an adaptation of the Cut'N'Run library preparation protocol [14], which is a one-tube library preparation protocol with no intermediate cleanup steps. However, using the original protocol led to various artifacts. We thus introduced an extra size selection and cleanup step after the end-repair and A-tailing (ERA) reaction to reduce those and further enrich for small DNA fragments. The adapters and enrichment primers used in this protocol add dual indexes to each fragment and thus enable two steps of multiplexing, but the protocol should be compatible with every T-overhang adapter. This part of the protocol can be completed in a day, but there are possible stopping steps for example before or after the amplification PCR.

3.3.1 End-Repair and A-Tailing (ERA) (1.5 h, see **Note 11**)

1. Prepare ERA mix as indicated below and add 5.4 μL ERA mix per sample into a fresh well of a 96-well PCR plate. Keep on ice.
   ERA mix per sample (5.4 μL in total), mix thoroughly after adding all components:

NEB 10× T4 DNA ligase Buffer	2 μL
10 mM dNTPs	1 μL
10 mM ATP	0.5 μL
50% PEG 4000	1.1 μL
T4 DNA Pol (5 U/uL)	0.1 μL
Taq Pol (5 U/uL)	0.1 μL
T4 PNK (10 U/uL)	0.6 μL
Total ERA mix	**5.4 μL**

2. Add 14.6 μL of each sample from Subheading 3.2.3, **step 14** to the wells containing ERA master mix and mix well. One ChEC reaction can be used for two library preparations.
3. Incubate with the following program (Table 1) in a thermocycler with the lid heated to 60 °C.

3.3.2 Cleanup and Size Selection (1 h, see **Note 11**)

1. Add 0.5× volumes of SPRI beads (10 μL) to each sample, mix thoroughly by pipetting, and wait for 3 min to allow for binding of large DNA fragments.

Table 1
ERA reaction PCR program

12 °C for 20 min	End repair by T4 DNA Polymerase
37 °C for 15 min	Phosphorylation by T4 PNK
58 °C for 45 min	Deactivation of T4 DNA Polymerase and A-tailing by Taq polymerase
Pause at 8 °C	Pause until the next step

2. Move the 96-well plate to a magnet stand and wait 3 min for the beads to settle.
3. Transfer the supernatant (~30 μL) to a new well of the PCR plate. Remove the plate from the magnet stand.
4. Add a 1.3× original volume (26 μL) of SPRI beads and 5.4× original volume (108 μL) of isopropanol to the supernatant. Mix thoroughly by pipetting and wait for 3 min to allow for the binding of remaining small DNA fragments.
5. Move the plate back to the magnet stand and wait 3 min for beads (loaded with the small DNA fragments) to settle.
6. Remove the supernatant with a pipette.
7. Wash the beads twice with 160 μL 85% EtOH.
8. Aspirate the remaining EtOH carefully, remove the plate from the magnet stand, and air-dry the beads for 3–5 min.
9. In the meantime, thaw the barcoded Y-shaped adapters and ligase buffer on ice.
10. Resuspend the beads in 18 μL of 0.1× TE buffer and wait for 3 min for DNA elution.
11. Put the cleanup plate back on the magnetic stand, wait 3 min for beads to settle, and move eluate (~16.4 μL, containing small DNA fragments) to a fresh 96-well plate. Keep on ice.

3.3.3 Barcoded Adapter Ligation and Reaction Cleanup (2 h, see Note 11)

1. Prepare ligation master mix on ice.
   Add the following reagents per sample:

2× Quick ligase buffer	20 μL
Quick ligase (2000 U/μL)	2 μL
Total Ligation Mix	**22 μL**

2. Add 1.6 μL of a differently barcoded adapter and 22 μL of the ligation master mix to each sample (eluate from Subheading 3.3.2, **step 11**). Mix thoroughly by pipetting.
3. Incubate the ligation reaction at 20 °C for 15–20 min in a thermocycler with a heated lid at 25 °C.
4. In the meantime, prepare 50 μL of HXP buffer per sample (*see* Subheading 2.3.3).
5. After ligation, add 1.2× volume of SPRI beads (~48 μL) to each sample and incubate for 3 min to bind adapter-ligated DNA fragments.
6. Move the plate to a magnetic stand, wait for 3 min for beads to settle, and remove the supernatant with a pipette.
7. Wash the beads twice with ~160 μL of 80% EtOH.
8. Aspirate all EtOH, remove the plate from the magnetic stand, and air-dry beads for 3 min.
9. Resuspend beads in 30 μL 0.1× TE buffer and wait for 3 min.
10. Add 1.3× volumes (39 μL) of HXP buffer to the resuspend beads and wait for 3 min to rebind DNA.
11. Move the plate to a magnetic stand, and wait for 3 min for beads to settle. Remove the supernatant with a pipette.
12. Wash the beads twice with ~160 μL of 80% EtOH.
13. Aspirate all EtOH, remove the plate from the magnetic stand, and air-dry beads for 3 min.
14. In the meantime, thaw 2× HiFi PCR master mix, P5 and P7 primer stocks for Subheading 3.3.4, **step 1**.
15. Resuspend the dried beads in 25 μL 0.1× TE buffer and wait 3 min for the elution.
16. After eluting, put the plate back on the magnetic stand, wait 3 min for the beads to settle, and transfer eluate (23 μL) to a fresh well.

*3.3.4 Enrichment PCR and Final SPRI Cleanup (1.5 h, see **Note 11**)*

1. Prepare 2 μL primer mix per sample. The primer mix can be stored at −20 °C for several months. For example, for 55 samples (110 μL) prepare:

100 μM P5 primer	8.8 μL
100 μM P7 primer	8.8 μL
dd H_2O	92.4 μL
Total Primer mix	**110 μL**

Table 2
PCR program for library amplification

98 °C for 45 s	Activation	
98 °C for 15 s 60 °C for 15 s	Strand separation Annealing + Extension	X 16 cycles
72 °C for 1 min	Final Extension	

2. Prepare the PCR master mix for each sample as follows:

2× HiFI PCR Master mix	25 μL
Primer mix	2 μL
Total PCR mix	**27 μL**

3. Add 27 μL of PCR master mix to each sample (eluate from Subheading 3.3.3, **step 16**) and mix thoroughly by pipetting.
4. Start the enrichment PCR with the following settings (Table 2). The lid is heated to 105 °C.
5. After enrichment PCR, add 1.1× volumes (55 μL) of SPRI beads to each well, mix thoroughly by pipetting, and wait 3 min for DNA binding.
6. Move the plate to a magnetic stand and wait 3 min for beads to settle and remove the supernatant.
7. Wash the beads twice with 160 μL 80% EtOH.
8. Aspirate the remaining EtOH with a pipette or multi-well vacuum aspirator, remove the plate from the magnet stand, and air-dry beads for 3–5 min.
9. Resuspend the beads in 15 μL 0.1× TE buffer, and incubate for 2 min for elution.
10. Place the plate back on a magnetic stand, wait 2 min for the beads to settle, and collect the eluate containing the NGS library.
11. Measure the DNA concentration with a fluorometer such as Qubit and analyze the fragment size distribution with a Tape station or similar. Library concentrations >1 ng/μL are desired, but also those with lower concentrations can be sequenced and give reliable profiles (just with a lower complexity). *See* Fig. 3 for exemplary profiles (*see* **Notes 17** and **18**).
12. Perform paired-end sequencing with at least 25 bases per read and both indexes with around three million reads/sample. However, fewer reads per sample (e.g., 200,000) are, in most cases, sufficient for reliable profiling (Fig. 4, *see* **Note 19**).

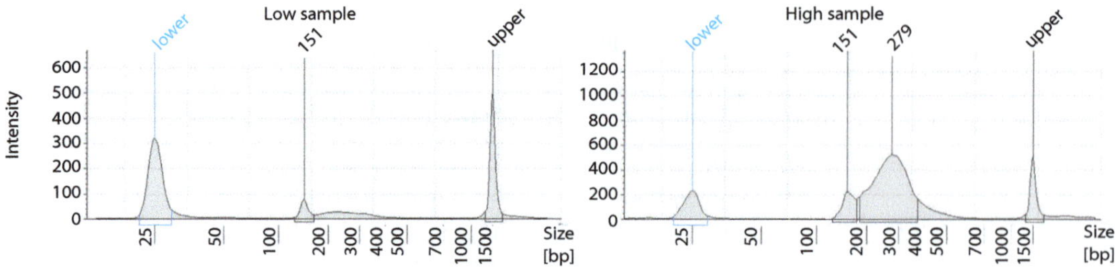

Fig. 3 Tape station profile of two exemplary successful ChEC-seq samples. Shown are the tape station profiles of two ChEC libraries, one of low (low sample) and one of high concentration (high sample). Both libraries had reliable profiles after the sequencing. The profiles are shown after the final PCR reaction and cleanup. The indicated peaks at ~150 bp and ~260 bp correspond to primer dimers (150 bp) and the adapter-ligated ChEC fragments (260 bp), respectively. The large fragments above the upper limit contribute to the Qubit concentration measurement but are inert during the sequencing

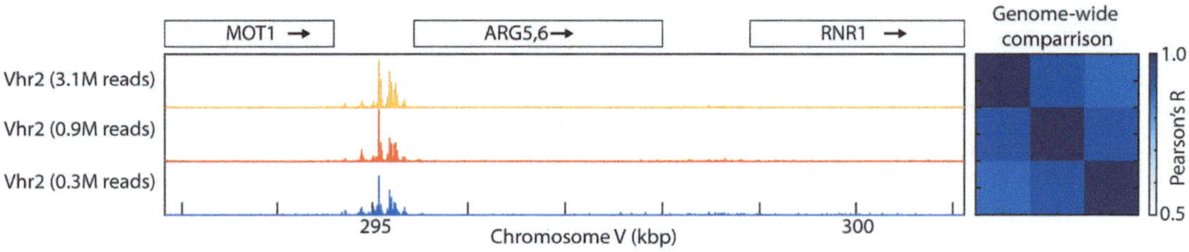

Fig. 4 Vhr2 binding profiles with different sequencing depth. *Left:* Normalized binding profile of three Vhr2 samples with indicated read depths around the Vhr2 target gene *Arg5,6*. Of note is the very similar peak position and height in all three samples and the slightly higher noise in the sample with the lowest number of reads. *Right:* Heatmap showing the pair-wise correlation between the different samples along the whole genome. The signal was smoothed with a 150 bp moving average

3.4 Bioinformatic Processing Pipeline to Generate Genome-Wide Binding Profiles

After obtaining and demultiplexing the raw data from the next-generation sequencing (NGS) run, we usually run them through a simple pipeline using published bioinformatic tools, but there are of course other ways to analyze the data. A few considerations that we made when designing our pipeline were as follows: (1) The small fragment size of the library makes it difficult to remove all adapter dimers via size selection; thus, we bioinformatically filter out those reads looking like adapter dimers. (2) While ChEC should theoretically only result in very small DNA fragments (~20–50 bp) consisting of the protected DNA-binding site, in practice many DNA fragments are bigger (up to ~150–200 bp) when they encompass the actual binding site and a nearby nucleosome protecting the DNA fragment from cleavage. We use all these fragments for our analysis. (3) In contrast to ChIP-seq, where the DNA-binding site, e.g., DNA motif, in case of transcription factors, can be observed across the DNA fragment; for ChEC-seq, we usually found them close to the end of the fragments. Thus, when generating the profile, we only consider a single base at the 5'end of both reads for each fragment.

The following steps show the crude processing pipeline:

1. Filter adapter dimers and library preparation artifacts using Cutadapt.
2. Aligning paired-end reads to the *S. cerevisiae* reference genome (R64-1-1) using Bowtie 2.
3. Calculate genome coverage from properly paired reads, only considering the 5′ end of each read (e.g., the beginning and end of the fragment) using `genomeCoverage` from Bedtools.

4 Notes

1. For non-library CRISPR-based genome editing, the guide RNA must be specific to the N- or C-terminus of the selected protein.
2. For non-library gene editing, the L1/L2 or L3/L4 overhangs need to be replaced by gene-specific overhangs, which also include a mutation of the CRISPR site.
3. NaCl, EDTA, and ddH_2O can be mixed beforehand and stored at 4 °C; the rest should be added freshly.
4. Buffer should be kept at room temperature (18–22 °C) and used within 1 h after preparation as part of the ingredients are unstable.
5. Before first use, adapter stocks should be prepared by mixing 1 Vol of forward primer, 1 Vol of reverse primer (both at 100 μM), and 2 Volumes of ultrapure water. Heat primer mixture to 100 °C and then slowly cool down to room temperature (18–22 °C) and store at −20 °C. Working solutions of 0.18 μM are prepared from the stocks.
6. The barcode sequences in the reverse oligo will be read as Index1 during Illumina sequencing. Up to 96 oligos with different barcodes can be ordered from vendors such as IDT to enable multiplexing.
7. The P5 barcode will be read out as Index2 during the Illumina sequencing. Therefore, different P5 primers distinguish different samples and allow a second level of multiplexing.
8. There are two possible sources of false positives, i.e., colonies without MNase repair: (i) background mutations in the CRISPR target sequence and (ii) inefficient cleavage/targeting by the CRISPR guide. We have verified that this guide efficiently targets the URA3 locus with 0–10 false positive colonies.

9. We usually get >200 colonies on the repair plate and <10 on the control plate indicating >90% true positive colonies, i.e., with MNase integration.
10. Replica plating on hygromycin plates is also possible.
11. Working with low-retention filter tips throughout this part is recommended to maximize yield and reduce contamination.
12. For absolute quantification between samples of the same experiment, a fixed number of spike-in cells can be added to each strain at the start of the ChEC procedure. For this purpose, we usually use 0.5 mL of OD600 = 4 of a yeast strain where a transcription factor with a sufficiently different binding profile is tagged with MNase. In our experience, Aro80 serves as a good spike-in as it has only a few binding sites.
13. Cross-linking of the protein of interest is possible but must be tested in a pilot study. In a pilot study with Msn2, it was possible to use cross-linked cells as an input at Subheading 3.2.1, **step 2**. While 1% formaldehyde cross-linking resulted in profiles, footprinting profiles of cells cross-linked with 0.1% formaldehyde are much closer to the in vivo profiles without cross-linking. Although −80 °C storage of cells is possible, it has a negative effect on the footprinting profiles.
14. As the amount and thus activity of MNase depends on the specific protein tagged, it seems possible that more abundant proteins require less time and might give unspecific fragments faster. Thus, it might be worth checking different incubation times 10, 30, and 90 s for a starter and see if there is any difference in the footprinting profiles.
15. Work inside a fume hood, wear protective clothing, and follow local safety regulations.
16. Isopropanol lowers the size threshold for SPRI beads [25].
17. If too many (>50%) adapter dimers with a fragment size of ~150 bp are observed, it is possible to perform another cleanup step with 0.9× SPRI beads, but this also leads to the loss of informative small fragments.
18. Big discrepancies between DNA concentration measured with fragment analyzers and Qubit might result from large fragments that are inert during next-generation sequencing.
19. The number of reads is based on our experience with transcription factors that only bind a relatively small number of loci in the genome (<1000). More than 200,000 reads are necessary for proteins that bind more DNA locations, i.e., histones.

References

1. Rossi MJ, Lai WKM, Pugh BF (2018) Simplified ChIP-exo assays. Nat Commun 9(1):2842
2. He Q, Johnston J, Zeitlinger J (2015) ChIP-nexus enables improved detection of in vivo transcription factor binding footprints. Nat Biotechnol 33(4):395–401
3. Schmid M, Durussel T, Laemmli UK (2004) ChIC and ChEC; genomic mapping of chromatin proteins. Mol Cell 16(1):147–157
4. Zentner GE, Kasinathan S, Xin B et al (2015) ChEC-seq kinetics discriminates transcription factor binding sites by DNA sequence and shape in vivo. Nat Commun 6(1):8733
5. Wang H, Mayhew D, Chen X et al (2012) "Calling cards" for DNA-binding proteins in mammalian cells. Genetics 190(3):941–949
6. van Steensel B, Henikoff S (2000) Identification of in vivo DNA targets of chromatin proteins using tethered dam methyltransferase. Nat Biotechnol 18(4):424–428
7. Saleh MM, Tourigny JP, Zentner GE (2021) Genome-wide profiling of protein-DNA interactions with chromatin endogenous cleavage and high-throughput sequencing (ChEC-Seq). Methods Mol Biol 2351:289–303
8. Grünberg S, Zentner GE (2017) Genome-wide mapping of protein-DNA interactions with ChEC-seq in Saccharomyces cerevisiae. J Vis Exp 124:55836
9. Gera T, Jonas F, More R et al (2022) Evolution of binding preferences among whole-genome duplicated transcription factors. elife 11: e73225
10. Jonas F, Carmi M, Krupkin B et al (2023) The molecular grammar of protein disorder guiding genome-binding locations. Nucleic Acids Res 51(10):4831–4844
11. Kumar DK, Jonas F, Jana T et al (2023) Complementary strategies for directing in vivo transcription factor binding through DNA binding domains and intrinsically disordered regions. Mol Cell 83(9):1462–1473.e1465
12. Anand RP, Memisoglu G (2017) Cas9-mediated gene editing in Saccharomyces cerevisiae. Protocol Exchange
13. Yofe I, Weill U, Meurer M et al (2016) One library to make them all: streamlining the creation of yeast libraries via a SWAp-Tag strategy. Nat Methods 13(4):371–378
14. Skene PJ, Henikoff S (2017) An efficient targeted nuclease strategy for high-resolution mapping of DNA binding sites. elife 6:e21856
15. Strobino M, Wenda JM, Padayachy L et al (2020) Loss of histone H3.3 results in DNA replication defects and altered origin dynamics in. Genome Res 30(12):1740–1751
16. Foss EJ, Tripathy S, Gatbonton-Schwager T et al (2021) Chromosomal Mcm2-7 distribution and the genome replication program in species from yeast to humans. PLoS Genet 17(9):e1009714
17. Yofe I, Schuldiner M (2014) Primers-4-Yeast: a comprehensive web tool for planning primers for Saccharomyces cerevisiae. Yeast 31(2): 77–80
18. Blecher-Gonen R, Barnett-Itzhaki Z, Jaitin D et al (2013) High-throughput chromatin immunoprecipitation for genome-wide mapping of in vivo protein-DNA interactions and epigenomic states. Nat Protoc 8(3): 539–554
19. Martin M (2011) Cutadapt removes adapter sequences from high-throughput sequencing reads. EMBnet J 17(1):10
20. Langmead B, Salzberg SL (2012) Fast gapped-read alignment with Bowtie 2. Nat Methods 9(4):357–359
21. Quinlan AR, Hall IM (2010) BEDTools: a flexible suite of utilities for comparing genomic features. Bioinformatics 26(6):841–842
22. Weill U, Yofe I, Sass E et al (2018) Genome-wide SWAp-Tag yeast libraries for proteome exploration. Nat Methods 15(8):617–622
23. Dubreuil B, Sass E, Nadav Y et al (2019) YeastRGB: comparing the abundance and localization of yeast proteins across cells and libraries. Nucleic Acids Res 47(D1):D1245–D1249
24. Gietz RD, Schiestl RH (2007) High-efficiency yeast transformation using the LiAc/SS carrier DNA/PEG method. Nat Protoc 2(1):31–34
25. Fishman A, Light D, Lamm AT (2018) QsRNA-seq: a method for high-throughput profiling and quantifying small RNAs. Genome Biol 19(1):113

Index

A

Adapter 49, 52, 55, 65, 66, 92, 94, 98, 99, 103–105, 114, 135, 137, 148, 154, 159–163, 166, 181, 183, 186, 192, 195, 196, 206, 209, 216, 218, 222, 226–228, 234, 236, 237, 239, 270, 276–278, 280–282
Adipocytes 35–45
Adipose tissues 14, 35–45
Automated workflow 48

B

Bioinformatics 47, 48, 65, 67, 137, 138, 191–212, 218, 265, 270, 280
Biotin-streptavidin system 124
Brain 14, 138, 139, 141, 149
Bone marrow-derived macrophages (BMDMs) 22, 31, 93, 96–100, 103–105

C

ChEC-seq 192, 263–282
ChIP-exo 91, 92, 264
ChIP-nexus 92, 94, 96, 98, 102–105, 264
ChIP-seq 2, 45, 47–61, 63–87, 91, 92, 104, 109, 134–138, 151, 152, 164, 169, 191–193, 200, 201, 203, 209, 211, 212, 216, 264, 280
Chromatin accessibility 182, 207, 244, 248, 249, 256, 257
Chromatin biology 47
Chromatin remodeling 47, 133, 170, 181
Cleavage under targets and release using nuclease (CUT&RUN) 2, 133–149, 151–166, 169–179
Cleavage under targets and tagmentation (CUT&Tag) 2, 181–189, 191–212
Cofactors 17, 133
Conda 48, 68, 85, 193, 223, 234

D

Dexamethasone 19, 22, 97, 105, 161
DiffBind 64, 69–76, 78, 79, 83, 86, 87
Differential ChIP-Seq analysis 76
DNA-binding 17, 22, 24, 25, 35–45, 93, 97, 99, 101, 103, 109, 133, 134, 158, 169, 170, 172, 191, 192, 211, 264, 265, 279, 280
DNA probes 124
DNA-protein interactions 31, 109, 123, 191
DNase-seq 244, 248, 249, 256, 258, 259

E

Epigenetics 152, 169–179, 182, 192, 203, 215, 216
Epigenome 63, 181–189, 192
Estrogen receptors 75
Exonuclease digestion 95, 98, 101

F

FACS sorting 221, 225
FastQC 49, 52, 55, 82, 193–197, 219, 224, 236

G

Gene-centered approach 123, 124
Genome coverage 53, 55–59, 61, 160, 281
Genome-wide 1–15, 59, 61, 63, 104, 109, 151, 192, 200, 209, 240, 244, 263–282
Genomic footprinting 244
Glucocorticoid receptor (GR) 19, 22, 25, 28–31, 91–93, 96–98, 104, 105, 133, 159, 161

H

High-throughput 63, 65, 152, 170, 191, 201, 243, 244, 265, 266, 270–272
High-throughput sequencing 193, 198, 249, 263
Histone modifications 1–15, 17, 44, 45, 47, 56, 57, 59, 61, 64, 72, 74, 79, 82, 86, 151, 159, 162–164, 170, 175, 177, 181, 182, 192, 203, 209, 215, 216, 220, 221, 223, 224, 235, 238–240
Hybridization 124–126, 128–131, 232

Franziska Greulich (ed.), *Chromatin Immunoprecipitation: Methods and Protocols*, Methods in Molecular Biology, vol. 2846, https://doi.org/10.1007/978-1-0716-4071-5,

I

Immune system 164
In-vivo binding 134

L

Library preparation 6, 10–13, 38, 43, 55, 65, 92, 95–96, 98, 102–104, 110, 112–116, 134, 135, 145, 146, 152, 155, 158–160, 162, 172, 173, 176, 177, 188, 192, 206, 264, 265, 269–270, 276–281

M

Macrophages 91–106, 152, 153, 155, 161, 165
Mass spectrometry 124, 129
Micrococcal nuclease (MNase) 1, 4, 7, 9, 10, 13, 14, 56, 134, 135, 148, 152, 162, 164, 169–171, 174, 175, 178, 221, 264–267, 270–274, 281, 282

N

Native protein-DNA interactions 1
Next-generation sequencing (NGS) 10, 12, 14, 50, 64, 65, 103, 109, 110, 134, 137, 149, 177, 191, 264, 269–270, 276–280, 282
Nuclear isolation/isolation of nuclei 37, 39, 124–127, 139, 141, 182, 184–185
Nuclei preparation 2–4, 6–9, 14, 184, 185, 258

P

Peak calling 53, 57, 59, 65–70, 86, 87, 137, 177, 193, 198–202, 208, 209
Plant 124, 126, 129
Polymerase chain reaction (PCR) 2, 5, 6, 9, 11–15, 17–34, 38, 40, 41, 43, 52, 53, 56, 57, 60, 61, 65, 84, 94–98, 102, 104, 106, 111, 113–115, 126, 129, 137, 154, 158–160, 164, 172, 177, 179, 181, 183, 187, 192, 205, 206, 218, 221–223, 226–228, 231–233, 237, 240, 268–272, 275–280
Protein co-localization 109
Protein-DNA binding 63–87, 92
Protein-protein interactions 135

Q

Quality controls 12, 36, 38, 43, 45, 47, 49–52, 55, 56, 65, 71, 76, 96, 103, 141, 146, 155, 159, 160, 166, 193–195, 201, 219, 223, 236, 247, 254
Quantitative PCR (qPCR) 2, 6, 11, 14, 15, 19, 22, 27, 28, 33, 38, 43, 45, 110, 126, 155, 160, 161, 164, 166, 247, 254, 259, 260, 264

R

Regulatory elements 109, 209, 212
Reproducibility 47, 48, 105, 162, 194, 201, 202, 210, 211
RNA sequencing (RNA-seq) 45, 71, 72, 137, 191–212

S

Saccharomyces cerevisiae 263–282
Sequencing data 51, 55, 82, 194, 195, 198, 206, 258
Sequential chromatin immunoprecipitation 110
Short-read sequencing 269–270, 276–279
Single-cell ChIC-seq technique 215–240
Sonication 1, 19, 32, 99, 100, 105, 124, 125, 127, 130, 131, 134, 152, 165, 169, 191, 237

T

Tissue 1–15, 35–45, 70, 72, 75, 119, 126, 134, 136, 138–141, 147, 149, 170, 181–189, 192, 193, 195, 216, 218, 235, 258
Tn5 transposase 135
Transcription factor footprinting 91, 243–260
Transcription factors (TFs) 2, 17, 19, 35–45, 47, 63, 71, 72, 74, 78, 80, 82, 83, 86, 91, 109, 123, 124, 134, 147, 151, 159, 163–165, 170, 172, 175, 178, 181, 191, 192, 195, 203, 207, 208, 211, 240, 244, 272, 280, 282

W

Workflow 30, 47–61, 65–70, 75, 84–86, 137, 191–212, 219, 234–236

MIX
Papier aus verantwortungsvollen Quellen
Paper from responsible sources
FSC® C105338

If you have any concerns about our products,
you can contact us on
ProductSafety@springernature.com

In case Publisher is established outside the EU,
the EU authorized representative is:
Springer Nature Customer Service Center GmbH
Europaplatz 3, 69115 Heidelberg, Germany

Printed by Libri Plureos GmbH
in Hamburg, Germany